Lecture Note2238

Edited by G. Goos

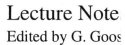

6000514470

D1328769

Springer
*Berlin
Heidelberg
New York
Barcelona
Hong Kong
London
Milan
Paris
Tokyo*

Gregory D. Hager   Henrik I. Christensen
Horst Bunke   Rolf Klein (Eds.)

Sensor Based
Intelligent Robots

International Workshop
Dagstuhl Castle, Germany, October 15-20, 2000
Selected Revised Papers

 Springer

Series Editors

Gerhard Goos, Karlsruhe University, Germany
Juris Hartmanis, Cornell University, NY, USA
Jan van Leeuwen, Utrecht University, The Netherlands

Volume Editors

Gregory D. Hager
Johns Hopkins University, Department of Computer Science
E-mail: hager@cs.jhu.edu

Henrik Iskov Christensen
Royal Institute of Technology, Center for Autonomous Systems, CVAP
E-mail: hic@nada.kth.se

Horst Bunke
University of Bern, Department of Computer Science
E-mail: bunke@iam.unibe.ch

Rolf Klein
University of Bonn, Institute for Computer Science I
E-mail: rolf.klein@uni-bonn.de

Cataloging-in-Publication Data applied for

Die Deutsche Bibliothek - CIP-Einheitsaufnahme

Sensor based intelligent robots : international workshop, Dagstuhl Castle,
Germany, October 15 - 20, 2000 ; selected revised papers / Gregory D. Hager
... (ed.). - Berlin ; Heidelberg ; New York ; Barcelona ; Hong Kong ; London ;
Milan ; Paris ; Tokyo : Springer, 2002
  (Lecture notes in computer science ; Vol. 2238)
  ISBN 3-540-43399-6

CR Subject Classification (1998): I.4, I.2.9, I.2, I.6, H.5.2

ISSN 0302-9743
ISBN 3-540-43399-6 Springer-Verlag Berlin Heidelberg New York

Springer-Verlag Berlin Heidelberg New York
a member of BertelsmannSpringer Science+Business Media GmbH

http://www.springer.de

© Springer-Verlag Berlin Heidelberg 2002
Printed in Germany

Typesetting: Camera-ready by author, data conversion by Olgun Computergrafik
Printed on acid-free paper   SPIN: 10845834   06/3142   5 4 3 2 1 0

Preface

Robotics is a highly interdisciplinary research topic, that requires integration of methods for mechanics, control engineering, signal processing, planning, graphics, human-computer interaction, real-time systems, applied mathematics, and software engineering to enable construction of fully operational systems. The diversity of topics needed to design, implement, and deploy such systems implies that it is almost impossible for individual teams to provide the needed critical mass for such endeavors. To facilitate interaction and progress on sensor-based intelligent robotics inter-disciplinary workshops are necessary through which in-depth discussion can be used for cross dissemination between different disciplines.

The Dagstuhl foundation has organized a number of workshops on Modeling and Integration of Sensor Based Intelligent Robot Systems. The Dagstuhl seminars take place over a full week in a beautiful setting in the Saarland in Germany. The setting provides an ideal environment for in-depth presentations and rich interaction between the participants.

This volume contains papers presented during the fourth workshop held October 15–20, 2000. All papers were submitted by workshop attendees, and were reviewed by at least one reviewer. We wish to thank all of the reviewers for their invaluable help in making this a high-quality selection of papers. We gratefully acknowledge the support of the Schloss Dagstuhl Foundation and the staff at Springer-Verlag. Without their support the production of this volume would not have been possible.

February 2002

G.D. Hager
H.I. Christensen
H. Bunke
R. Klein

Table of Contents

Sensing

Robotics

Intelligence

Generic Model Abstraction from Examples

Yakov Keselman[1] and Sven Dickinson[2]

[1] Department of Computer Science
Rutgers University
New Brunswick, NJ 08903, USA
[2] Department of Computer Science
University of Toronto
Toronto, Ontario M5S 3G4, Canada

Abstract. The recognition community has long avoided bridging the representational gap between traditional, low-level image features and generic models. Instead, the gap has been eliminated by either bringing the image closer to the models, using simple scenes containing idealized, textureless objects, or by bringing the models closer to the images, using 3-D CAD model templates or 2-D appearance model templates. In this paper, we attempt to bridge the representational gap for the domain of model acquisition. Specifically, we address the problem of automatically acquiring a generic 2-D view-based class model from a set of images, each containing an exemplar object belonging to that class. We introduce a novel graph-theoretical formulation of the problem, and demonstrate the approach on real imagery.

1 Introduction

1.1 Motivation

The goal of generic object recognition is to recognize a novel exemplar from a known set of object classes. For example, given a generic model of a coffee cup, a generic object recognition system should be able to recognize "never before seen" coffee cups whose local appearance and local geometry vary significantly. Under such circumstances, traditional CAD-based recognition approaches (e.g., [25,29,22]) and the recently popular appearance-based recognition approaches (e.g., [44,31,27]) will fail, since they require a priori knowledge of an imaged object's exact geometry and appearance, respectively. Unfortunately, progress in generic object recognition has been slow, as two enormous challenges face the designers of generic object recognition systems: 1) creating a suitable generic model for a class of objects; and 2) recovering from an image a set of features that reflects the coarse structure of the object. The actual matching of a set of salient, coarse image features to a generic model composed of similarly-defined features is a much less challenging problem.

The first challenge, which we will call generic model acquisition, has been traditionally performed manually. Beginning with generalized cylinders (e.g., [4, 1,33,6]), and later through superquadrics (e.g., [34,20,42,24,43,28]) and geons

G.D. Hager et al. (Eds.): Sensor Based Intelligent Robots, LNCS 2238, pp. 1–24, 2002.

(e.g., [3, 12, 14, 13, 2, 37, 47, 18, 10, 5]), 3-D generic model acquisition required the designer to not only identify what features were common to a set of object exemplars belonging to a class, but to construct a model, i.e., class prototype, in terms of those features. The task seems quite intuitive: most cups, for example, have some kind of handle part attached to the side of a larger container-like part, so choose some parameterized part vocabulary that can accommodate the within-class part deformations, and put the pieces together. Although such models are generic (and easily recognizable[1]), such intuitive, high-level representations are extremely difficult (under the best of conditions) to recover from a real image.

The generic object recognition community has long been plagued by this representational gap between features that can be robustly segmented from an image and the features that make up a generic model. Although progress in segmentation, perceptual grouping, and scale-space methods have narrowed this gap somewhat, generic recognition is as elusive now as it was in its prime in the 1970's. Back then, those interested in generic object recognition eliminated the gap by bringing the objects they imaged closer to their models, by removing surface markings and structural detail, controlling lighting conditions, and reducing scene clutter. Since then, the recognition community has eliminated the gap by steadily bringing the models closer to the imaged objects, first resulting in models that were exact 3-D reproductions (CAD-based templates) of the imaged objects, followed by today's 2-D appearance-based templates.

Interestingly enough, both approaches to eliminating this gap are driven by the same limiting assumption: there exists a one-to-one correspondence between a "salient" feature in the image (e.g., a long, high-contrast line or curve, a well-defined homogeneous region, a corner or curvature discontinuity or, in the case of an appearance-based model, the values of a set of image pixels) and a feature in the model. This assumption is fundamentally flawed, for saliency in the image does not equal saliency in the model. Under this assumption, object recognition will continue to be exemplar-based, and generic recognition will continue to be contrived.

Returning to our two challenges, we first seek a (compile-time) method for automatically acquiring a generic model that bridges the representational gap between the output of an image segmentation module and the "parts" of a generic model. Next, from an image of a real exemplar, we seek a (run-time or recognition-time) method that will recover a high-level "abstraction" that contains the coarse features that make up some model. In this paper, we address the first challenge – that of generic model acquisition.

1.2   An Illustrative Example

Assume that we are presented with a collection of images, such that each image contains a single exemplar, all exemplars belong to a single known class, and that

---

[1] Take a look at the object models of Marr and Nishihara [30], Nevatia and Binford [33], Brooks [6], Pentland [34], or Biederman [3], and you will easily recognize the classes represented by these models.

the viewpoint with respect to the exemplar in each image is similar. Fig. 1(a) illustrates a simple example in which three different images, each containing a block in a similar orientation, are presented to the system. Our task is to find the common structure in these images, under the assumption that structure that is common across many exemplars of a known class must be definitive of that class. Fig. 1(b) illustrates the class "abstraction" that is derived from the input examples. In this case, the domain of input examples is rich enough to "intersect out" irrelevant structure (or appearance) of the block. However, had many or all the exemplars had vertical stripes, the approach might be expected to include vertical stripes in that view of the abstracted model.

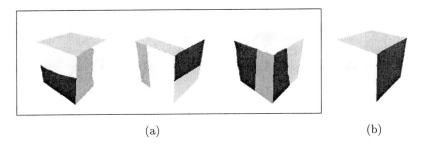

(a)   (b)

Fig. 1. Illustrative Example of Generic Model Acquisition: (a) input exemplars belonging to a single known class; (b) generic model abstracted from examples.

Any discussion of model acquisition must be grounded in image features. In our case, each input image will be region-segmented to yield a region adjacency graph. Similarly, the output of the model acquisition process will yield a region adjacency graph containing the "meta-regions" that define a particular view of the generic model. Other views of the exemplars would similarly yield other views of the generic model. The integration of these views into an optimal partitioning of the viewing sphere, or the recovery of 3-D parts from these views (e.g., see [12, 14, 13]) is beyond the scope of this paper. For now, the result will be a collection of 2-D views that describe a generic 3-D object. This collection would then be added to the view-based object database used at recognition time.

1.3   Related Work

Automatic model acquisition from images has long been associated with object recognition systems. One of the advantages of appearance-based modeling techniques, e.g., [44, 31, 27, 7] is that no segmentation, grouping, or abstraction is necessary to acquire a model. An object is simply placed on a turntable in front of a camera, the viewing sphere is sampled at an appropriate resolution, and the resulting images (or some clever representation thereof) are stored in a database. Others have sought increased illumination-, viewpoint-, or occlusion-invariance by extracting local features as opposed to using raw pixel values, e.g., [36, 38,

32, 45]. Still, the resulting models are very exemplar-specific due to the extreme locality at which they extract and match features (e.g., one pixel or at best, a small neighborhood around one pixel). The resulting models are as far from generic as one can get.

In the domain of range images, greater success has been achieved in extracting coarse models. Generic shape primitives, such as restricted generalized cylinders, quadrics, and superquadrics have few parameters and can be robustly recovered from 3-D range data [35, 42, 24, 21, 43, 11]. Provided the range data can be segmented into parts or surfaces, these generic primitives can be used to approximate structural detail not belonging to the class. Unlike methods operating on 2-D data, these methods are insensitive to perceived structure in the form of surface markings or texture.

In the domain of generating generic models from 2-D data, there has been much less work. The seminal work of Winston [46] pioneered learning descriptions of 3-D objects from structural descriptions of positively or negatively labeled examples. Nodes and edges of graph-like structures were annotated with shapes of constituent parts and their relations. As some shapes and relations were abstractions and specializations of others, the resulting descriptions could be organized into specificity-based hierarchy. In the 2-D shape model domain, Ettinger learned hierarchical structural descriptions from images, based on Brady's curvature primal sketch features [17, 16]. The technique was successfully applied to traffic sign recognition and remains one of the more elegant examples of feature abstraction and generic model acquisition.

1.4   What's Ahead

In the following sections, we begin by presenting a detailed formulation of our problem and conclude that its solution is computationally intractable. Next, we proceed to reformulate our problem by focusing on deriving abstractions from pairs of input images through a top-down procedure that draws on our previous work in generic 2-D shape matching. Given a set of pairwise abstractions, we present a novel method for combining them to form an approximation to the solution of our original formulation. We demonstrate the approach by applying it to subsets of images belonging to a known class, and conclude with a discussion of the method's strengths and weaknesses, along with a glimpse of where we're heading.

2   Problem Formulation

Returning to Fig. 1, let us now formulate our problem more concretely. As we stated, each input image is processed to form a region adjacency graph (we employ the region segmentation algorithm of Felzenzwalb and Huttenlocher [19]). Let us now consider the region adjacency graph corresponding to one input image. We will assume, for now, that our region adjacency graph represents an

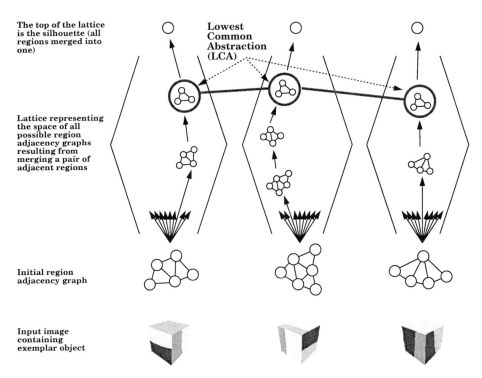

Fig. 2. The Lowest Common Abstraction of a Set of Input Exemplars

oversegmentation of the image (later on, we will discuss the problem of under-segmentation, and how our approach can accommodate it). Under this assumption, the space of all possible region adjacency graphs formed by any sequence of merges of adjacent regions will form a lattice, as shown in Fig. 2. The lattice size is exponential in the number of regions obtained after initial oversegmentation[2].

Each of the input images will yield its own lattice. The bottom node in each lattice will be the original region adjacency graph. In all likelihood, if the exemplars have different shapes (within-class deformations) and/or surface markings, the graphs forming the bottom of their corresponding lattices may bear little or no resemblance to each other. Clearly, similarity between the exemplars cannot be ascertained at this level, for there does not exist a one-to-one correspondence between the "salient" features (i.e., regions) in one graph and the salient features in another. On the other hand, the top of each exemplar's lattice, representing a silhouette of the object (where all regions have been merged into one region), carries little information about the salient surfaces of the object. Following some preliminary definitions, our problem can be stated as follows:

---

[2] Indeed, considering the simplest case of a long rectangular strip subdivided into $n + 1$ adjacent rectangles, the first pair of adjacent regions that can be merged can be selected in n ways, the second in $n - 1$, and so on, giving a lattice size of $n!$. The lattice is even larger for more complex arrangements of regions.

Definitions: Given N input image exemplars, $E_1, E_2, \ldots, E_N$, let $L_1, L_2, \ldots,$ L_N be their corresponding lattices, and for a given lattice, L_i, let $L_i n_j$ be its constituent nodes, each representing a region adjacency graph, g_{ij}. We define a *common abstraction*, or CA, as a set of nodes (one per lattice) $L_1 n_{j_1}, L_2 n_{j_2}, \ldots,$ $L_N n_{j_N}$ such that for any two nodes $L_p n_{j_p}$ and $L_q n_{j_q}$, their corresponding graphs g_{pj_p} and g_{qj_q} are isomorphic. Thus, the root (silhouette) of each lattice is a common abstraction. We define the *lowest common abstraction*, or LCA, as the common abstraction whose underlying graph has maximal size (in terms of number of nodes). Note that there may be more than one LCA.

Problem Definition: For N input image exemplars, find the LCA.

Intuitively, we are searching for a node (region segmentation) that is common to every input exemplar's lattice and that retains the maximum amount of structure. Unfortunately, the presence of a single, heavily undersegmented exemplar (a single-node silhouette in the extreme case) will drive the LCA towards the trivial silhouette CA. In a later section, we will relax our LCA definition to make it less sensitive to region undersegmentation.

3   The Lowest Common Abstraction of Two Examples

3.1   Overview

For the moment, we will focus our attention on finding the LCA of two lattices; in the next section, we will accommodate any number of lattices. Since the input lattices are exponential in the number of regions, actually computing the lattices is intractable. Our approach will be to restrict the search for the LCA to the *intersection* of the lattices, which is much smaller than either lattice, and leads to a tractable search space. But how do we generate this new "intersection" search space without enumerating the lattices?

Our solution is to work top-down, beginning with a node known to be in the intersection set – the root node. If one or both of the roots have no children in the lattice, i.e., the original region segmented image was already a silhouette, then the process stops and the LCA is simply the root. However, in most cases, each root (silhouette) has many possible decompositions, or *specializations*. We will restrict ourselves to the space of specializations of each root into two component regions, and attempt to find a specialization that is common to both lattices. Again, there may be multiple 2-region specializations that are common to both lattices; each is a member of the intersection set.

Assuming that we have some means for ranking the matching *common specializations* (if more than one exists), we pick the best one (the remainder constituting a set of backtracking points), and recursively apply the process to each pair of isomorphic subregions. The process continues in this fashion, "pushing" its way down the intersection lattice, until no further common specializations are found. This lower "fringe" of the search space represents the LCA of the original two lattices. In the following subsections, we will formalize this process.

3.2 The Common Specialization of Two Abstraction Graphs

One of the major components of our algorithm for finding the LCA of two examples is finding the common specialization of two abstraction graphs. In this subsection, we begin by formulating the problem as a search for a pair of corresponding cuts through the two abstraction graphs. Next, we reformulate the problem as the search for a pair of corresponding paths in the dual graph representations of the two abstraction graphs. Finally, high search complexity motivates our transformation of the problem into the search for a shortest path in a product graph of the two dual graphs. In the following subsections, we elaborate on each of these steps.

Problem Definition. Our *specialization* problem can be formulated as follows: Given a pair of isomorphic graphs G_1 and G_2 in L_1 and L_2, find a pair of isomorphic specializations of G_1 and G_2, denoted by $H_1 \in L_1$ and $H_2 \in L_2$, if such a pair exists. Two decompositions (in general, two region adjacency graphs) are isomorphic if their corresponding regions have similar shapes and similar relations. For corresponding regions, it is imperative that we define a similarity metric that accounts for coarse shape similarity. Since the exemplars are all slightly different, so too are the shapes of their abstracted regions. To compute the coarse shape distance between two regions, we draw on our previous work in generic 2-D object recognition [40, 39, 41], in which distance is a weighted function of a region's part structure and part geometry. For relational (or arc) similarity, we must check the relational constraints imposed on pairs of corresponding regions. Such constraints can take the form of relative size, relative orientation, or degree of boundary sharing. We implicitly check the consistency of these pairwise constraints by computing the shape distance (using the same distance function referred to above) between the union of the two regions forming one pair (i.e., the union of a region and its neighbor defined by the arc) and the union of the two regions forming the other. If the constraints are satisfied, the distance will be small.

The Search for Corresponding Graph Cuts. The decomposition of a region into two subregions defines a cut in the original region adjacency subgraph defining the region. Unfortunately, the number of possible 2-region decompositions for a given region may be large, particularly for nodes higher in the lattice. To reduce the computational complexity of finding a pair of corresponding cuts, we will restrict ourselves to cuts that generate regions that are simply connected in the topological sense, i.e., they have no internal "holes". Despite this restriction, the complexity is still prohibitive, and we need to take further measures to simplify our formulation.

The Search for Corresponding Paths in a Dual Graph. To find a pair of corresponding cuts, we first define a *dual graph* to be any graph with the property that a cut in the original graph can be generated by a path in its dual graph[3]. Thus, finding a pair of corresponding cuts in the original graphs reduces to finding a pair of corresponding paths in their dual graphs. Moreover, our restriction of open cuts (preventing holes) will result in a corresponding restriction (preventing cycles) on the paths in the dual graphs. Before we discuss how to generate a dual graph, which we visit in a later section, we will assume that the dual graphs exist and proceed to find a pair of corresponding paths in the dual graphs.

The Product Graph of Two Dual Graphs. Our transformation to the dual graph has not affected the complexity of our problem, as there could be an exponential number of paths in each dual graph, leading to an even larger number of possible pairs of paths (recall our checkerboard example). Rather than enumerating the paths in each dual graph and then enumerating all pairs, we will generate the pairs directly using a heuristic that will generate more promising pairs first. To accomplish this, we define the *product graph* of two dual graphs, such that each path in the product graph corresponds to a pair of paths in the dual graphs. Moreover, with suitably defined edge weights in the product graph, we can ensure that paths resulting in nearly optimal values of an appropriately chosen objective function will correspond to more promising pairs of paths in the dual graphs.

The product graph $G = G_1 \times G_2 = (V, E)$ of graphs $G_1 = (V_1, E_1)$, $G_2 = (V_2, E_2)$ is defined as follows:

$$V = V_1 \times V_2 = \{(v_1, v_2) : v_1 \in V_1, v_2 \in V_2\}$$
$$E = \{((u_1, u_2), (v_1, v_2)) : (u_1, v_1) \in E_1, (u_2, v_2) \in E_2\} \quad \cup$$
$$\{((v_1, u_2), (v_1, v_2)) : v_1 \in V_1, (u_2, v_2) \in E_2\} \quad \cup$$
$$\{((u_1, v_2), (v_1, v_2)) : (u_1, v_1) \in E_1, v_2 \in V_2\}$$

Hence, a simple path $(u_1, v_1) \to (u_2, v_2) \to \cdots \to (u_n, v_n)$ in the product graph corresponds to two sequences of nodes in the initial dual graphs, $u_1 \to u_2 \to \cdots \to u_n$ and $v_1 \to v_2 \to \cdots \to v_n$ which, after the elimination of successive repeated nodes, will result in two simple paths (whose lengths may be different) in the initial dual graphs.

Algorithm for Finding the Common Specialization. Notice that a path that is optimal with respect to an objective function defined in terms of edge weights of the product graph may result in unacceptable partitions. Therefore, we will evaluate several near-optimal paths in terms of similarity of regions resulting from the partitions. Our generic algorithm for finding the common specialization of two abstraction nodes is shown in Algorithm 1. In the following sections, we will elaborate on a number of components of the algorithm, including the choice of a dual graph, edge weights, and the objective function.

---

[3] While our definition agrees in principle with the usual notion of the dual graph for planar graphs [23], we have chosen a more flexible definition which does not prescribe its exact form.

Algorithm 1 A Generic Algorithm for Finding a Common Specialization

---

1: Let A_1, A_2 be subgraphs of the original region adjacency graphs that correspond to isomorphic vertices of the abstraction graphs.
2: Let G_1, G_2 be dual graphs of A_1, A_2.
3: Form the product graph $G = G_1 \times G_2$, as described above.
4: Choose an objective function f, compute edge weights w_i, and select a threshold $\varepsilon > 0$.
5: Let P_f be the optimal path with respect to $(f, \{w_i\})$ with value $F(P_f)$.
6: Let $P = P_f$
7: **while** $|f(P) - f(P_f)| < \varepsilon$ **do**
8:     Let P_1 and P_2 be the paths in G_1, G_2 corresponding to P.
9:     Let (V_1, W_1) and (V_2, W_2) be the resulting cuts in A_1, A_2
10:     **if** region V_1 is similar to region V_2, and region W_1 is similar to region W_2, and arcs (V_1, U_1^i), (V_2, U_2^i) are similar for all isomorphic neighbors U_1^i, U_2^i of V_1, V_2 respectively, and arcs (W_1, U_1^i), (W_2, U_2^i) are similar for all isomorphic neighbors U_1^i, U_2^i of W_1, W_2 respectively **then**
11:         **output** decompositions (V_1, W_1) and (V_2, W_2).
12:         **return**
13:     **end if**
14:     Let P be the next optimal path with respect to $(f, \{w_i\})$.
15: **end while**
16: **output** "no non-trivial specialization is found".

Choosing a Dual Graph. We now turn to the problem of how to define a dual graph of an abstraction graph. In this section, we will present two alternatives. However, before describing these alternatives, we must first establish some important definitions, in conjunction with Fig. 3.

- A region pixel is a *boundary pixel* if its 8-connected neighborhood contains pixels belonging to one or more other regions.
- At a given step, our algorithm for computing the LCA of two examples will focus on a connected subgraph of the input region adjacency graph. A region belonging to this subgraph is called a *foreground region*, while any region not belonging to this subgraph is called a *background region*.
- A boundary pixel of a region is *exterior* if it is adjacent to a background region; it is *interior* otherwise.
- A *boundary segment* of a region is a contiguous set of its boundary pixels, and is *interior* if all its points (except possibly for the endpoints) are such.
- A *junction* is a point whose neighborhood of a fixed radius r includes pixels from at least 3 different regions[4]. If one or more of the regions are background, the junction is *exterior*; otherwise, it is *interior*.

---

[4] The radius r is a parameter of the definition and should be chosen so as to counteract the effects of image noise. In the ideal case, the immediate neighbors of the junction point will belong to 3 different regions, which may not be true if the boundaries of the regions are noisy.

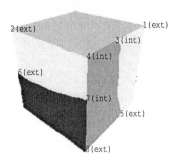

Fig. 3. Illustration of Basic Definitions. All the regions belonging to the cube are foreground. Junctions are labeled as either interior or exterior. The boundary segments between junctions 2 and 4, and 4 and 7 are interior, while those between 1 and 5, and 5 and 8 are exterior.

Our initial choice for a dual graph representation was a junction graph:

- The *junction graph* of a region adjacency graph is a graph whose nodes represent region junctions and whose arcs are those internal boundary segments that connect junctions[5].

Fig. 4 illustrates a junction graph. Since a path between two exterior junctions along internal boundary segments separates regions into two disjoint groups, a path in the junction graph between nodes corresponding to external junctions generates a cut in the region adjacency graph. Thus, the junction graph is dual to the region adjacency graph.

Despite the simplicity and intuitive appeal of the junction graph, it poses some limitations. For example, consider the different types of edges that arise in the product graph. Looking back at its definition, we notice that some edges are of type "edge-edge", corresponding to the $((u_1, u_2), (v_1, v_2))$ terms, while others are of type "node-edge", corresponding to the $((v_1, u_2), (v_1, v_2))$ and $((u_1, v_2), (v_1, v_2))$ terms. Since edge weights in the product graph will be based on node and edge data in the dual graphs, a definition of an edge weight in the junction graph that will result in acceptable paths will require sufficiently precise geometric alignment of the object silhouettes, such that junctions and internal boundary segments are closely aligned[6]. However, such a close alignment is unlikely, with both the shapes and locations of the boundary segments varying across exemplars. As boundary segments possess not only positional but also local shape information, below we define a graph whose edges *and nodes* carry information about internal boundary segments.

If we define a graph on undirected boundary segments, i.e., take boundary segments as nodes and define node adjacency according to the adjacency of

---

[5] There can be multiple arcs between two nodes corresponding to different internal boundary segments connecting the two junctions.

[6] A simpler, watershed-like method, where blurred dark region boundaries are overlaid on top of each other, and darkest curves are found, might also be applicable.

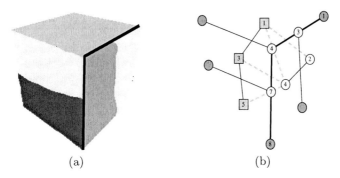

(a)   (b)

Fig. 4. The Dual Graph of a Region Adjacency Graph. (a) Segmented image with a path between external junctions shown in black. (b) Junction graph overlaid on the region adjacency graph, with the corresponding path in the junction graph shown in black. The path edges cut the region adjacency graph (at the dashed edges) into two parts, one whose nodes are square and shaded and one whose nodes are circular and unshaded.

the corresponding boundary segments, then many paths in the dual graph will not result in cuts in the region adjacency graph due to the presence of 3-cycles occurring at junction points. Although these unacceptable paths can be modified to produce cuts in the region adjacency graph, they may still be generated among near-optimal paths, thus decreasing the overall efficiency. Orienting boundary segments, i.e., splitting each undirected boundary segment into two directed copies, eliminates the problem, leading to the following alternative dual graph:

– The *boundary segment graph* of a region adjacency graph has directed internal boundary segments of the regions as its node set (two nodes per undirected internal boundary segment), and an edge from boundary segment b_1 to b_2 if the ending point of b_1 coincides with the starting point of b_2, unless b_1 and b_2 are directed versions of the same undirected boundary segment, in which case they are not adjacent. Nodes of the graph are attributed with the corresponding boundary segments, while an edge is attributed with the union of the adjacent boundary segments corresponding to the nodes it spans.

The directed boundary segment graph possesses the interesting property that for each path starting at a node v_1 and ending at a node v_2, there is a "reverse directed" path starting at the duplicate of v_2 and ending at the duplicate of v_1, which is characteristic of undirected graphs. An example of a boundary segment graph is given in Fig. 5.

Assigning Edge Weights to the Product Graph. Consider now the task of defining edge weights so that optimal paths will result in regions of similar shape. The minimal requirement that shape similarity imposes on the paths is that their shapes are similar and that they connect "corresponding points" on the object's silhouettes. Despite the fact that corresponding contours are

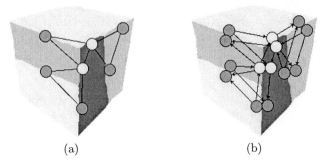

(a)   (b)

Fig. 5. Two Possible Boundary Segment Graphs. Nodes representing boundary segments are placed on or near the segments. A path between any two green nodes in
both graphs corresponds to a cut in the region adjacency graph. Notice that in the
undirected version, (a), cycles have no meaningful interpretation in terms of curves in
the original image. The directed acyclic version of the same graph, (b), in which each
boundary segment is duplicated and directed, eliminates this problem.

unlikely to be closely aligned, their proximity in both shape and position can
be taken into account. We therefore define edge weights as a combination of
metrics reflecting the amount of rigid displacement and the amount of non-rigid
transformation required to align one boundary segment with another. Due to the
special type of the initial boundary segment graphs, whose nodes and edges are
both attributed with boundary segments, it is sufficient to define distances on
pairs of boundary segments from different images. We employ a simple Hausdorff-
like distance between the two boundary segments, yielding a local approximation
to the global similarity of the regions.

Choosing an Objective Function. In our dual graph, smaller edge weights
correspond to pairs of more similar boundary segments. This leads to a number
of very natural choices for an objective function, if we interpret edge weights
as edge lengths. The total path length, $tl(p) = \sum_{p_i \in p} l(p_i)$, is a well-studied
objective function [9]. Fast algorithms for generating successive shortest and
simple shortest paths are given in [15, 26]. However, the above objective function
tends to prefer shorter paths over longer ones, assuming path edges are of equal
average length. For our problem, smaller paths will result in smaller regions being
cut off, which is contrary to our goal of finding the lowest common abstraction[7].

To overcome this problem, we turn to a different objective function that measures the maximum edge weight on a path, $ml(p) = \max_{p_i \in p} l(p_i)$. A well-known
modification[8] of Dijkstra's algorithm [9] finds paths of minimal maximum edge
weight (minmax paths) between a chosen node and all other graph nodes, and

[7] A small region is unlikely to be common to many input exemplars.
[8] Instead of summing up edge weights when determining the distance to a node, it
takes their maximum.

has the same complexity, $O(|E| + |V| \log |V|)$, as the original algorithm. How-
ever, efficient algorithms for finding successive minmax paths are not readily
available. Leaving development of such an algorithm for the future, we will employ a mixed strategy. Namely, we find pairs of nodes providing near-optimal
values of the objective function, and along with the minmax path between the
nodes we also generate several successive shortest paths between them. For this,
we use Eppstein's algorithm [15], which generates k successive shortest paths
between a chosen pair of nodes in $O(|E| + |V| \log |V| + k \log k)$ time. The mixed
strategy, whose overall complexity is $O(|V|(|E| + |V| \log |V|))$ for small k, has
proven to be effective in preliminary empirical testing.

3.3   Algorithm

Now that we have fully specified our algorithm for finding a common specialization of two abstraction graphs, we will embed it in our solution to the problem of
finding the LCA of two examples. Recall that our solution to finding the LCA of
two examples computes the intersection of the respective lattices in a top-down
manner. Beginning with the two root nodes (the sole member of the initialized
intersection set), we recursively seek the "best" common specialization of these
nodes, and add it to the intersection set. The process is recursively applied to
each common specialization (i.e., member of the intersection set) until no further
common specializations are found. The resulting set of "lowest" common specializations represents the LCA of the two lattices. The procedure is formalized
in Algorithm 2.

Algorithm 2 Finding the maximal common abstraction of two region adjacency
graphs.

1: Let A_1, A_2 be the initial region adjacency graphs.
2: Let G_1, G_2 denote abstraction graphs belonging to abstraction lattices, L_1 and L_2
   respectively.
3: Let G_1^0, G_2^0 be the topmost nodes of the lattices.
4: Let $G_1 = G_1^0$, $G_2 = G_2^0$.
5: **while** there are unexplored non-trivial isomorphic nodes $u_1 \in G_1$, $u_2 \in G_2$ **do**
6:    Let U_1 and U_2 be the corresponding subgraphs of A_1, A_2.
7:    **if** there is a *common specialization* $U_1 = V_1 \cup W_1$ and $U_2 = V_2 \cup W_2$ **then**
8:       Split the nodes $u_1 \in G_1$, $u_2 \in G_2$ by forming the *specialization graphs* $H_1 =$
         $(G_1 - \{u_1\}) \cup \{v_1, w_1\}$, $H_2 = (G_2 - \{u_2\}) \cup \{v_2, w_2\}$ with edges established
         using A_1, A_2.
9:       Let $G_1 = H_1$, $G_2 = H_2$, and **goto** 5.
10:   **end if**
11: **end while**
12: **output** G_1, G_2.

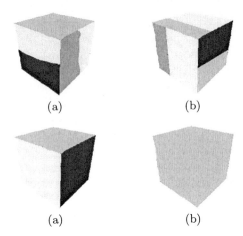

(a)   (b)

(a)   (b)

Fig. 6. The straightforward computation of the Lowest Common Abstraction of exemplars (a)-(d) gives the exemplar in (d). However, (c) is the Lowest Common Abstraction of exemplars (a)-(c), and therefore is more representative.

4   The LCA of Multiple Examples

So far, we've addressed only the problem of finding the LCA of two examples. How then can we extend our approach to find the LCA of multiple examples? Furthermore, when moving towards multiple examples, how do we prevent a "noisy" example, such as a single, heavily undersegmented silhouette from derailing the search for a meaningful LCA? To illustrate this effect, consider the inputs (a)-(d) shown in Fig. 6. If the definition of the pairwise LCA is directly generalized, thus requiring the search for an element common to all abstraction lattices, the correct answer will be the input (d). However, much useful structure is apparent in inputs (a)-(c); input (d) can be considered to be an outlier.

To extend our two-exemplar LCA solution to a robust (to outliers), multi-exemplar solution, we begin with two important observations. First, the LCA of two exemplars lies in the intersection of their abstraction lattices. Thus, both exemplar region adjacency graphs can be transformed into their LCA by means of sequences of region merges. Second, the total number of merges required to transform the graphs into their LCA is minimal among all elements of the intersection lattice. Our solution begins by relaxing the first property. We will define the LCA of a set of region adjacency graphs to be that element in the intersection of two or more abstraction lattices that minimizes the total number of edit operations (merges or splits) required to obtain the element from *all* the given exemplars. As finding the desired abstraction according to this definition would still involve the construction of many abstraction lattices, whose complexity is intractable, we will pursue an approximation method.

Consider the closure of the set of the original region adjacency graphs under the operation of taking pairwise LCA's. In other words, starting with the initial region adjacency graphs, we find their pairwise LCA's, then find pairwise LCA's

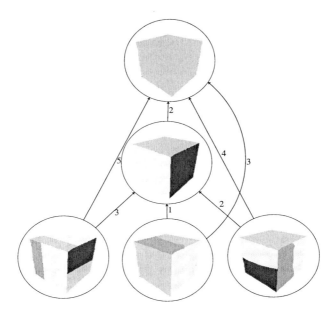

Fig. 7. Embedding Region Adjacency Graphs and their Pairwise LCA's in a Weighted Directed Acyclic Graph. The center node is the median, as its distance sum value is $3 + 1 + 2 + 2 = 8$, while the sum is $5 + 3 + 4 + 0 = 12$ for the topmost node.

of the resulting abstraction graphs, and so on (note that duplicate graphs are removed). We take all graphs, original and LCA, to be nodes of a new *closure* graph. If graph H was obtained as the LCA of graphs G_1 and G_2, then directed arcs go from nodes corresponding to G_1, G_2 to the node corresponding to H in the closure graph.

Although a graph may not be directly linked to all of its abstractions, if H is an abstraction of G, there is a directed path between the nodes corresponding to G and H. Thus, any abstraction is reachable from any of its specializations by a directed path. Each edge in the closure graph is assigned a weight equal to the merge edit distance that takes the specialization to the abstraction. The edit distance is simply the difference between the numbers of nodes in the specialization graph and the abstraction graph. As a result, we obtain a weighted directed acyclic graph. An example of such a graph, whose edges are shown directed from region adjacency graphs to their LCA's, is given in Fig. 7.

Given such a graph, the robust LCA of *all* inputs will be that node that minimizes the sum of shortest path distances from the initial input region adjacency graphs. In other words, we are looking for the "median" of the graph. Note that the resulting solution is bound to lie in the constructed graph, and therefore may be only an approximation to the true answer. To find a possibly better solution, one must consider a supergraph of the closure graph. Algorithm 3 computes the LCA for a set of input examples.

To prove correctness of the algorithm, we must prove that the distance sum for the output node is minimal. Adopting the convention that edges are directed

---

Algorithm 3 Finding the median of the closure graph

1: Let the *sink node*, s, be the topmost node in the closure graph.
2: Solve the "many-to-one" directed shortest path problem on the graph with the
   source nodes being the original adjacency graphs and with the specified edge
   weights. Find the distance sum, $DS(s)$, for the sink node.
3: Similarly, find distance sums, $DS(s_i)$, for all unexplored $s_i \in N(s)$.
4: **if** $min_i(DS(s_i)) \geq DS(s)$ **then**
5:     **return** s
6: **else**
7:     Let $s = \arg\min_i DS(s_i)$.
8:     **goto** 2.
9: **end if**

---

towards the current sink node, we denote $R_1, \ldots, R_k$ to be those nodes whose
outgoing edges point directly to the sink, and denote $L_1, \ldots, L_k$ to be subgraphs
with nodes $R_1, \ldots, R_k$ as their roots (L_i will consist of all nodes having R_i as
their abstraction). It suffices to prove that if $DS(\text{sink}) > DS(R_i)$ and $DS(R_i) \leq$
$DS(R_j)$ for all j, then a solution lies in L_i.

Consider the original problem restricted to L_i's and call the restricted objective function of the i-th problem DS_i. The values of the objective functions
approximately satisfy the following relations:

$$DS(R_i) = \sum_{j=1}^{k} DS_j(R_j) + 2(k-1) - F(i) \qquad (1)$$

Here $2(k-1)$ comes from the fact that to get from an element of L_j, that is
not in L_i, to R_i, we must necessarily pass through the two edges connecting the
sink node to L_i, L_j. $F(i)$ is the sum of shortest path distances from the original
nodes to R_i taken over the nodes that belong to both L_i and L_j (for some
$i \neq j$). Under the assumption $DS(R_i) \leq DS(R_j)$, we have that $F(i) \geq F(j)$,
which means that L_i wholly contains at least as many shortest paths as any
other L_j. This, in turn, implies that a node minimizing the original objective
function will lie in L_i.

To analyze the complexity of the algorithm, notice that the first step, i.e.,
finding the distance sum to the topmost node, can be performed in linear time
in the graph size, since the closure graph is a directed acyclic graph, and the
single source shortest path problem in such graphs can be solved in $O(|V|+|E|)$
time [9]. Since the algorithm can potentially examine a constant fraction of the
graph nodes (consider the case of a line graph), the total running time can be
as high as $O(|V|(|V|+|E|))$. The complexity can be somewhat reduced by using
the relations (1). The average case complexity will depend on the particular
distribution of the initial data and is beyond the scope of this paper. In practice,
the algorithm stops after a few iterations.

The possibility that the size of the generated graph is exponential in the size
of the initial region adjacency graphs cannot be ruled out. This could happen,
for example, when the images are segmented too finely, and different pairs of

region adjacency graphs are abstracted to similar but unequal graphs. We hope to address this issue in the future. Alternatives include resegmenting the images when the size of the generated closure graph exceeds a threshold value, and subsampling the graph in a randomized fashion.

5   Experiments

In this section, we begin by showing two results of our approach applied to synthetic image data, and follow that with two results using real image data. As mentioned before, images were region segmented using the approach of Felzenzwalb and Huttenlocher [19]. For the experimental results shown below, considerable parameter tuning in the region segmentation algorithm was performed to avoid region oversegmentation.

5.1   Synthetic Imagery

The first input exemplar set to our program consisted of four synthetic images of books, whose segmentations are shown as rectangular nodes with black, bold outline in the closure graph, shown in Fig. 8. The other two nodes are the LCA's of each pair of nodes just below them. Edges of the graph are labeled with the merge edit distance between the nodes, while nodes are labeled with the distance sum value. Although the three upper nodes are labeled with the same minimal value, 9, the upper node is optimal according to the algorithm. Our computed LCA agrees with our intuition given the input set.

The next input set to our program consists of four images of bottles, whose segmented versions are shown as rectangular nodes with black, bold outline in the closure graph, shown in Fig. 9. The fifth, upper node is the LCA of the two nodes just below it. Again, edges of the graph are labeled with the merge edit distance between the nodes, while nodes are labeled with the distance sum value. The center node, with rounded corners and boldest outline, is optimal according to the algorithm.

On the positive side, the computed LCA preserved all features present in the majority of the inputs. For example, the four rectangular stripes, into which the region segmentation algorithm segmented the label of the bottle, are preserved, as is the region corresponding to the cork. However, on the negative side, the subdivision of the label into four stripes is undesirable, as it does not correspond to a partition of the object into meaningful structure based on coarse shape. This is more a limitation of the original exemplar set. As was pointed out earlier, our algorithm finds the finest-level common structure present in the input region adjacency graphs, which may not correspond to the desired shape-based structure. If additional examples of bottles were available, in which the label was not segmented into several pieces or was segmented very differently, a more appealing answer would likely have resulted.

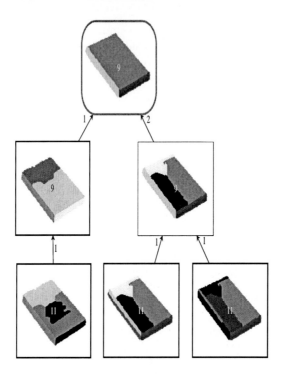

Fig. 8. The LCA of Multiple Books. The upper node is optimal. See text for details.

5.2 Real Imagery

We now turn to two examples of our approach applied to real imagery. In the figures below, each of the two rows corresponds to one entry, the first column containing the original intensity image, the second column containing its region segmentation, and the third column containing the LCA of the two region segmentations (although each lattice's LCA is shown for completeness, either one could be chosen for inclusion in a database).

In the first example, we compute the LCA of two input image exemplars, as shown in Fig. 10. The computed LCA captures the coarse structure with one exception. The ideal LCA would assign the handle two regions rather than one. The reason the "hole" was left out of the handle is that the shape matching module found that the "handle silhouette" regions (i.e., the union of the handles with their holes) were sufficiently similar in shape, while the two-region (handle,hole) representation was found to exceed the shape dissimilarity threshold. This is a deficiency of the current implementation of the shape similarity module, and it will be eliminated in the near future.

Our second example, again computing the LCA of two exemplars, is shown in Fig. 11. In this case, the results deviate more significantly from our expectation. For example, the handle of the second cup was not segmented from the body of the cup. This illustrates the need for a region splitting operation, which we

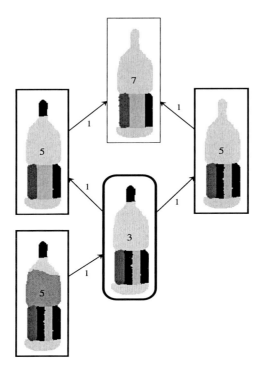

Fig. 9. The LCA of Multiple Bottles. The node in the center, with rounded corners, is optimal. See text for details.

discuss in the next section. Notice, however, that even if the second cup was better segmented, the segmentation of the first cup would have prevented the algorithm from obtaining the correct abstraction, since the handle of the first cup is directly attached to the top portion, while in the second cup, it is attached to the body portion. In this case, a split would be required on the first cup. The other undesirable feature computed in the LCA is the stripe at the bottom of the LCA. As has been already mentioned, this stripe would disappear if there were other examples, most of whom lacked the stripe.

6   Limitations

There are a number of limitations of our proposed approach that deserve mentioning. As we discussed earlier in the paper, we have assumed that a lattice can be formed from an input region adjacency graph through a sequence of adjacent region merge operations. However, this assumes that the input region adjacency graph is *oversegmented*. This is a limiting assumption, for lighting effects and segmentation errors can lead to region *undersegmentation*[9]. Granted, our algo-

---

[9] Frequent undersegmentation of our input data was the main reason we did not use the segmentation algorithm of Comaniciu and Meer [8]

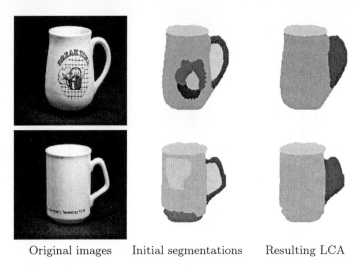

Original images   Initial segmentations   Resulting LCA

Fig. 10. The LCA of Two Cups.

Original images   Initial segmentations   Resulting LCA

Fig. 11. The LCA of Two Cups.

rithm for computing the LCA of multiple examples, through its search for the median of the closure graph, can avoid the influence of an undersegmented example, such as a silhouette. Nevertheless, a more direct approach to recovering from region undersegmentation would be appropriate. Such an approach might decompose selected regions in each input exemplar into a set of subregions which could remain intact, merge with each other, or merge with other neighbors. Although discretion must be exercised in selecting which regions should be split (undue oversegmentation would result in a much more complex and possibly ineffective input graph), we are fortunate in that the shock graph representation of regions specifies a finite number of splits (branches in the shock graph, which can be severed).

A second limitation involves the method by which a common specialization of two graphs is found. Recall that this consists of finding corresponding paths through two graphs that yield matching subregions. In order to reduce the complexity of examining a possibly exponential number of corresponding paths (not to mention application of the shape similarity function), we reformulated the problem as a search for the shortest path in a product graph. The limitation arises from the fact that in order to make the search tractable, our edge weights capture local information. Thus, although the algorithm may find a global minimum path, such a path may not represent the best cut. We are investigating the incorporation of additional information into the edge weights, including more global shape information.

Finally, more experimentation is needed to better understand the performance of the framework as a function of the number and nature of the input exemplars. For example, although the algorithm for finding the LCA of multiple examples (i.e., the median of the closure graph) can theoretically handle any number of input exemplars, our experiments to date have included only a few input exemplars. Experiments using larger numbers of input exemplars are necessary to establish the performance of the algorithm. In addition, although our experiments to date have included some undersegmentation and oversegmentation, we have not evaluated the performance of the entire framework as a function of degree of segmentation error. A more thorough set of experiments, parameterized in terms of number of exemplars and degree of segmentation error, is essential before the approach can be fully evaluated.

7   Conclusions

The quest for generic object recognition hinges on an ability to generate abstract, high-level descriptions of input data. This process is essential not only at runtime, for the recognition of objects, but at compile time, for the automatic acquisition of generic object models. In this paper, we address the latter problem – that of generic model acquisition from examples. We have introduced a novel formulation of the problem, in which the model is defined as the lowest common abstraction of a number of segmentation lattices, representing a set of input image exemplars. To manage the intractable complexity of this formulation, we focus our search on the intersection of the lattices, reducing complexity by first considering pairs of lattices, and later combining these local results to yield an approximation to the global solution. We have shown some very preliminary results that compute a generic model from a set of example images belonging to a known class. Although these results are encouraging, further experimentation is necessary and a number of limitations need to be addressed.

Our next major step is the actual recognition of the derived models from a novel exemplar. Our efforts are currently focused on the analysis of the conditions under which two regions are merged. If we can derive a set of rules for the perceptual grouping of regions, we will be able to generate abstractions from images. Given a rich set of training data derived from the model acquisition process (recall that the LCA of two examples yields a path of region merges), we are applying machine learning methods to uncover these conditions. Combined with our model acquisition procedure, we can close the loop on a system for generic object recognition which addresses a representational gap that has been long ignored in computer vision.

Acknowledgements

The authors gratefully acknowledge the support of the National Science Foundation, the Army Research Office, and the National Sciences and Engineering Research Council of Canada.

References

1. G. Agin and T. Binford. Computer description of curved objects. *IEEE Transactions on Computers*, C-25(4):439–449, 1976.
2. R. Bergevin and M. D. Levine. Part decomposition of objects from single view line drawings. *CVGIP: Image Understanding*, 55(1):73–83, January 1992.
3. I. Biederman. Human image understanding: Recent research and a theory. *Computer Vision, Graphics, and Image Processing*, 32:29–73, 1985.
4. T. Binford. Visual perception by computer. In *Proceedings, IEEE Conference on Systems and Control*, Miami, FL, 1971.
5. D. Borges and R. Fisher. Class-based recognition of 3d objects represented by volumetric primitives. *Image and Vision Computing*, 15(8):655–664, 1997.
6. R. Brooks. Model-based 3-D interpretations of 2-D images. *IEEE Transactions on Pattern Analysis and Machine Intelligence*, 5(2):140–150, 1983.
7. O. Camps, C. Huang, and T Kanungo. Hierarchical organization of appearance based parts and relations for object recognition. In *IEEE Conference on Computer Vision and Pattern Recognition*, pages 685–691, Santa Barbara, CA, 1998.
8. D. Comaniciu and P. Meer. Robust analysis of feature spaces: Color image segmentation. In *IEEE Computer Society Conference on Computer Vision and Pattern Recognition*, pages 750–755, 1997.
9. T. H. Cormen, C. E. Leiserson, and R. L. Rivest. *Introduction to Algorithms*, chapter 25. The MIT Press, 1993.
10. S. Dickinson, R. Bergevin, I. Biederman, J.-O. Eklundh, A. Jain, R. Munck-Fairwood, and A. Pentland. Panel report: The potential of geons for generic 3-D object recognition. *Image and Vision Computing*, 15(4):277–292, April 1997.
11. S. Dickinson, D. Metaxas, and A. Pentland. The role of model-based segmentation in the recovery of volumetric parts from range data. *IEEE Transactions on Pattern Analysis and Machine Intelligence*, 19(3):259–267, March 1997.
12. S. Dickinson, A. Pentland, and A. Rosenfeld. A representation for qualitative 3-D object recognition integrating object-centered and viewer-centered models. In K. Leibovic, editor, *Vision: A Convergence of Disciplines*. Springer Verlag, New York, 1990.

13. S. Dickinson, A. Pentland, and A. Rosenfeld. From volumes to views: An approach to 3-D object recognition. *CVGIP: Image Understanding*, 55(2):130–154, 1992.
14. S. Dickinson, A. Pentland, and A. Rosenfeld. 3-D shape recovery using distributed aspect matching. *IEEE Transactions on Pattern Analysis and Machine Intelligence*, 14(2):174–198, 1992.
15. David Eppstein. Finding the k shortest paths. *SIAM J. Computing*, 28(2):652–673, 1999.
16. G. Ettinger. Large hierarchical object recognition using libraries of parameterized model sub-parts. In *Proceedings, IEEE Conference on Computer Vision and Pattern Recognition*, pages 32–41, Ann Arbor, MI, 1988.
17. G.J. Ettinger. Hierarchical object recognition using libraries of parameterized model sub-parts. Technical Report 963, MIT Artificial Intelligence Laboratory, 1987.
18. R. Fairwood. Recognition of generic components using logic-program relations of image contours. *Image and Vision Computing*, 9(2):113–122, 1991.
19. P. Felzenszwalb and D. Huttenlocher. Image segmentation using local variation. In *IEEE Conference on Computer Vision and Pattern Recognition*, pages 98–104, Santa Barbara, CA, 1998.
20. F. Ferrie, J. Lagarde, and P. Whaite. Darboux frames, snakes, and superquadrics. In *Proceedings, IEEE Workshop on Interpretation of 3D Scenes*, pages 170–176, 1989.
21. F. Ferrie, J. Lagarde, and P. Whaite. Darboux frames, snakes, and super-quadrics: Geometry from the bottom up. *IEEE Trans. Pattern Analysis and Machine Intelligence*, 15(8):771–784, 1993.
22. D. Forsyth, J.L. Mundy, A. Zisserman, C. Coelho, A. Heller, and C. Rothwell. Invariant descriptors for 3d object recognition and pose. *IEEE PAMI*, 13:971–992, October 1991.
23. R. Gould. *Graph Theory*, pages 170–172. The Benjamin/Cummings Publishing Company, Inc., 1988.
24. A. Gupta. Surface and volumetric segmentation of 3D objects using parametric shape models. Technical Report MS-CIS-91-45, GRASP LAB 128, University of Pennsylvania, Philadelphia, PA, 1991.
25. D. Huttenlocher and S. Ullman. Recognizing solid objects by alignment with an image. *International Journal of Computer Vision*, 5(2):195–212, 1990.
26. N Katoh, T. Ibaraki, and H. Mine. An efficient algorithm for k shortest simple paths. *Networks*, 12:411–427, 1982.
27. A. Leonardis and H. Bischoff. Dealing with occlusions in the eigenspace approach. In *Proceedings, IEEE Conference on Computer Vision and Pattern Recognition*, pages 453–458, San Francisco, CA, June 1996.
28. A. Leonardis, F. Solina, and A. Macerl. A direct recovery of superquadric models in range images using recover-and-select paradigm. In *Proceedings, Third European Conference on Computer Vision (Lecture Notes in Computer Science, Vol 800)*, pages 309–318, Stockholm, Sweden, May 1994. Springer-Verlag.
29. D. Lowe. *Perceptual Organization and Visual Recognition*. Kluwer Academic Publishers, Norwell, MA, 1985.
30. D. Marr and H. Nishihara. Representation and recognition of the spatial organization of three-dimensional shapes. *Royal Society of London*, B 200:269–294, 1978.
31. H. Murase and S. Nayar. Visual learning and recognition of 3-D objects from appearance. *International Journal of Computer Vision*, 14:5–24, 1995.

32. R. Nelson and A. Selinger. A cubist approach to object recognition. In *Proceedings, IEEE International Conference on Computer Vision*, Bombay, January 1998.

33. R. Nevatia and T. Binford. Description and recognition of curved objects. *Artificial Intelligence*, 8:77–98, 1977.

34. A. Pentland. Perceptual organization and the representation of natural form. *Artificial Intelligence*, 28:293–331, 1986.

35. A. Pentland. Automatic extraction of deformable part models. *International Journal of Computer Vision*, 4:107–126, 1990.

36. A. Pope and D. Lowe. Learning object recognition models from images. In *Proceedings, IEEE International Conference on Computer Vision*, pages 296–301, Berlin, May 1993.

37. N. Raja and A. Jain. Recognizing geons from superquadrics fitted to range data. *Image and Vision Computing*, 10(3):179–190, April 1992.

38. C. Schmid and R. Mohr. Combining greyvalue invariants with local constraints for object recognition. In *Proceedings, IEEE Conference on Computer Vision and Pattern Recognition*, pages 872–877, San Francisco, CA, June 1996.

39. A. Shokoufandeh and S. Dickinson. Applications of bipartite matching to problems in object recognition. In *Proceedings, ICCV Workshop on Graph Algorithms and Computer Vision (web proceedings: http://www.cs.cornell.edu/iccv-graph-workshop/papers.htm)*, September 1999.

40. A. Shokoufandeh, S. Dickinson, K. Siddiqi, and S. Zucker. Indexing using a spectral encoding of topological structure. In *IEEE Conference on Computer Vision and Pattern Recognition*, pages 491–497, Fort Collins, CO, June 1999.

41. K. Siddiqi, A. Shokoufandeh, S. Dickinson, and S. Zucker. Shock graphs and shape matching. *International Journal of Computer Vision*, 30:1–24, 1999.

42. F. Solina and R. Bajcsy. Recovery of parametric models from range images: The case for superquadrics with global deformations. *IEEE Transactions on Pattern Analysis and Machine Intelligence*, 12(2):131–146, 1990.

43. D. Terzopoulos and D. Metaxas. Dynamic 3D models with local and global deformations: Deformable superquadrics. *IEEE Transactions on Pattern Analysis and Machine Intelligence*, 13(7):703–714, 1991.

44. M. Turk and A. Pentland. Eigenfaces for recognition. *Journal of Cognitive Neuroscience*, 3(1):71–86, 1991.

45. Markus Weber, M. Welling, and Pietro Perona. Unsupervised learning of models for recognition. In *European Conference on Computer Vision*, volume 1, pages 18–32, 2000.

46. P.H. Winston. Learning structural descriptions from examples. In *The Psychology of Computer Vision*, chapter 5, pages 157–209. McGraw-Hill, 1975.

47. K. Wu and M. Levine. Recovering parametric geons from multiview range data. In *Proceedings, IEEE Conference on Computer Vision and Pattern Recognition*, pages 159–166, Seattle, WA, June 1994.

Tracking Multiple Moving Objects
in Populated, Public Environments

Boris Kluge

Research Institute for Applied Knowledge Processing,
Helmholtzstr. 16, 89081 Ulm, Germany
kluge@faw.uni-ulm.de

Abstract. The ability to cope with rapidly changing, dynamic environments is an important requirement for autonomous robots to become valuable assistants in everyday life. In this paper we focus on the task of tracking moving objects in such environments. Objects are extracted from laser range finder images and correspondence between successive scan images is established using network flow algorithms. This approach is implemented on a robotic wheelchair and allows it to follow a guiding person and to perform some simplistic reasoning about deliberate obstructions of it, these two applications requiring robust tracking of humans in the robot's vicinity.

1 Introduction

Today most existing robot systems are not designed to cope with rapidly changing, dynamic environments. Usually they are programmed once and then perform the same motion many thousand times. If there is an unforeseen situation, like a human entering the working cell of an assembly robot or people crossing a mobile robot's desired path, at best the robot stops and possibly tries to evade this obstacle in order not to hurt anybody. On the other hand, such a robot might be easily obstructed by a person playing jokes on it. So this lack of awareness is a serious obstacle to a widespread use of such systems, and a basic requirement to this awareness is continuous observation of objects surrounding the robot.

1.1 Related Work

Tracking human motion with computer vision is an active field of research [7], most approaches using video image sequences. Range images are used, e.g., in intelligent vehicles or driver assistance systems [6, 14]. An object tracking system using a laser range finder has previously been implemented at our institute [10].

There is a correspondence problem similar to ours in stereo vision, where features from two camera images are to be matched. As an alternative to the widely used dynamic programming approach Roy and Cox [13] use maximum-flow computations in graphs to solve this problem, yielding considerably better results, but unfortunately at increased computational costs.

G.D. Hager et al. (Eds.): Sensor Based Intelligent Robots, LNCS 2238, pp. 25–38, 2002.
© Springer-Verlag Berlin Heidelberg 2002

Matching multiple moving point objects by computing minimum cost matchings in bipartite graphs is mentioned in [1] as an application example. Our network optimization approach is inspired by this application example and the relationship between matchings and flows in bipartite graphs [1, 4].

1.2   Overview

Tracking of objects around the robot is performed by repeated execution of the following steps. At the beginning of each cycle a scan image of the environment is taken. This scan is segmented (Sect. 2.1) and further split into point sets representing objects (Sect. 2.2). As our matching approach is based on graph theory some notions are reviewed (Sect. 3.1) before the algorithms themselves are presented (Sect. 3.2). As applications we present detection of deliberate obstructions by humans (Sect. 4.1) and accompanying a guiding person (Sect. 4.2). Results of our experiments are shown (Sect. 5) and possible further work is pointed out (Sect. 6) before concluding the paper (Sect. 7).

2   Finding Objects

A laser scan image is given by a sequence $S = (s_1, \ldots, s_L)$ of samples $s_i \in \mathbb{R}^{\geq 0}$ representing the distances from the range finder to the closest opaque obstacles in the horizontal half plane in front of the robot. Let θ_i be the direction in which distance sample s_i is measured, and without loss of generality $\theta_1 = 0$, $\theta_i \leq \theta_j$ if $i \leq j$, and $\theta_L < 2\pi$. Thus we may associate each distance sample s_i to the point $p_i = (s_i \cos \theta_i, s_i \sin \theta_i) \in \mathbb{R}^2$.

Objects are extracted from laser scan images by two heuristic steps. At first the scan is segmented into densely sampled parts. In the second step these parts are split into subsequences describing "almost convex" objects.

2.1   Scan Segmentation

Consider two adjacent distance samples s_i and s_{i+1}. Intuitively, if they are taken from the same obstacle, their values will be similar. On the other hand, if one of these samples is taken from a different obstacle, which is located in front of or behind the other sample's obstacle, we will encounter a significant difference $|s_i - s_{i+1}|$ in those two distances. Thus it is plausible to use such distance gaps as split positions for the sequence of samples in order to find distinct objects. A threshold value δ_{max} is chosen in advance for the maximal allowed distance gap. The result of the scan segmentation is a finite sequence $((s_1, \ldots, s_{i_1-1}), (s_{i_1}, \ldots, s_{i_2-1}), \ldots, (s_{i_p}, \ldots, s_L))$ of subsequences of S where indices $i_1, i_2, \ldots, i_p$ denote split positions such that $|s_k - s_{k-1}| > \delta_{max}$ iff $k \in \{i_1, i_2, \ldots, i_p\}$.

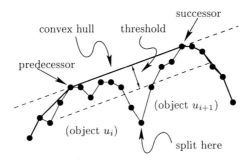

Fig. 1. Using the convex hull to find split points

2.2   Object Extraction

The subsequences of sample points yielded by the preceding step are divided further, as we would otherwise encounter problems with objects situated close to other objects, for example humans leaning against or walking close to a wall or other humans. We assume that most objects of interest in the robot's environment are either almost convex or can be decomposed into almost convex subobjects in a reasonable way.

Thus our approach is to compute the visible part of the convex hull for each of the given subsequences (see Fig. 1). For each sample point on the interior side of the convex hull its distance to the line defined by the next preceding and succeeding sample points on the hull is computed. If there is a point whose distance exceeds a threshold value (i.e. the shape is not even "almost convex"), the sequence is divided at a point with a maximum distance, and this split procedure is applied again in a recursive manner. Note that the visible part of the convex hull can be computed efficiently in linear time, since we have an angular ordering of the points around the range finder. The algorithm is an adaptation of Graham's scan [12].

The result of the object extraction step is a set $U = \{u_1, \ldots, u_n\}$ of objects and associated indices $begin(u_i)$ and $end(u_i)$ such that object u_i refers to the scan points $\{p_{begin(u_i)}, \ldots, p_{end(u_i)}\}$.

3   Object Correspondence

As we intend to track objects in a dynamic environment, we have to compare information about objects from successive scans. This is done by a combination of graph algorithms. Before presenting our approach itself, some notions in graph theory are briefly reviewed below.

3.1   Notions in Graph Theory

A *(directed) graph* is an ordered pair (V, E) with *vertices* $V \neq \emptyset$ and *edges* $E \subseteq V \times V$. An edge $e = (u, v)$ is said to have *source node* u, *target node* v, and to be *incident* to u and v. The *degree* of a vertex v is the number of edges incident to it.

Let $G = (V, E)$ a graph, $V' \subseteq V$. The *subgraph of G induced by V'* is the graph $G' = (V', E \cap (V' \times V'))$.

A graph $G = (V, E)$ is *bipartite* if there is a partitioning of the nodes $V = S \cup T$, $S \cap T = \emptyset$, such that $E \cap (S \times S) = \emptyset$ and $E \cap (T \times T) = \emptyset$.

A *matching* M in a graph $G = (V, E)$ is a subset of its edges $M \subseteq E$, such that every node $u \in V$ is incident to at most one edge $e \in M$ of the matching. A matching M is *perfect* if for each node $v \in V$ there is an edge $e \in M$ such that e is incident to v.

Let $pred(v) = \{u \in V \mid (u, v) \in E\}$ the set of *predecessors* of v and $succ(v) = \{w \in V \mid (v, w) \in E\}$ the set of *successors* of v.

Let $e : V \to \mathbb{R}$ a node label, the *excess* of a node. Let $u : E \to \mathbb{R}^{\geq 0}$ and $c : E \to \mathbb{R}^{\geq 0}$ edge labels, the *capacity bound* and the *cost* of an edge. Now we can ask for an edge label $f : E \to \mathbb{R}^{\geq 0}$, a *flow*, which complies with the *mass balance condition*

$$\sum_{u \in pred(v)} f((u, v)) + e(v) = \sum_{w \in succ(v)} f((v, w)) \tag{1}$$

for each node $v \in V$ and the capacity bound

$$f(e) \leq u(e) \tag{2}$$

for each edge $e \in E$. Such a label f is called a *feasible flow* in G. Often only a special case is considered, where there is exactly one node $s \in V$ with $e(s) > 0$, the *source node*, and exactly one node $t \in V$ with $e(t) < 0$, the *sink node*. A well known problem is the question for a *maximum flow* in this special case, that is, how large may the amount of excess $e(s) = -e(t)$ be such that there is still a feasible flow.

If there is any feasible flow in a given labeled graph, we can search for a feasible flow f^* with minimal cost

$$\sum_{e \in E} f^*(e) \cdot c(e) = \min_{f \text{ is a feasible flow}} \left(\sum_{e \in E} f(e) \cdot c(e) \right) \tag{3}$$

among all feasible flows f.

There are several algorithms which solve these problems efficiently. Implementations are available for example via the LEDA library [5]. For more details on graph theory and network optimization see [1, 4, 8].

3.2   Finding Object Matchings

From the previous and the current scan two sets of objects $U = \{u_1, \ldots, u_n\}$ and $V = \{v_1, \ldots, v_m\}$ are given. The goal is to find for objects $u_i \in U$ from the previous scan corresponding objects $v_j \in V$ from the current scan. This can be seen as a matching in the bipartite graph $(U \cup V, U \times V)$.

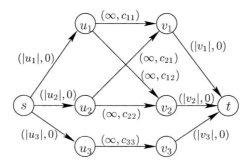

Fig. 2. Example graph for object matching

To find a matching representing plausible assignments between objects at successive points of time in the real world, we start by computing a maximum flow with minimal cost in a graph $G = (\{s, t\} \cup U \cup V, E)$ as illustrated by Fig. 2 and Alg. 1. Each edge is labeled by its capacity and its costs, where $|u|$ denotes the shape length of object u, and c_{ij} denotes the "dissimilarity" of objects u_i and v_j as described below. We have

$$E = \{ (s, u) \mid u \in U \} \cup$$
$$\{ (v, t) \mid v \in V \} \cup \qquad (4)$$
$$\{ (u, v) \mid P(u, v) \}$$

for some predicate $P : U \times V \to \{true, false\}$ of reasonably low computational complexity. We could use the constant predicate $true$ here (i.e. accept all edges), but for practical reasons P is chosen such that $P(u, v)$ is true if the distance between the centers of gravity of u and v does not exceed a threshold value. This is equivalent to the assumption of a speed limit for objects in our environment. The finite scanning frequency does not allow tracking of arbitrary fast objects, anyway. Thus the size of the graph is reduced without imposing a further restriction, resulting in faster computations of minimum cost maximum flows.

Finally an object matching is deduced by retaining only edges conveying large amounts of this minimum cost maximum flow, i.e. we compute a maximum weight matching. Details on capacity, cost, and weight labels as well as on computational complexities are given below.

Capacity Labels. For each object $u_i \in U$ we compute the length

$$|u_i| = \sum_{i=begin(u_i)}^{end(u_i)-1} d_2(p_i, p_{i+1}) \qquad (5)$$

of the polygonal chain induced by its scan points. This length is taken as capacity label $u(e)$ for the edge (s, u_i) (see Alg. 2). Capacities of edges (v_j, t), $v_j \in V$, are assigned analogously. Edges $(u_i, v_j) \in U \times V$ are assigned infinite capacity.

Algorithm 1   OBJECT MATCHING
___
1: **input:** objects $U = \{u_1, \ldots, u_n\}$ from the previous scan
2: **input:** objects $V = \{v_1, \ldots, v_m\}$ from the current scan
3: let $E_{su} = \{(s, u) \mid u \in U\}$
4: let $E_{uv} = \{(u, v) \in U \times V \mid P(u, v)\}$ for some predicate P
5: let $E_{vt} = \{(v, t) \mid v \in V\}$
6: let $E = E_{su} \cup E_{uv} \cup E_{vt}$
7: define graph $G = (\{s, t\} \cup U \cup V, E)$
8: compute a maximum flow $\hat{f}$ in G from source node s to sink node t w.r.t. capacity labels as defined by Alg. 2.
9: compute node excess labels

$$excess(w) = \begin{cases} \sum_{u \in U} \hat{f}(s, u) & \text{if } w = s, \\ \sum_{v \in V} -\hat{f}(v, t) & \text{if } w = t, \\ 0 & else \end{cases}$$

10: compute a minimum cost flow f^* in G w.r.t. capacity and cost labels as defined by Alg.s 2 and 3, and excess labels as defined above.
11: let $w = f^*|_{E_{uv}}$ a weight label
12: compute a maximum weight matching M^* in $G[E_{uv}] = (U \cup V, E_{uv})$ w.r.t. weight label w
13: **return** matching $M^* \subseteq U \times V$
___

Intuitively we try to assign as much object shape length as possible from one scan to the next by computing a maximum flow. This is reasonable if we assume small changes of this length for each object between two successive scans.

Cost Labels. Edges (s, u_i) incident to the source node and edges (v_j, t) incident to the target node are assigned zero costs. So now consider edges (u_i, v_j) incident only to nodes representing objects. These edges will be assigned costs that reflect the similarities in shape and position of these objects in the real world, rendering less plausible object matchings more expensive than plausible ones. Our approach to compute these cost labels is to roughly estimate the work needed to transform one object into the other. Note that for a resting point of mass the work that is necessary to move it by a certain distance in a constant period of time is proportional to the square of this distance.

So for each object $u_i \in U$ a constant number of evenly spread sample points $U_i = \{u_1^i, \ldots, u_N^i\} \subseteq \mathbb{R}^2$ are taken from its shape (i.e. the polygonal chain induced by its scan points) as an approximation of it. Analogously for each object $v_j \in V$ points $V_j = \{v_1^j, \ldots, v_N^j\} \subseteq \mathbb{R}^2$ are taken from its shape. Using these points we construct for each edge $(u_i, v_j) \in E \cap (U \times V)$ of graph G the bipartite graph $H_{ij} = (U_i \cup V_j, U_i \times V_j)$ and an edge label $d_{ij} : U_i \times V_j \to \mathbb{R}$ for H_{ij} with $d_{ij}(u_k^i, v_l^j) = (d_2(u_k^i, v_l^j))^2$, where d_2 denotes the Euclidean distance in the plane. Since we do not want to make an assumption about the maintenance of the order of points on an object shape between successive scans, we follow a

---

Algorithm 2   EDGE CAPACITIES $capacity : E \to \mathbb{R}$

1: **input:** edge $(p, q) \in E = E_{su} \cup E_{uv} \cup E_{vt}$ (edge set E as in Alg. 1)
2: let $c \in \mathbb{R}$
3: **if** $(p, q) = (s, u_i)$ for some $u_i \in U$ **then**
4:      **return** (length of the polygonal chain representing object u_i)
5: **else if** $(p, q) = (v_j, t)$ for some $v_j \in V$ **then**
6:      **return** (length of the polygonal chain representing object v_j)
7: **else**
8:      **return** $+\infty$
9: **end if**

Algorithm 3   EDGE COSTS $costs : E \to \mathbb{R}$

1: **input:** edge $(p, q) \in E = E_{su} \cup E_{uv} \cup E_{vt}$ (edge set E as in Alg. 1)
2: let $c \in \mathbb{R}$
3: let $N \in \mathbb{N}$ a constant number
4: **if** $p = s$ or $q = t$ **then**
5:      **return** 0
6: **else**
7:      $(p, q) = (u_i, v_j)$ for some i, j
8:      let $C_{u_i} \subset \mathbb{R}^2$ the polygonal chain representing object u_i.
9:      let $U_i = \{u_1^i, \ldots, u_N^i\} \subset C_{u_i}$ a set of N points evenly spread along chain C_{u_i}
10:      let $C_{v_j} \subset \mathbb{R}^2$ the polygonal chain representing object v_j.
11:      let $V_j = \{v_1^j, \ldots, v_N^j\} \subset C_{v_j}$ a set of N points evenly spread along chain C_{v_j}
12:      let M^* a minimum cost perfect matching in the bipartite graph $H_{ij} = (U_i \cup V_j, U_i \times V_j)$ w.r.t. squared Euclidean distance d_2^2 in the plane as edge costs.
13:      **return** $\sum_{(u_k^i, v_l^j) \in M^*} d_2^2(u_k^i, v_l^j)$
14: **end if**

---

least effort approach and compute a minimum cost perfect matching M_{ij}^* in H_{ij}. The total costs $c_{ij} = \sum_{e \in M_{ij}^*} d_{ij}(e)$ of this matching are taken as a rough estimate for the necessary work and are assigned as the value of the cost label to the according edge (u_i, v_j) of our prior graph G. This approach is also described in Alg. 3.

Object Matching. The computed flow gives an idea of the motion in the environment of the robot but does not yet induce a unique matching. There may be objects that split and rejoin in successive scan images (consider a human and his arm) and thus induce a flow from one node to two successors and reversely. As the length of the shape of an object varies there may be a small flow reflecting these changes as well. Thus it is a good idea to focus attention on edges with a large amount of flow. Consequently the final step in our approach is to compute a matching of maximum weight in the bipartite subgraph of G induced by the two object sets U and V, using the flow labels computed in the previous step as weight labels (see Alg. 1 again). We finally have unique assignments between objects from two successive scans by this matching.

Computational Complexity. We now examine the time complexity of the presented object matching approach. The size of the input is described by the number L of samples per laser scan and the sizes of the sets $U \neq \emptyset$ and $V \neq \emptyset$ of objects to be matched. Clearly, the graph G as defined in line 7 of Alg. 1 contains $|U| + |V| + 2$ nodes and at most $|U| + |V| + |U| \cdot |V|$ edges.

Proposition 1. *Computing cost labels for all edges $e \in G$ requires no more than $\mathcal{O}(L + (|U| \cdot |V|))$ computational steps.*

Proof. Algorithm 3 uses a set of N evenly spread points per object, where N is a constant number. These point sets can be computed by two subsequent sweeps over a scan point sequence S, where object shape lengths are determined during the first sweep and point sets are chosen during the the second sweep. This is accomplished within $\mathcal{O}(L)$ computational steps.

Next, we have to compute a minimum cost perfect matching in a bipartite graph with a constant number of nodes and edges for each edge of G. Clearly this requires a constant amount of time for each edge of G, so this step is accomplished within $\mathcal{O}(|U| \cdot |V|)$ computational steps.

Finally, edges with source s or target t are assigned zero costs, which is easily accomplished within $\mathcal{O}(|U| + |V|)$ computational steps. Summing up establishes the claimed time complexity. □

Proposition 2. *Computing capacity labels capacity(e) for all edges $e \in G$ requires at most $\mathcal{O}(L + (|U| \cdot |V|))$ computational steps.*

Proof. As shown in the proof of proposition 1 computation of object shape lengths can be accomplished within $\mathcal{O}(L)$ computational steps. The remaining effort per edge is constant. As there are $\mathcal{O}(|U| \cdot |V|)$ edges in G, the claimed time complexity is proven. □

Proposition 3. *Algorithm 1 can be implemented such that it terminates after at most $\mathcal{O}(L + (|U| + |V|)^3 \log((|U| + |V|) \cdot C))$ computational steps, where $C = \max_{e \in G} costs(e)$ denotes the maximum edge cost in G.*

Proof. The construction of graph G in line 7 of Alg. 1 together with cost and capacity labels requires at most $\mathcal{O}(L + (|U| \cdot |V|))$ computational steps as shown above. The computation of a maximum flow $\hat{f}$ in line 8 requires at most $\mathcal{O}((|U| + |V|)^3)$ computational steps if we use the FIFO preflow push algorithm [1]. The calculation of the node excess labels in line 9 consumes another $\mathcal{O}(|U|+|V|)$ steps. The computation of a minimum cost flow f^* in line 10 requires at most $\mathcal{O}((|U| + |V|)^3 \log((|U|+|V|) \cdot C))$ steps, if we use the cost scaling algorithm [1]. The maximum weight matching M^* in line 10 can be computed within at most $\mathcal{O}((|U| + |V|) \cdot (|U| \cdot |V| + (|U|+|V|) \log(|U|+|V|))) \subseteq \mathcal{O}((|U|+|V|)^3)$ steps [5]. Summing up establishes the claimed time complexity. □

In other words for a set of $n = |U| + |V|$ objects and a constant laser scan size L we may need up to $\mathcal{O}(n^3 \log(nC))$ computational steps in order to find an object matching following the approach presented above.

There are several improvements for bipartite network flows [2]. However they require the network to be unbalanced in order to substantially speed up the algorithms, i.e. either $|U| \ll |V|$ or $|U| \gg |V|$, which is not the case in our context.

The complexity of finding an optimal (minimum or maximum weight) matching might be reduced if the cost label is also a metric on the node set of the underlying graph. For example if the nodes of the graph are points in the plane and the cost label is the L_1 (*manhattan*), L_2 (*Euclidean*) or L_∞ metric there are lower time complexity bounds for the problem of finding a minimum weight perfect matching [15] than in the general case. However it is not obvious if (and if so, how) this can be applied to the given object correspondence problem.

4   Applications

Fast and robust tracking of moving objects is a versatile ability, which we use in two applications: detection of obstructions and motion coordination with a guiding person. In these examples the motion of the robot is controlled by a reactive system that enables it to avoid collisions with static and dynamic obstacles. More details on this system and the underlying "velocity obstacle" paradigm can be found in [3, 9, 11].

4.1   Obstruction Detection

A goal beyond the scope of this paper is to enable a mobile system to recognize certain situations in its environment. As a first situation we address deliberate obstructions by humans, people who attempt to stop or bother the mobile system. This situation has a certain importance to real world applications of mobile robots, since systems like these may attract passers-by to experiment on how the system reacts and controls its motion.

Our detection of obstructions is based on the definition of a supervised area in front of the robot and three counters assigned to each tracked object. The supervised area represents the space that the robot needs for its next motion steps. We call an object *blocking* if it is situated inside this area but does not move at a sufficient speed in the robot's desired direction. The three counters represent the following values:

1. Number of entrances into supervised area. This is the number of transitions from non-blocking to blocking state of a tracked object.
2. Duration of current blocking. If the considered object is blocking, the elapsed time since the last entrance event of this object is counted.
3. Overall duration of blocking. The total time spent blocking by the considered object is counted.

If any of these counters exceeds a threshold value, the corresponding object is considered obstructing. Following a dynamic window approach, these counters forget blockings after a period of time. Note that a passively blocking object is evaded by the control scheme, i.e. its relative position leaves the supervised area. Therefore this counter approach in conjunction with the used control scheme detects active and deliberate obstructions fairly well.

The same object may be considered obstructing several times. Each time the response of the system is increased. At first the system informs the object that it has been tracked and that it interferes with the robot's motion plan. Next, the robot stops, asks the obstructor to let it pass by and waits for a short period of time, hopefully losing the obstructor's focus of interest. Finally, the robot might choose an alternative path to its goal to evade this persistent object. In our current implementation, however, it just gives up.

4.2   Motion Coordination

Objects in the environment are not always opponent to the vehicle. In our second application one object is accepted as a guide that the robot has to accompany. This idea is inspired by the implementation on a robotic wheelchair. Typical applications are a walk in a park, shopping mall or pedestrian area together with a disabled person. Ideally, there is no need for continuous steering maneuvers, it is sufficient to indicate the guiding person. This ability to accompany a moving object is realized by a modification to the underlying velocity obstacle approach. We give the basic idea here, for further details see [9].

At each step of time the set RV of dynamically reachable velocities of the vehicle is computed, velocity referring to speed and direction. Omitting velocities CV causing collisions with moving and static obstacles, an avoidance velocity v is selected for the next cycle from the set $RAV = RV \backslash CV$ of reachable avoidance velocities. In the original velocity obstacle work, v is selected in order to reduce the distance to a goal position. In our case the goal is not a position in the environment but a velocity v_g which is composed of the velocity of the guide person, and an additional velocity vector in order to reach a desired position relative to the guide person. Thus a velocity v is selected from RAV for the next cycle such that the difference $v - v_g$ is sufficiently small. So this approach is of beautiful simplicity but yet powerful enough to enable a robotic vehicle to accompany a human through natural, dynamic environments, as shown by our experiments.

5   Experiments

The described system is implemented on a robotic wheelchair equipped with a SICK laser range finder and a sonar ring [11]. Computations are performed on an on-board PC (Intel Pentium II, 333 MHz, 64 MB RAM). The laser range finder is used to observe the environment, whereas the sonar ring helps to avoid collisions with obstacles that are invisible to the range finder.

The tracking system has been tested in our lab environment and in the railway station of Ulm. It proved to perform considerably more robust than its predecessor [10] which is based on a greedy nearest neighbor search among

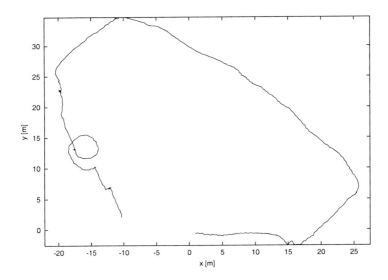

Fig. 3. Outdoor trajectory

the objects' centers of gravity. The number of objects extracted from a scan typically ranges from ten to twenty, allowing cycle times of about 50 milliseconds using the hardware mentioned above. However, the performance of the serial communication link to the range finder imposes a restriction to three cycles per second.

Figure 3 shows the trajectory of a guide walking outside on the parking place in front of our lab. The guide has been tracked and accompanied for 1073 cycles (more than five minutes), until he finally became occluded to the range finder. The wheelchair moved counterclockwise. The small loop is caused by the guide walking around the wheelchair.

Figure 4 shows trajectories of numerous objects tracked in the railway station of Ulm. The wheelchair moved from $(0,0)$ to about $(30, -10)$ accompanying a guide. Pedestrians' trajectories crossing the path of the robot or moving parallel can be seen as well as static obstacles, apparently moving as their centers of gravity slowly move due to change of perspective and dead reckoning errors. This scene has been observed for 247 cycles (82 seconds).

6   Further Work

Unfortunately, the tracking system still loses tracked objects occasionally. One obvious cause is occlusion. It is evident that invisible objects cannot be tracked by any system. But consider an object occluded by another object passing between the range finder and the first object. Such an event cancelled the tracking shown in Fig. 3, where the guide was hidden for exactly one scan. Hence a future system should be enabled to cope at least with short occlusions.

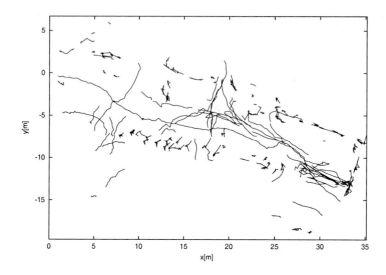

Fig. 4. Trajectories of objects tracked in the railway station of Ulm

But tracked objects get lost occasionally even if they are still visible. This might happen for example if new objects appear and old objects disappear simultaneously, as the visual field of the range finder is limited. To illustrate this, imagine a linear arrangement of three objects. Now delete the leftmost object and insert an object next to the rightmost. A flow computed as described above induces a false assignment, that is a shift to the right. This problem is partially dealt with by restriction to a local search for correspondents as presented in Sect. 3.2. It might be further improved if we do not assign any old object to new objects that become visible by a change of perspective due to the robot's motion.

In some cases object extraction fails to properly split composed objects. If these objects are recognized separately in the previous scan, either of them is lost. But this situation may be recognized by looking at the minimum cost flow in the graph, if there is a significant flow into one node from two predecessors. This might give a hint to split the newly extracted object.

As object extraction is error-prone, one might follow the idea to compute the flow based on the scan points before extracting objects by searching for proximity groups of parallel edges carrying flow. However this might be computationally infeasible, since the sizes of the graphs involved in the computations of the flows are heavily increased.

Information about the motion of an object drawn from previous scan images could be used to compute an approximation of its current position and thus direct the search for corresponding points. A first implementation of this regarding the motion of centers of gravity shows poor behaviour in some environments, for example considering walls moving as their visible part grows. Another bad effect of position prediction is its tendency to create errors by a chain effect, as even a single incorrect object assignment results in incorrect prediction of future positions and therefore may result in further incorrect assignments.

7   Conclusion

In this paper we presented an object tracking system based on laser range finder images and graph algorithms. The basic idea of our tracking approach is to represent the motion of object shapes in successive scan images as flows in bipartite graphs. By optimization (maximum flow, minimum cost, maximum weighted matching) we get plausible assignments of objects from successive scans. Furthermore, we briefly presented an approach to detect deliberate obstructions of a robot and a method for motion coordination between a human and a robot. However, a more robust object extraction and the relatively high computational complexity of the network flow algorithms remain open problems.

Acknowledgements

This work was supported by the German Department for Education and Research (BMB+F) under grant no. 01 IL 902 F6 as part of the project MORPHA.

References

1. R. K. Ahuja, T. L. Magnati, and J. B. Orlin. *Network Flows: Theory, Algorithms, and Applications.* Prentice Hall, 1993.
2. R. K. Ahuja, J. B. Orlin, C. Stein, and R. E. Tarjan. Improved algorithms for bipartite network flow. *SIAM Journal on Computing*, 23(5):906–933, Oct. 1994.
3. P. Fiorini and Z. Shiller. Motion planning in dynamic environments using velocity obstacles. *International Journal of Robotics Research*, 17(7):760–772, July 1998.
4. E. L. Lawler. *Combinatorial optimization: networks and matroids.* Rinehart and Winston, New York, 1976.
5. K. Mehlhorn and S. Näher. *LEDA—A Platform for Combinatorial and Geometric Computing.* Cambridge University Press, 1999.
6. E. B. Meier and F. Ade. Object detection and tracking in range image sequences by separation of image features. In *IEEE International Conference on Intelligent Vehicles*, pages 280–284, 1998.
7. T. B. Moeslund. Computer vision-based human motion capture – a survey. Technical Report LIA 99-02, University of Aalborg, Mar. 1999.
8. H. Noltemeier. *Graphentheorie: mit Algorithmen und Anwendungen.* de Gruyter, 1976.
9. E. Prassler, D. Bank, B. Kluge, and M. Strobel. Coordinating the motion of a human and a mobile robot in a populated, public environment. In *Proc. of Int. Conf. on Field and Service Robotics (FSR)*, Helsinki, Finland, June 2001.
10. E. Prassler, J. Scholz, M. Schuster, and D. Schwammkrug. Tracking a large number of moving objects in a crowded environment. In *IEEE Workshop on Perception for Mobile Agents*, Santa Barbara, June 1998.

11. E. Prassler, J. Scholz, and M. Strobel. Maid: Mobility assistance for elderly and disabled people. In *Proc. of the 24th Int. Conf. of the IEEE Industrial Electronics Soc. IECON'98*, Aachen, Germany, 1998.

12. F. P. Preparata and M. I. Shamos. *Computational geometry : an introduction*. Springer Verlag, 1988.

13. S. Roy and I. J. Cox. A maximum-flow formulation of the n-camera stereo correspondence problem. In *Proceedings of the International Conference on Computer Vision*, Bombai, Jan. 1998.

14. K. Sobottka and H. Bunke. Vision-based driver assistance using range imagery. In *IEEE International Conference on Intelligent Vehicles*, pages 280–284, 1998.

15. P. M. Vaidya. Geometry helps in matching. *SIAM Journal on Computing*, 18(6): 1201–1225, Dec. 1989.

Omnidirectional Vision
for Appearance-Based Robot Localization

B.J.A. Kröse, N. Vlassis, and R. Bunschoten

Real World Computing Partnership, Novel Functions Laboratory SNN,
Department of Computer Science, University of Amsterdam,
Kruislaan 403, NL-1098 SJ Amsterdam, The Netherlands
{krose,vlassis,bunschot}@science.uva.nl

Abstract. Mobile robots need an internal representation of their environment to do useful things. Usually such a representation is some sort of geometric model. For our robot, which is equipped with a panoramic vision system, we choose an appearance model in which the sensoric data (in our case the panoramic images) have to be modeled as a function of the robot position. Because images are very high-dimensional vectors, a feature extraction is needed before the modeling step. Very often a linear dimension reduction is used where the projection matrix is obtained from a Principal Component Analysis (PCA). PCA is optimal for the reconstruction of the data, but not necessarily the best linear projection for the localization task. We derived a method which extracts linear features optimal with respect to a risk measure reflecting the localization performance. We tested the method on a real navigation problem and compared it with an approach where PCA features were used.

1   Introduction

An internal model of the environment is needed to navigate a mobile robot optimally from a current state toward a desired state. Such models can be topological maps, based on labeled representations for objects and their spatial relations, or geometric models such as polygons or occupancy grids in the task space of the robot.

A wide variety of probabilistic methods have been developed to obtain a robust estimate of the location of the robot given its sensory inputs and the environment model. These methods generally incorporate some observation model which gives the probability of the sensor measurement given the location of the robot and the parameterized environment model. Sometimes this parameter vector describes *explicit* properties of the environment (such as positions of landmarks [8] or occupancy values [4]) but can also describe an *implicit* relation between a sensor pattern and a location (such as neural networks [6], radial basis functions [10] or look-up tables [2]).

Our robot is equipped with a panoramic vision system. We adopt the implicit model approach: we are not going to estimate the parameters of some sort of

G.D. Hager et al. (Eds.): Sensor Based Intelligent Robots, LNCS 2238, pp. 39–50, 2002.

CAD model but we model the relation between the images and the robot location directly (*appearance modeling*).

In section 2 we describe how this model is used in a Markov localization procedure. Then we discuss the problem of modeling in a high dimensional image space and describe the standard approach for linear feature extraction by Principal Component Analysis (PCA). In order to evaluate the method we need a criterion, which is discussed in section 5. The criterion can also be used to find an alternative linear projection: the supervised projection. Experiments on real robot data are presented in sections 6 and 7 where we compare the two linear projection methods.

2   Probabilistic Appearance-Based Robot Localization

Let $\mathbf{x}$ be a stochastic vector (e.g., 2-D or 3-D) denoting the robot position in the workspace. Similar to [1] we employ a form of *Markov localization* for our mobile robot. This means that at each point in time we have a belief where the robot is indicated by a probability density $p(\mathbf{x})$. Markov localization requires two probabilistic models to maintain a good position estimate: a *motion* model and an *observation* model.

The motion model describes the effect a motion command has on the location of the robot and can be represented by a conditional probability density

$$p(\mathbf{x}_t | u, \mathbf{x}_{t-1}) \tag{1}$$

which determines the distribution of $\mathbf{x}_t$ (the position of the robot after the motion command u) if the initial robot position is $\mathbf{x}_{t-1}$.

The observation model describes the relation between the observation, the location of the robot, and the parameters of the environment. In our situation the robot takes an omnidirectional image $\mathbf{z}$ at position $\mathbf{x}$. We consider this as a realization of a stochastic variable $\mathbf{z}$. The *observation model* is now given by the conditional distribution

$$p(\mathbf{z} | \mathbf{x}; \boldsymbol{\theta}), \tag{2}$$

in which the parameter vector $\boldsymbol{\theta}$ describes the distribution and reflects the underlying environment model.

Using the Bayes' rule we can get an estimate of the position of the robot after observing $\mathbf{z}$:

$$p(\mathbf{x} | \mathbf{z}; \boldsymbol{\theta}) = \frac{p(\mathbf{z} | \mathbf{x}; \boldsymbol{\theta}) p(\mathbf{x})}{\int p(\mathbf{z} | \mathbf{x}; \boldsymbol{\theta}) p(\mathbf{x}) d\mathbf{x}}. \tag{3}$$

Here $p(\mathbf{x})$ gives the probability that the robot is at $\mathbf{x}$ before observing $\mathbf{z}$. Note that $p(\mathbf{x})$ can be derived using the old information and the motion model $p(\mathbf{x}_t | u, \mathbf{x}_{t-1})$ repeatedly. If both models are known we can combine them and decrease the motion uncertainty by observing the environment again.

In this paper we will focus on the observation model (2). In order to estimate this model we need a dataset consisting of positions $\mathbf{x}$ and corresponding observations $\mathbf{z}$ [1]. We are now faced with the problem of modeling data in a high-dimensional space, particularly since the dimensionality of $\mathbf{z}$ (in our case the omnidirectional images) is high. Therefore the dimensionality of the sensor data has to be reduced. Here we restrict ourselves to linear projections, in which the image can be described as a set of linear features. We will start with a linear projection obtained from a Principal Component Analyis (PCA), as is usually done in appearance modeling [5]

3 Principal Component Analysis

Let us assume that we have a set of N images $\{\mathbf{z}_n\}$, $n = 1, \ldots, N$. The images are collected at respective 2-dimensional robot positions $\{\mathbf{x}_n\}$. Each image consists of d pixels and is considered as a d-dimensional data vector. In a Principal Component Analysis (PCA) the eigenvectors of the covariance matrix of an image set are computed and used as an orthogonal basis for representing individual images. Although, in general, for perfect reconstruction all eigenvectors are required, only a few are sufficient for visual recognition. These eigenvectors constitute the $q, (q < d)$ dimensions of the *eigenspace*. PCA projects the data onto this space in such a way that the projections of the original data have uncorrelated components, while most of the variation of the original data set is preserved.

First we subtract from each image the average image over the entire image set, $\bar{\mathbf{z}}$. This ensures that the eigenvector with the largest eigenvalue represents the direction in which the variation in the set of images is maximal. We now stack the N image vectors to form the rows of an $N \times d$ image matrix $\mathbf{Z}$. The numerically most accurate way to compute the eigenvectors from the image set is by taking the singular value decomposition [7] $\mathbf{Z} = \mathbf{ULV}^T$ of the image matrix $\mathbf{Z}$, where $\mathbf{V}$ is a $d \times q$ orthonormal matrix with columns corresponding to the q eigenvectors $\mathbf{v}_i$ with largest eigenvalues λ_i of the covariance matrix of $\mathbf{Z}$ [3].

These eigenvectors $\mathbf{v}_i$ are now the linear features. Note that the eigenvectors are vectors in the d-dimensional space, and can be depicted as images: the *eigenimages*. The elements of the $N \times q$ matrix $\mathbf{Y} = \mathbf{ZV}$ are the projections of the original d-dimensional points to the new q-dimensional eigenspace and are the q-dimensional feature values.

4 Observation Model

The linear projection gives us a feature vector $\mathbf{y}$, which we will use for localization. The Markov localization procedure, as presented in Section 2, is used on the feature vector $\mathbf{y}$:

---

[1] In this paper we assume we have a set of positions and corresponding observations: our method is *supervised*. It is also possible to do a simultaneous localization and map building (SLAM). In this case the only available data is a stream of data $\{\mathbf{z}^{(1)}, u^{(1)}, \mathbf{z}^{(2)}, u^{(2)}, \ldots, \mathbf{z}^{(T)}, u^{(T)}\}$ in which u is the motion command to the robot. Using a model about the uncertainty of the motion of the robot it is possible to estimate the parameters from these data [8].

$$p(\mathbf{x}|\mathbf{y};\boldsymbol{\theta}) = \frac{p(\mathbf{y}|\mathbf{x};\boldsymbol{\theta})p(\mathbf{x})}{\int p(\mathbf{y}|\mathbf{x};\boldsymbol{\theta})p(\mathbf{x})d\mathbf{x}}, \tag{4}$$

where the denominator is the marginal density over all possible $\mathbf{x}$. We now have to find a method to estimate the observation model $p(\mathbf{y}|\mathbf{x};\boldsymbol{\theta})$ from a dataset $\{\mathbf{x}_n, \mathbf{y}_n\}, n = 1, \ldots, N$.

We used a kernel density estimation or Parzen estimator. In a Parzen approach the density function is approximated by a sum of kernel functions over the N data points from the training set. Note that in a strict sense this is not a 'parametric' technique in which the parameters of some pre-selected model are estimated from the training data. Instead, the training points themselves as well as the chosen kernel width may be considered as the parameter vector $\boldsymbol{\theta}$. We write $p(\mathbf{y}|\mathbf{x};\boldsymbol{\theta})$ as

$$p(\mathbf{y}|\mathbf{x};\boldsymbol{\theta}) = \frac{p(\mathbf{y},\mathbf{x};\boldsymbol{\theta})}{p(\mathbf{x};\boldsymbol{\theta})} \tag{5}$$

and represent each of these distribution as a sum of kernel functions:

$$p(\mathbf{x},\mathbf{y};\boldsymbol{\theta}) = \frac{1}{N}\sum_{n=1}^{N}g_y(\mathbf{y}-\mathbf{y}_n)g_x(\mathbf{x}-\mathbf{x}_n) \tag{6}$$

$$p(\mathbf{x};\boldsymbol{\theta}) = \frac{1}{N}\sum_{n=1}^{N}g_x(\mathbf{x}-\mathbf{x}_n). \tag{7}$$

where

$$g_y(\mathbf{y}) = \frac{1}{(2\pi)^{q/2}h^q}\exp\left(-\frac{||\mathbf{y}||^2}{2h^2}\right) \quad \text{and} \quad g_x(\mathbf{x}) = \frac{1}{2\pi h^2}\exp\left(-\frac{||\mathbf{x}||^2}{2h^2}\right) \tag{8}$$

are the q- and two-dimensional Gaussian kernel, respectively. For simplicity in our experiments we used the same width h for the g_x and g_y kernels.

5   Feature Representation

As is made clear in the previous sections, the performance of the localization method depends on the linear projection, the number of kernels in the Parzen model, and the kernel widths. First we discuss two methods with which the model can be evaluated. Then we will describe how a linear projection can be found using the evaluation.

5.1   Expected Localization Error

A model evaluation criterion can be defined by the average error between the true and the estimated position. Such a risk function for robot localization has been proposed in [9]. Suppose the difference between the true position $\mathbf{x}^*$ of the robot and the the estimated position by $\mathbf{x}$ is denoted by the loss function

$L(\mathbf{x}, \mathbf{x}^*)$. If the robot observes $\mathbf{y}^*$, the expected localization error $\varepsilon(\mathbf{x}^*, \mathbf{y}^*)$ is (using Bayes' rule) computed as

$$\varepsilon(\mathbf{x}^*, \mathbf{y}^*) = \int_{\mathbf{x}} L(\mathbf{x}, \mathbf{x}^*) p(\mathbf{x}|\mathbf{y}^*) d\mathbf{x}$$
$$= \int_{\mathbf{x}} L(\mathbf{x}, \mathbf{x}^*) \frac{p(\mathbf{y}^*|\mathbf{x})p(\mathbf{x})}{p(\mathbf{y}^*)} d\mathbf{x}. \qquad (9)$$

To obtain the total risk for the particular model, the above quantity must be averaged over all possible observations $\mathbf{y}^*$ obtained from $\mathbf{x}^*$ and all possible $\mathbf{x}^*$ to give

$$R_L = \int_{\mathbf{x}^*} \int_{\mathbf{y}^*} \varepsilon(\mathbf{x}^*, \mathbf{y}^*) p(\mathbf{y}^*, \mathbf{x}^*) d\mathbf{y}^* d\mathbf{x}^* \qquad (10)$$

The *empirical* risk is computed when estimating this function from the data:

$$\hat{R}_L = \frac{1}{N} \sum_{n=1}^{N} \varepsilon(\mathbf{x}_n, \mathbf{y}_n)$$
$$= \frac{1}{N} \sum_{n=1}^{N} \frac{\sum_{l=1}^{N} L(\mathbf{x}_l, \mathbf{x}_n) p(\mathbf{y}_n|\mathbf{x}_l)}{\sum_{l=1}^{N} p(\mathbf{y}_n|\mathbf{x}_l)}. \qquad (11)$$

This risk penalizes position estimates that appear far from the true position of the robot. A problem with this approach is that if at a few positions there are very large errors (for example two distant locations have similar visual features and may be confused), the average error will be very high.

5.2 Measure of Multimodality

An alternative way of evaluating the linear projection from $\mathbf{z}$ to $\mathbf{y}$ is to consider the average degree of *modality* of $p(\mathbf{x}|\mathbf{y})$ [11]. The proposed measure is based on the simple observation that, for a given observation $\mathbf{z}_n$ which is projected to $\mathbf{y}_n$, the density $p(\mathbf{x}|\mathbf{y} = \mathbf{y}_n)$ will always exhibit a mode on $\mathbf{x} = \mathbf{x}_n$. Thus, an approximate measure of multimodality is the *Kullback-Leibler* distance between $p(\mathbf{x}|\mathbf{y} = \mathbf{y}_n)$ and a unimodal density sharply peaked at $\mathbf{x} = \mathbf{x}_n$, giving the approximate estimate $-\log p(\mathbf{x}_n|\mathbf{y} = \mathbf{y}_n)$ plus a constant. Averaging over all points $\mathbf{y}_n$ we have to minimize the risk

$$\hat{R}_K = -\frac{1}{N} \sum_{n=1}^{N} \log p(\mathbf{x}_n|\mathbf{y} = \mathbf{y}_n) \qquad (12)$$

with $p(\mathbf{x}|\mathbf{y} = \mathbf{y}_n)$ computed with kernel smoothing as in (5). This risk can be regarded as the negative average log-likelihood of the data given a model defined by the kernel widths and the specific projection matrix. The computational costs of this is $O(N^2)$, in contrast with the $O(N^3)$ for $\hat{R}_L$.

5.3   Supervised Projection

We use the risk measure to find a linear projection $\mathbf{y} = \mathbf{W}^T\mathbf{z}$ alternative to the PCA projection. The smooth form of the risk $\hat{R}_K$ as a function of the projection matrix $\mathbf{W}$ allows the minimization of the former with nonlinear optimization. For constrained optimization we must compute the gradient of $\hat{R}_K$ and the gradient of the constraint function $\mathbf{W}^T\mathbf{W} - \mathbf{I}_q$ with respect to $\mathbf{W}$, and then plug these estimates in a constrained nonlinear optimization routine to optimize with respect to $\hat{R}_K$. We followed an alternative approach which avoids the use of constrained nonlinear optimization. The idea is to parameterize the projection matrix $\mathbf{W}$ by a product of Givens (Jacobi) rotation matrices and then optimize with respect to the angle parameters involved in each matrix (see [12] for details). Such a *supervised projection* method must give better results than an unsupervised one like PCA (see experiments in this paper and [11]). In the next sections we will experimentally test both methods.

6   Experiments Using PCA Features

First we want to know how good the localization is when using PCA features. In particular we investigate how many features are needed.

6.1   Datasets and Preprocessing

We tested our methods on real image data obtained from a robot moving in an office environment, of which an overview is shown in figure 1. We made use of the MEMORABLE robot database. This database is provided by Tsukuba Research Center, Japan, for the Real World Computing Partnership and contains a dataset of about 8000 robot positions and associated measurements from sonars, infrared sensors and omni-directional camera images. The measurements in the database were obtained by positioning the robot (a Nomad 200) on the grid-points of a virtual grid with distances between the grid-points of 10 cm. One of the properties of the Nomad 200 robot is that it moves around in its environment while the sensor head maintains a constant orientation. Because of this, the state of the robot is characterized by the position $\mathbf{x}$ only.

The omni-directional imaging system consists of a vertically oriented standard color camera and a hyperbolic mirror mounted in front of the lens. This results in images as depicted in figure 2. Using the properties of the mirror we transformed the omni-directional images to 360 degrees panoramic images. To reduce the dimensionality we smoothed and subsampled the images to a resolution of 64×256 pixels (figure 3). A set of 2000 images was randomly selected from the total set to derive the eigenimages and associated eigenvalues. We found that for 80% reconstruction error we needed about 90 eigenvectors. However, we are not interested in the reconstruction of images, but in the use of the low-dimensional representation for robot localization.

Fig. 1. The environment from which the images were taken.

Fig. 2. Typical image from a camera with a hyperbolic mirror

6.2   Observation Model

In section 4 we described the kernel estimator as a way to represent the observation model $p(\mathbf{y}|\mathbf{x};\boldsymbol{\theta})$. In such a method usually all training points are used for the modeling. In our database we have 8000 points we can use. If we use this whole dataset this means that in the operational stage we should calculate the distance to all 8000 points for the localization, which, even though the dimensions of $\mathbf{x}$ and $\mathbf{y}$ are low, is computationally too slow. We are therefore interested in taking only a part of these points in the kernel density estimation model. In the following sections a set of about 300 images was selected as a training set. These images were taken from robot positions on the grid-points of a virtual grid with distances between the grid-points of 50 cm.

Another issue in the kernel method is a sensible choice for the width of the Gaussian kernel. The optimal size of the kernel depends on the real distribution (which we do not know), the number of kernels and the dimensionality of the

Fig. 3. Panorama image derived with the omnidirectional vision system.

problem. When modeling $p(\mathbf{x}, \mathbf{y})$ for a one-dimensional feature vector $\mathbf{y}$ with our training set we found that $h \approx 0.1$ maximized the likelihood of an independent test set. The test set consisted of 100 images, randomly selected from the images in the database not designated as training images. When using more features for localization (a higher dimensional feature vector $\mathbf{y}$) the optimal size of the kernel was found to be higher. We used these values in our observation model.

6.3   Localization

In a Markov localization procedure, an initial position estimate is updated by the features of a new observation using an observation model. The initial position estimate is computed using the motion model, and gives an informed prior in the Bayes rule. Since we are only interested in the performance of the observation model, we assume a flat prior distribution on the admissible positions $\mathbf{x}$.

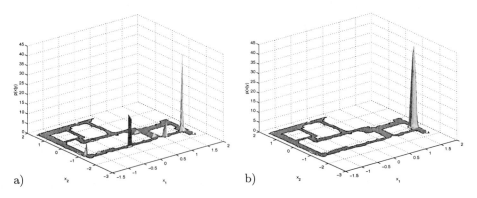

Fig. 4. An image at position (1.74, -0.96) is taken. The figure depicts the probability distribution over the learned locations. a) The first eigenvector (with the highest eigenvalue) is used as feature. b) The first 5 eigenvectors are used.

In the current experiments we studied how many of the principal components are needed for good localization In figure 4 we see the distribution $p(\mathbf{x}|\mathbf{y})$ for an image which was taken at position (1.74, -0.96), for two different number of features: five eigenimages or one eigenimage. We observe that the distribution

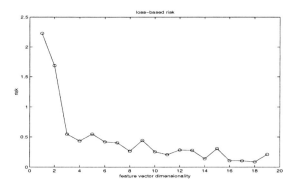

Fig. 5. Performance for different number features.

when using a single feature has multiple peaks, indicating multiple hypotheses for the position. This is solved if more features are taken into account. In both situations the maximum *a posteriori* value is close to the real robot position. This illustrates that the model gives a good prediction of the robot's real position. In some cases we observed a maximal value of the distribution at an erroneous location if too few features were used. So we need sufficient features for correct localization. The effect of the number of features is depicted in figure 5 where we plotted localization risk for different number of features. We see that for this dataset (300 positions) the performance levels out after about 10-15 features.

7 Comparing PCA with a Supervised Projection

We also compared the localization of the robot when using PCA features and when using supervised projection features. Here we used data collected in our own laboratory. The Nomad robot follows a predefined trajectory in our mobile robot lab and the adjoining hall as shown in Fig. 6. The data set contains 104 omnidirectional images (320×240 pixels) captured every 25 centimeters along the robot path. Each image is transformed to a panoramic image (64×256) and these 104 panoramic images together with the robot positions along the trajectory constitute the training set of our algorithm. A typical panoramic image shot at the position A of the trajectory is shown in Fig. 7.

In order to apply our supervised projection method, we first sphered the panoramic image data. Sphering is a normalization to zero-mean and identity covariance matrix of the data. This is always possible through PCA. We kept the first 10 dimensions explaining about 60% of the total variance. The robot positions were normalized to zero mean and unit variance. Then we applied the supervised projection method (see [12] for details of optimization) projecting the sphered data points from 10-D to 2-D.

In Fig. 8 we plot the resulting two-dimensional projections using (a) PCA, and (b) our supervised projection method. We clearly see the advantage of the proposed method over PCA. The risk is smaller, while from the shape of the

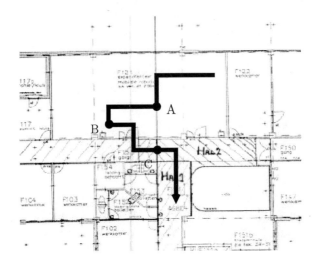

Fig. 6. The robot trajectory in our building.

Fig. 7. A panoramic snapshot from position A in the robot trajectory.

projected manifold we see that taking into account the robot position during projection can significantly improve the resulting features: there are fewer self-intersections of the projected manifold in our method than in PCA which, in turn, means better robot position estimation on the average (smaller risk). In [11] we also show that localization is more accurate when using the supervised projection method.

Finally, the two feature vectors (directions in the image space on which the original images are projected) that correspond to the above two solutions are shown in Fig. 9. In the PCA case these are the familiar first two eigenimages of the panoramic data which, as is normally observed in typical data sets, exhibit low spatial frequencies. We see that the proposed supervised projection method yields very different feature vectors than PCA, namely, images with higher spatial frequencies and distinct local characteristics.

8   Discussion and Conclusions

We showed that appearance-based methods give good results on localizing a mobile robot. In the experiments with the PCA features, the average expected

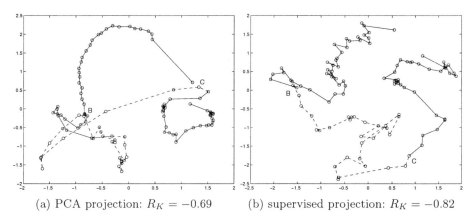

(a) PCA projection: $R_K = -0.69$ (b) supervised projection: $R_K = -0.82$

Fig. 8. Projection of the panoramic image data from 10-D. (a) Projection on the first two principal components. (b) Supervised projection optimizing the risk R_K. The part with the dashed lines corresponds to projections of the panoramic images captured by the robot between positions B and C of its trajectory.

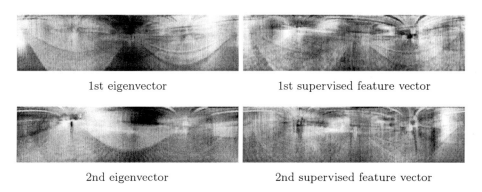

1st eigenvector   1st supervised feature vector

2nd eigenvector   2nd supervised feature vector

Fig. 9. The first two feature vectors using PCA (left) and our supervised projection method (right).

localization error from our test set is about 40cm if around 15 features are used and the environment is represented with 300 training samples. Note that we studied the worst-case scenario: the robot has no prior information about its position (the 'kidnapped robot' problem), and combined with a motion model the localization accuracy should be better. A second observation is that the environment can be represented by only a small number of parameters. For the 300 15-dimensional feature vectors the storage capacity is almost negligible and the look-up can be done very fast.

The experiments with the supervised projection showed that this method resulted in a lower risk, and therefore a better expected localization. In [11] we describe an experiment where we used the full Markov procedure to localize the robot. The supervised projection method gave significantly better results than the PCA features.

Both experiments were carried out with extensive data sets, with which we were able to get good estimates on the accuracy of the method. However, the data were obtained in a static environment, with constant lighting conditions. Our current research in this line focuses on investigating which features are most important if changes in the illumination will take place.

Acknowledgment

We would like to thank the people in the Real World Computing Partnership consortium and the Tsukuba Research Center in Japan for providing us with the MEMORABLE robot database.

References

1. W. Burgard, A. Cremers, D. Fox, G. Lakemeyer, D. Hähnel, D. Schulz, W. Steiner, Walter, and S. Thrun. The interactive museum tour-guide robot. In A. P. Press, editor, *Proceedings of the Fifteenth National Conference on Artificial Intelligence*, 1998.
2. J. L. Crowley, F. Wallner, and B. Schiele. Position estimation using principal components of range data. In *Proc. IEEE Int. Conf. on Robotics and Automation*, Leuven, Belgium, May 1998.
3. I. Jolliffe. *Principal Component Analysis*. Springer-Verlag, New York, 1986.
4. K. Konolige and K. Chou. Markov localization using correlation. In *Proc. International Joint Conference on Artificial Intelligence*, pages 1154–1159. Morgan Kauffmann, 1999.
5. H. Murase and S. K. Nayar. Visual learning and recognition of 3-d objects from appearance. *Int. Jrnl of Computer Vision*, 14:5–24, 1995.
6. S. Oore, G. E. Hinton, and G. Dudek. A mobile robot that learns its place. *Neural Computation*, 9:683–699, 1997.
7. W. H. Press, S. A. Teukolsky, B. P. Flannery, and W. T. Vetterling. *Numerical Recipes in C*. Cambridge University Press, 2nd edition, 1992.
8. S. Thrun, W. Burgard, and D. Fox. A probabilistic approach to concurrent mapping and localization for mobile robots. *Machine Learning*, 31:29–53, 1998.
9. S. Thrun. Bayesian landmark learning for mobile robot localization. *Machine Learning*, 33(1), 1998.
10. N. Vlassis and B. Kröse. Robot environment modeling via principal component regression. In *IROS'99, Proceedings of 1999 IEEE/RSJ International Conference on Intelligent Robots and Systems*, pp 677–682, 1999.
11. N. Vlassis, R. Bunschoten, and B. Kröse. Learning task-relevant features from robot data. In *IEEE International Conference on Robotics and Automation*, Seoul, Korea, May 2001. pp 499–504, 2001.
12. N. Vlassis, Y. Motomura and B. Kröse. Supervised Dimension Reduction of Intrinsically Low-dimensional Data. *Neural Computation*, to appear, 2001.

Vision for Interaction

H.I. Christensen, D. Kragic, and F. Sandberg

Centre for Autonomous Systems
Numerical Analysis and Computer Science
Royal Institute of Technology
SE-100 44 Stockholm, Sweden

1 Introduction

Society is experiencing a significant aging over the next few decades [1]. This will result in an increase by 30% more elderly and retired people and an increase of 100% in the number of people above 85 years of age. This increase in age will require significant new services for managed care and new facilities for providing assistance to people in their homes to maintain a reasonable quality of life for society in general and elderly and handicapped in particular. There are several possible solutions to the aging problem and the delivery of the needed services. One of the potential solutions is use of robotic appliances to provide services such as cleaning, getting dressed, mobility assistance, etc. In addition to providing assistance to elderly it can further be envisaged that such robotic appliances will be of general utility to humans both at the workplace and in their homes, for many different functions.

At the same time the basic methods for navigation and operation in a domestic environment is gradually becoming available. Today there are methods available for (semi-) autonomous mapping and navigation in domestic settings [2, 3]. For service robots to be truly useful they must include facilities for interaction with the environment, to be able to pick-up objects, change controls, support humans, etc. Interaction can be implemented using dedicated actuator systems such as lift and simple grippers as found on many research platforms. The ultimate facility for interaction is of course a light-weight mobile manipulator that can support a range of different tasks.

Interaction with objects requires facilities for recognition of objects, grasping and manipulation. For the instruction of a robot to carry out such tasks there is also a need for flexible interaction with human users. The by far most flexible sensory modality that provides methods for both recognition, action interpretation, and servoing is computational vision. In this paper the issues of computational vision for human-computer interaction and visual servoing for manipulation will be discussed and example solutions for use of vision in the context of the service robot systems: The Intelligent Service Robot (ISR) [4] will be presented. The ISR system is a service robot demonstrator aimed at operation in a natural domestic setting for fetch and carry type tasks. The system is to be used by regular people for operation in an unmodified setting, which implies that it must rely on sensory information for navigation, interaction and instructions. It is in

G.D. Hager et al. (Eds.): Sensor Based Intelligent Robots, LNCS 2238, pp. 51–73, 2002.

general assumed that the users have no special training which imposes special requirements in terms of interaction.

The paper is organised in the following manner: In Section 2 the use of vision for gesture interpretation is presented, Section 3 presents a systems for visual servoing for object manipulation, while Section 4 provides a summary and directions for future research.

2   Human-Robot Interaction

2.1   Structure of Interaction System

Interaction with humans can take on a number of different forms. The most typical forms are outlined below in terms of output and input;

Output

Screen. Output in terms of text or graphics on a computer screen is quite common for human computer interaction and widely studied in HCI. Mounting a computer screen on the robot requires that the user is situated in front of the screen, which is a rather strong requirement.

Speech. Spoken feedback is becoming more and more common as the quality of speech synthesis is improving. Excellent examples includes the Festival systems from University of Edinburgh and commercial systems from Microsoft and IBM. The advantage of spoken feedback is that it can be made omni directional and thus gives added flexibility in terms of position with respect to the robot. Unfortunately the spoken feedback might be challenging for elderly (with reduced hearing) and in noisy environments.

Embodied agent. An alternative to graphics and speech is use of an embodied agent, where a small character is used for feedback. An example of such an agent is show in figure 1. The agent has four degrees of freedom. The head can perform pan and tilt movements, corresponding to nodding and shaking of the head. Each of the arms have a single degree of rotational freedom, i.e. lifting of arms. Using this methodology it is possible to nod the head as a "yes" response and shake the head as a "no" response. In addition the agent can raise both hands to indicate surrender, i.e. giving up as part of a dialogue ("I did not understand what you said"). This is an intuitive modality for feedback and provides a good complement to speech and/or graphics.

Input

Keyboard. Keyboard and mouse are well known modalities for interaction with GUIs but in the context of a mobile robot it is not immediately obvious where to put the keyboard and/or the mouse. Wireless keyboards are available, but one would have to carry such a unit around. This modality is thus not very useful for mobile platforms.

Fig. 1. Example of an embodied agent (Developed by the Industrial Designer: Erik Espmark)

Touch screen. Some commercial systems such as the HelpMate use touch screens for interaction with users. This is a flexible way of providing input, provided that all possible commend alternatives can be enumerated on the screen. Unfortunately the modality requires that the user is situated in front of the screen to enable interaction.

Speech Input. Gradually speech interpretation systems with large vocabularies are becoming available. Good examples include the IBM ViaVoice and Dragon Dictate. Both of these systems are unfortunately speaker specific and require extensive training before they can be deployed. For limited vocabularies it is now possible to use speaker independent software for input generation. An excellent example is the Esmeralda system developed by Fink et al. [5, 6]. A problem with speech systems is that the signal to noise ratio typically must be very good to obtain reasonable recognition rates. This either requires a good microphone in combination with a quiet setting or alternatively mounting of a microphone on the user (i.e. a wireless microphone). The second option is by far the most frequently used option. Use of specific vocabularies is well suited for specification of a sequence of commands such as "fetch the milk from the refrigerator". Unfortunately such a mode of interaction is only well suited for a set of prior known objects that can be uniquely identified. Thus a system would not be able to understand commands such as "pick this object and deliver it to Susan", or "please clean this area". Use of speech along is thus only suitable for well defined settings.

Gestures. Gestures are well suited for generation of instructions in terms of a specific set of commands. The sequencing of commands will typically require recognition of detailed timing which in most cases is difficult if not impossible. In addition a problem with gestures is that the user has to stand in front of the robot cameras and the action must be observable by the camera. This is a significant constraint. In addition vision has often had robustness problems for operation in domestic settings around the clock. Gestures are however well suited for specification of spatial locations and for simple instructions such as "follow me", "this area", "go left", "STOP!", etc.

Given the characteristics outlined one possible solution is to use a spoken dialogue systems in combination with a simple set of gestures and an embodied agent for basic level feedback. For processing of spoken input the Esmeralda system [6] is used. The spoken system provides a sequence of commands, while the gesture system may be used for generation of motion commands in combination with the speech systems. In the following the basic components of the gesture interpretation system are outlined.

2.2   Gesture Interpretation

Gestures have been widely studied in computer vision and human computer interaction as for example reported by [7, 8]. Gestures have also been used in a few robot applications as for example reported by [9, 10].

For interaction with a human operator it is necessary to consider the location of hands and the face of the users so as to allow both left and right handed people to interact with the system. In addition for some commands it is of interest to be able to use both hands. There is thus a need for the definition of a system that allows identification of the face and the two hands. Once the regions corresponding to these three components have been identified they must be tracked over time and finally the trajectories must be interpreted to allow recognition of the specific gestures. The process of gesture interpretation can thus be divided into three steps i) segmentation of hands and face, ii) tracking of regions and iii) interpretation of trajectories. Each of the three processes are described in detail below.

Colour Segmentation in Chromaticity Space. Recognition of hands and other skin coloured regions have been studied extensively in the literature in particular for face and gesture recognition. The by far most frequently used approach is based on colour segmentation. To achieve robustness to colour variation it is well known that the traditional RGB colour representation is not a good basis for segmentation. Frequently researchers such as Sterner [8] have used a Hue-Saturation-Intensity (HSI) representation as a basis for recognition. In this colour space Hue and Saturation are used for segmentation while the intensity component is ignored. Such approaches allow definition of methods that are more robust to light variation. Unfortunately not only intensity but also colour temperature varies over the day and is also dependent on the use of natural or artificial lighting. It is thus easy to illustrate that HSI based approaches will fail if the colour temperature is varied. A more careful consideration of the underlying physics reveals that the reflected light from an object such as the human skin can be modeled using a di-chromatic reflection model, i.e.:

$$L(\theta, \lambda) = m_s(\theta)L_s(\lambda) + m_b L_b(\lambda) \qquad (1)$$

where L is the light reflected as a function angle of incidence (θ) and the wavelength (λ). The reflected light is composed of two components one derived from surface reflection (L_s) and another dependent on light reflected from below the

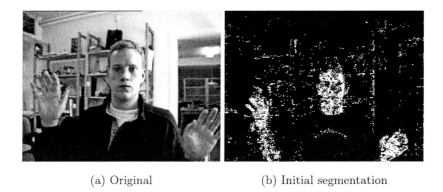

(a) Original   (b) Initial segmentation

Fig. 2. Initial segmentation of skin colour regions

surface (L_b). The skin colour and the content of melanin will determine the reflection coefficients (m_s and m_b). The total light reflected can be determined through integration over the full spectrum, using a spectral sensitivity function for the camera of $f_{RGB}(\lambda)$, i.e.:

$$C_{RGB} = \int_\lambda L(\theta,\lambda) f_{RGB}(\lambda) \, d\lambda \qquad (2)$$

The colour image is unfortunately still highly sensitive to illumination variations. It is however possible to change the image to Chromaticity coordinates where a more robust representation can be achieved. I.e.,

$$r = \frac{R}{R+G+B}$$
$$g = \frac{G}{R+G+B}$$
$$b = \frac{B}{R+G+B}$$

When processing images in chromaticity space normally only the $r - g$ components are used. The b component is redundant and in addition normally the blue channel of a camera is noisy in particular in low light conditions.

When considering images of human skin in $r - g$ space it is possible to identify a narrow region that contains all skin pixels. This region is often termed the skin locus [11]. Through simple box classification in $r - g$ space it is possible to perform a segmentation of skin regions with a high level of confidence. An example is shown in figure 2.

Through estimation of the density of skin coloured pixels in a local neighbourhood and subsequent thresholding it is possible to detect the three major skin coloured regions in an image. For images where a single person is the dominating figure this will result in reliable detection of the face and hands, as shown in figure 3:

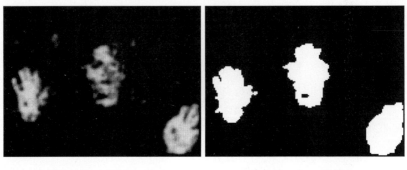

(a) Density of skin pixels   (b) Hands and Face

Fig. 3. Estimation of density of skin pixels and thresholding of the density image to extract hand and face regions

To further compensate for illumination variations it is possible to compute a histogram over the detected regions (in $r - g$ space) and feedback the histogram boundaries back to the estimation of the location of the skin locus box classifier to allow adaptation to potential variations in illumination.

Region Tracking. Once regions have been extracted the next problem is to track the regions over time to allow generation of a sequence of trajectories for interpretation of gestures. The tracking is here carried out using a conventional Kalman filter [12]. The regions are tracked directly in image space. For the tracking the center of gravity is used, i.e:

$$p = \begin{pmatrix} x \\ y \end{pmatrix} \tag{3}$$

For the tracking the position p and velocity $\dot{p}$ is used. This results in a system model, where the state is:

$$x = \begin{pmatrix} p \\ \dot{p} \end{pmatrix} \tag{4}$$

Under the assumption of a first order Taylor expansion the autonomy of the system is defined by:

$$x_{k+1} = \mathbf{A}x_k + \mathbf{C}w_k \tag{5}$$

$$\mathbf{A} = \begin{pmatrix} 1 & \Delta T \\ 0 & 1 \end{pmatrix} \tag{6}$$

$$\mathbf{C} = \begin{pmatrix} \Delta T^2/2 \\ \Delta T \end{pmatrix} \tag{7}$$

Where w_k models acceleration etc and is assumed to be Gaussian noise, with a variance σ_w^2. As only the position of regions are available in the image, the observation model is defined as:

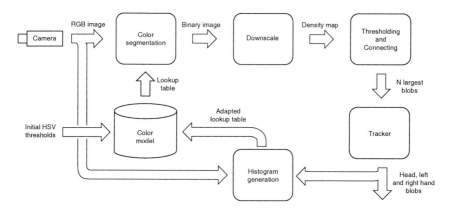

Fig. 4. The system for face and head region tracking

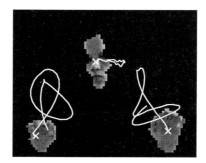

Fig. 5. Example trajectories extracted for face and hand tracking

$$z_k = \mathbf{H}x + n_k \tag{8}$$

$$\mathbf{H} = \begin{pmatrix} 1 & 0 \end{pmatrix} \tag{9}$$

Where z is the measurement vector and n_k is the measurement noise that is assumed to Gaussian noise with a variance σ_n^2. Using the standard Kalman updating model it is now possible to track regions over time. Details can be found in [12]. The matching between images is performed using a nearest neighbour algorithm, which is adequate when the algorithm is run at 25 Hz.

The full systems for tracking of hand and face regions is shown in figure 4.

An example result for tracking of regions is shown in figure 5. The trajectories illustrate how both hands and the face are tracked over a limited sequence.

The above mentioned system has been implemented on a standard 400 MHz Pentium-II computer running Linux. The system uses a standard Matrox framegrabber. The system is running in real-time. The execution time for the different phases of the algorithm are shown in Table 1 (for an 160x120 pixel image):

Interpretation of Gestures Using HMM. From the tracking module three sets of trajectories are available. These trajectories can be used for the interpre-

Table 1. CPU usage for segmentation and tracking of hand and face regions on a 400
MHz Pentium II computer running Linux

Phase	Time/Frame [ms]
Image Retrieval	4.3
Colour Segmentation	0.6
Density Estimation	2.1
Connected Components	2.1
Kalman filtering	0.3
Total	9.4

tation of gestures. For this purpose it is assumed that there is some minimum
motion between different gestures. I.e. the hands are almost stationary between
different gestures. Under this assumption velocity can be used to segment the
trajectories into independent parts that can be processed independent of each
other. Once separate trajectories have been extracted they are normalised to provide independence of size and position in the image. The normalized trajectory
segments are forwarded to a recognition module.

For the recognition of gestures a Hidden Markov Model is used [13]. A Hidden
Markov Model (HMM) is defined by a 5-tuple: $\lambda = (S, V, A, B, \pi)$ that is defined
as:

- States, $S = \{s_1, s_2, \ldots, s_N\}$. The state at time t is denoted q_t.
- Symbols, $V = \{v_1, v_2, \ldots, v_m\}$
- Transition probability distribution, $A = \{a_{ij}\}$, where
  $a_{ij} = P(q_{t+1} = s_j | q_t = s_i)$, $i, j \in \{1..N\}$
- Observation probability distribution, $B = \{b_j(k)\}$, where
  $b_j(k) = P(v_k \text{generated in state} j)$, $j \in \{1..N\}$, $k \in \{1..M\}$
- Initial state distribution, $\pi = \{\pi_i\}$ where
  $\pi_i = P(q_i = s_i)$, $i \in \{1..N\}$.

In this particular application the HMM is assumed to be a Markov chain,
and all trajectories are assumed to start in the first state, i.e.

$$\pi_i = \begin{cases} 0, i \neq 1 \\ 1 \ i = 1 \end{cases} \tag{10}$$

In many applications the observations are organised into a sequence, i.e. $O = \{x_1, x_2, \ldots x_K\}$. In this particular application the gestures are continuous rather
than discrete and the observation probability distribution is thus not discrete,
and is consequently replaced by a continuous Gaussian distribution:

$$b_j(x_t) = \frac{1}{\sqrt{(2\pi)^n |\Sigma_j|}} e^{-\frac{1}{2}(x_k - \mu_j)^T \Sigma_j^{-1}(x_t - \mu_j)} \tag{11}$$

where n is the dimension of the measurement space (here 2) and μ is the mean
and Σ_j is the covariance matrix for the observations in state j.

For the use of a HMM two problems have to be considered:

1. *Evaluation:* Determining $P(O|\lambda)$, the probability that a HMM λ generated a particular observation sequence. I.e. is a HMM for a particular gesture the best model for the observed trajectory.
2. *Training:* Given a sequence of test samples how does one determine the HMM that best "explains" the data set.

Let us consider each of these problem in the following.

Recognition/Evaluation: Recognition is performed by using the HMM that is the most like generator of a particular sequence. A simple algorithm to compute $P(O|\lambda)$ is through use of the Vieterbi algorithm [13]. Given:

$$\alpha_i(t) = P(\boldsymbol{x}_1, \boldsymbol{x}_2, \ldots, \boldsymbol{x}_t, q_t = s_i|t)$$

Initialize:

$$\alpha_i(1) = \pi_i b_i(\boldsymbol{x}_1), \ i \in \{1..N\} \tag{12}$$

Recursion

$$\alpha_j(t+1) = \left[\max_i \alpha_i(t)a_{ij}\right] b_j(\boldsymbol{x}_{t+1}), \ t \in \{1..T-1\}, \ j \in \{1..N\} \tag{13}$$

Termination

$$P(O|\lambda) = \max_i \alpha_i(K), \ i \in \{1..N\} \tag{14}$$

The Vieterbi algorithm generates a maximum likelihood estimate for the trajectory. To allow identification of the case "no gesture" it is necessary to introduce a threshold that ignores low probability trajectories.

Training The training phase is carried out by using a sequence of test trajectories that are recorded and subsequently processed in a batch process. The training is carried out in two phases: i) initialisation and ii) re-estimation.
    The initialisation of the HMM is carried out by simple statistics, i.e.

$$\boldsymbol{\mu}_i = \frac{1}{N_j} \sum_{\forall t, q_i = s_j} \boldsymbol{x}_t \tag{15}$$

and

$$\Sigma_j = \frac{1}{N_j} \sum_{\forall t, q_i = s_j} (\boldsymbol{x}_t - \boldsymbol{\mu}_j)(\boldsymbol{x}_t - \boldsymbol{\mu}_j)^T \tag{16}$$

where N_j is the number of observations in state j. The most likely trajectories are determined using the Vieterbi algorithm, and the state transition probabilities are approximated by relative frequencies:

$$a_{ij} = \frac{A_{ij}}{\sum_{k_1}^N A_{ik}} \tag{17}$$

The initialisation process is repeated until the estimates stabilize.

The initial parameters can be improved through use of the Baum-Welch forward-backward algorithm [14]. Let $L_j(t)$ denote the likelihood of being in state j at time t. Then the statistics can be revised to be:

$$\mu_j = \frac{\sum_{t=1}^{T} L_j(t)\boldsymbol{x}_t}{\sum_{t=1}^{T} L_j(t)} \tag{18}$$

and

$$\Sigma_j = \frac{\sum_{t=1}^{T} L_j(t)(\boldsymbol{x}_t - \boldsymbol{\mu}_j)(\boldsymbol{x}_t - \boldsymbol{\mu}_j)^T}{\sum_{t=1}^{T} L_j(t)} \tag{19}$$

The calculation of the state likelihoods, $L_j(t)$, is carried out using the forward-backward algorithm. The forward probability is defined by

$$\alpha_i(t) = P(\boldsymbol{x}_1, \boldsymbol{x}_2, \ldots, \boldsymbol{x}_t, q_t = s_i | \lambda) \tag{20}$$

Computed as mentioned in the recognition section (i.e., Eqs 12 – 14). The backward probability is defined as

$$\beta_i(t) = P(\boldsymbol{x}_{t+1}, \boldsymbol{x}_{t+2}, \ldots, \boldsymbol{x}_K | q(t) = i, \lambda) \tag{21}$$

The forward backward probabilities gives:

$$\alpha_i(t)\beta_i(t) = P(O, q_t = i | \lambda) \tag{22}$$

Which can be used for computation of the state probabilities and the state transition probabilities.

$$L_j(t) = \frac{1}{P(O|\lambda)} \alpha_j(t)\beta_j(t) \tag{23}$$

$$a_{ij} = \frac{\frac{1}{P(O|\lambda)} \sum_{t=1}^{T-1} \alpha_i(t) b_j(\boldsymbol{x}_{t+1})\beta_j(t+1)}{\frac{1}{P(O|\lambda)} \sum_{t=1}^{T-1} \alpha_i(t)\beta_j(t)} \tag{24}$$

The forward backward process is then repeated until the probabilities stabilize by which the network is ready for application.

The HMM classifier has been implemented on the same PC as used for the blob tracking. The classifier run in 23.4 ms for each full gesture and combined the blob tracking and the recognizer is thus able to classify gestures in real-time.

Evaluation. The gesture system has been evaluated using two different gesture sets. For the initial evaluation of the system the graffiti language developed by Palm Computing was implemented. The graffiti language has without doubt been developed to ensure a minimum of errors in recognition and consequently it is a good dataset for testing of the algorithm. After quite some initial testing a Markov model with four states was selected as a model.

Table 2. Recognition results achieved for the Palm graffiti language

Recognition rates	Position	Velocity	Combined
Result [%]	96.6	87.7	99.5

Attention	Idle	Forward	Back	Left	Right

Turn left	Turn right	Faster	Slower	Stop

Fig. 6. The gesture set used for basic robot interaction

For the testing a total of 2230 test sequences were recorded. The dataset involved 5 different persons all of them operating in a living room scenario. From the dataset 1115 of the sequences were used for training of the HMM. Subsequently the remaining 1115 sequences were used for testing of the algorithm. For recognition tests were carried using image position p, image velocity $\dot{p}$ and position and velocity. The achieved recognition rates are summarised in table 2.

Subsequently a language suitable for simple instruction was designed to allow for interaction with the robot. The basic instructions are feed into the robot control system to allow real-time interaction. The set of gestures is shown in figure 6.

The gesture set was trained on 500 image sequences. Subsequently the gestures were tested on a new set of 4 people using a limited set of 320 sequences. A recognition rate of 78% was achieved for this dataset. The lower recognition rate is in part due to the fact that the gesture set has not been optimized. Another problem experienced with a data set that involves both hands is occasional occlusions between the hands, which easily can lead to errors in the tracking process. When combined with the speech system it is however possible to obtain an overall system for human interaction that is suitable for use by regular people.

3 Model Based Visual Manipulation

One of the basic requirements of a service robot is the ability to manipulate objects in a domestic environment. Given a task such as to fetch a package of milk from a refrigerator, the robot should safely navigate to the kitchen, open the refrigerator door, fetch and deliver the milk. Assuming a perfect positioning

in front of the refrigerator door, two tasks remain: opening the door and fetching the milk. The first task - opening of the door may benefit from integrated use of different sensors, a camera and a force-torque sensor. This problem has been addressed in our previous work, [15]. Here, we concentrate on "fetch–and–carry" tasks.

A "fetch" or a "pick-up" task where vision is used as an underlying sensor may in general be divided into the following subtasks:

Detection. Given a view of the scene, the agent should be able to provide a binary answer whether the desired object is in the field of view or not. This implies that the scene might both contain a few similar objects or no object the all. "Simple" approaches based on colour or texture may be used for moderately complex scenes. However, it is not straightforward to design a general tool for highly dynamic environments where spatial position of objects may change completely unexpected.

Segmentation. Providing that the detection was successful, the image must be searched for the object. The hypotheses should be made and the image has to be segmented for further verification and actual recognition of the object. In a trivial manner, an opposite approach may be used where we exclude those regions that are not likely to contain the object, e.g. regions of uniform colour or highly textures region (depending, of course, on the objects appearance).

Recognition. The recognition tool should recognize the object and provide certainty measure for the recognition. Depending on the implementation, this part may also provide partial/full pose of the object [16]. Object recognition is a long studied research issue in the field of computational vision [17]. A promising system, based on Support Vector Machines [18] has recently been developed locally and used to detect a moderate set of everyday objects.

Alignment. After the object has been located, the robotic manipulator/arm should be visually servoed to the vicinity of the object. Many of the existing visual servo systems neglect the first three issues and concentrate on the alignment task. The approaches differ in the number of cameras used (stand–alone vs. eye–in–hand) and underlying control space (image based [19], position based [20] and 2 1/2 D visual servoing [21]). To perform the alignment in a closed–loop manner, the systems should have the ability to track object features during the arm movements and update their position if needed. This is especially obvious if the objects are not static or if the alignment and grasping are performed while the robot is moving.

Grasping. After the alignment has been performed, the grasping sequence should start. For "simple" objects like boxes and cylinders (which most of the food items have) a set of predefined grasps may be learned and used depending on the current pose of the object. However, there are many everyday object which are far from having a "nice", easy-graspable shape. It is obvious that for those objects vision should be used together with other sensory modalities like force-torque or touch sensors.

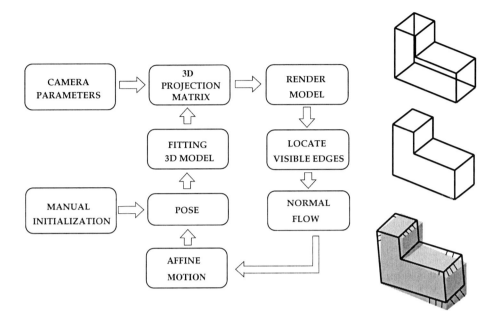

Fig. 7. Diagram of the model based tracking system.

Manipulation. In many cases the agent is required to deliver an object in a particular way, i.e. the object should be fit into or put onto some other object. This may additionally require of the agent to manipulate the object in the end–effector/hand before final delivery/placement. There therefore a need form the grasping tool to perform both a stable and a manipulable grasp, [22], [23] .

Each of the outlined issues have been widely studied in different research areas: computer vision, control, optimization, sensor integration, etc. However, their interaction and integration is still quite a young area.

Although somewhat neglected because of it computational complexity, the use of CAD models has been a quite popular approach to solving all of the outlined issues: pose estimation and tracking, [24, 25], alignment [26, 27] and grasping [28, 22]. This is also the approach taken in our current work. In this paper we address the issues of pose estimation, alignment (visual servoing) and tracking. For more information about the currently used recognition tool we refer to [18]. The diagram of the implemented tracking system is presented in figure 7. The approach is similar to the one proposed by Marchand in [26].

3.1   Pose Estimation

Pose estimation considers a computation of a rotation matrix (orientation) and a translation vector of the object (position), $\mathbf{H}(\mathbf{R}, \mathbf{t})$. Different researchers have

formulated closed form solutions to pose estimation with feature points in coplanar or non-coplanar configuration,[29–33]. It has been shown that the key to a
robust estimation is the use a larger number of feature points since the image
noise and measurement errors average out due to the feature redundancy. The
most straightforward method described by Roberts, [29], consists in retrieving
the 11 parameters of a projective transformation matrix as a solution to a linear system. Other notable methods were proposed by Tsai, [34], Lowe [24], and
Yuan, [35]. Mentioned techniques rely on the Newton-Raphson method which
requires the initial approximate pose as well as the computation of the Jacobian
matrix which is computationally expensive and usually not suitable for real-time
applications.

The method proposed by DeMenthon and Davis[36], relies on linear algebra
techniques. Although this method is also iterative, it does not require the $a-$
$priori$ knowledge of the initial pose nor the estimation of the Jacobian matrix.
At the first iteration step, the method computes the pose from orthography
and scaling with the assumption that the image projections of model points
were obtained under a scaled orthographic projection. Briefly, the method starts
by assuming a scaled orthographic projection and iteratively converges to a
perspective projection by minimizing the error between the original image points
and projected point using the current camera model. The method converges after
3-4 iterations which is suitable for real-time applications. In our implementation,
this step is followed by an extension of Lowe's [24] nonlinear approach proposed
in [37]. This step is called POSE in figure 7.

Once the pose of the object in the camera coordinate system is known, the
tracking is initiated. It involves 3 steps:

1. normal flow computation
2. fitting an affine/quadratic transformation model between two views
3. optimizing the pose with respect to spatial intensity gradients in the image

3.2   Model Based Tracking

After the pose estimation is obtained, the internal camera parameters are used
to project the model of the object onto the image plane. A simple rendering
technique is used to determine the visible edges of the object [38]. For each
visible edge, tracking nodes are assigned at regular intervals in image coordinates
along the edge direction. After that, a search is performed for the maximum
discontinuity (nearby edge) in the intensity gradient along the normal direction
to the edge. The edge normal is approximated with four directions: $0, 45, 90, 135$
degrees. This way we obtain a displacement vector:

$$d_i^{\perp} = \begin{pmatrix} \Delta x_i \\ \Delta y_i \end{pmatrix} \qquad (25)$$

representing the normal displacement field of visible edges.

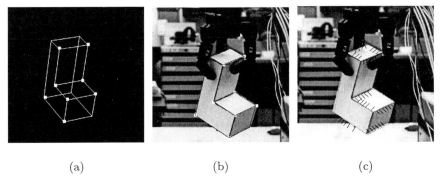

(a)   (b)   (c)

Fig. 8. Example images obtained during pose estimation and tracking: a) the model with points used in POSE to determine the pose, b) pose estimation (the model is overlaid in black), and c) normal flow estimation (the lines represent the direction and not the magnitude.

A 2D affine transformation is expressed by:

$$\begin{pmatrix} x_i^{t+1} \\ y_i^{t+1} \end{pmatrix} = \begin{pmatrix} a_1 \ a_2 \\ a_3 \ a_4 \end{pmatrix} \begin{pmatrix} x_i^t \\ y_i^t \end{pmatrix} + \begin{pmatrix} T_x \\ T_y \end{pmatrix} = \mathbf{W}(\boldsymbol{p}_i)\Theta \tag{26}$$

with $\Theta = (a_1, a_2, a_3, a_4, T_x, T_y)^T$, $\boldsymbol{p}_i^t = (x_i^t, y_i^t)^T$, $\boldsymbol{p}_i^{t+1} = \Psi_\Theta \boldsymbol{p}_i^t$ (where Ψ_Θ denotes affine transformation from (Eq. 26)) and

$$\mathbf{W}(\boldsymbol{p}) = \begin{pmatrix} x \ y \ 0 \ 0 \ 1 \ 0 \\ 0 \ 0 \ x \ y \ 0 \ 1 \end{pmatrix} \tag{27}$$

Displacement vector can be written as:

$$\boldsymbol{d}_i(\boldsymbol{p}_i) = \qquad\qquad \mathbf{W}(\boldsymbol{p}_i)\Theta' \tag{28}$$
$$= \mathbf{W}(\boldsymbol{p}_i)\left(\Theta - (1,0,0,1,0,0)^T\right) \tag{29}$$

From $(\boldsymbol{p}_i^t, \boldsymbol{d}_i)_{i=1...k}$, we estimate the 2D affine transformation Θ'. From Eq. 29, we have:

$$d_i^{\perp} = \boldsymbol{n}_i^T \boldsymbol{d}_i(\boldsymbol{p}_i) = \boldsymbol{n}_i^T \mathbf{W}(\boldsymbol{p}_i)\Theta' \tag{30}$$

where $\boldsymbol{n}_i$ is a unit vector orthogonal to the edge at a point $\boldsymbol{p}_i$. From Eq. 30 we can estimate the parameters of the affine model, $\widehat{\Theta}'$ using a M-estimator ρ:

$$\widehat{\Theta}' = \arg\min_{\Theta'} \sum_i \rho(d_i^{\perp} - \boldsymbol{n}_i^T \mathbf{W}(\boldsymbol{p}_i)\Theta') \tag{31}$$

Computed affine parameters give us new image positions of the points in time $t+1$. Thereafter, pose estimation step (POSE) is performed to obtain the pose of the object in the camera coordinate system, $\mathbf{H}(\mathbf{R}, \mathbf{t})_{init}$.

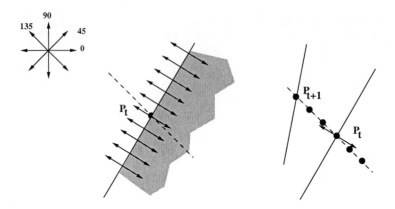

Fig. 9. Determining normal displacements for points on a contour in consecutive frames.

In certain cases, a six-parameter motion model is not sufficient. The following polynomial model may be used:

$$\begin{pmatrix} x_i^{t+1} \\ y_i^{t+1} \end{pmatrix} = \begin{pmatrix} a_0 \\ a_1 \end{pmatrix} + \begin{pmatrix} a_2 & a_3 \\ a_4 & a_5 \end{pmatrix} \begin{pmatrix} x_i^t \\ y_i^t \end{pmatrix} + \begin{pmatrix} a_6 & a_7 & 0 \\ 0 & a_6 & a_7 \end{pmatrix} \begin{pmatrix} x_i^{t^2} \\ x_i^t y_i^t \\ y_i^{t^2} \end{pmatrix} = \mathbf{W}(\boldsymbol{p}_i)\Theta \quad (32)$$

Fitting the CAD model to the spatial intensity gradients

Using the estimated affine parameters $\widehat{\Theta}'$ and positions of edge nodes at time t, we are able to compute their positions at time $t+1$ from:

$$\boldsymbol{p}_i^{t+1} = \Psi_\Theta \boldsymbol{p}_i^t \quad (33)$$

with

$$\widehat{\Theta} = \widehat{\Theta}' + (1, 0, 0, 1, 0, 0)^T \quad (34)$$

As already mentioned, the affine motion model does not completely account for the 3D motion and perspective effects that occur during the object motion. Therefore, the pose space, $\mathbf{H}(\mathbf{R}, \mathbf{t})$, should be searched for a best fit given the image data. As proposed in [26], the projection of the object model is fitted on the spatial intensity gradients in the image, using the $\mathbf{H}(\mathbf{R}, \mathbf{t})_{init}$ as the initialization:

$$\widehat{\mathbf{H}}(\mathbf{R}, \mathbf{t}) = \arg \min_{\mathbf{H}} \left[-\sum_{C_\mathbf{H}} \|\nabla \mathbf{G}(t+1)\| \right] \quad (35)$$

where $\nabla \mathbf{G}(t+1)$ is the intensity gradient along the projected model edges and $C_\mathbf{H}$ are the visible edges of the model given a pose $\mathbf{H}(\mathbf{R}, \mathbf{t})$.

The optimization method proposed in [26] uses a discrete hierarchical search approach with respect to to the pose parameters. However, the right discrete step of pose parameters is crucial in order to find the right minimum value. The affine motion model is particularly sensitive to the rotational motions of the object and large changes in rotation usually result with loss of tracking. For that reason we have extended the method proposed in [26] so that the step determination is dynamically changed based on the 3D object velocity.

This system has been used for three different tasks:

1. **Alignment and tracking.** Position based visual servoing approach [39] is used to align the robot hand with the object. In addition, the robot hand follows the object during its motion by keeping the constant pose between the objects and hands coordinate systems.
2. **Grasping.** Grasping is performed by using a set of pre–defined grasps depending on the objects pose.
3. **Visual servoing.** After the object is grasped, image based visual servoing [39] is used to guide the manipulator during the placement task.

To control the robot we have adopted both image based and position based visual servo approaches. Detailed description and characteristics of these approaches can be found in [39]. The basic idea is to minimize an error function usually defined as an image position or 3D pose difference between the current and desired objects set of tracked features. Detailed description of the outlined tasks are presented in following section together with the experimental setup.

3.3   Experimental SetUp

The dextrous robot hand used in this example is the Barrett Hand. It is an eight–axis, three–fingered mechanical hand with each finger having two joints. One finger is stationary and the other two can spread synchronously up to 180 degrees about the palm (finger 3 is stationary and fingers 1 and 2 rotate about the palm). The hand is controlled by four motors and has 4 degrees of freedom: 1 for the spread angle of the fingers, and 3 for the angles of the proximal links. Our current model of the hand uses slightly simplified link geometries, but the kinematics of the model match the real hand exactly. The hand is attached to Puma560 arm which operates in a simple workcell. The vision system uses a standard CCD camera (Sony XC-77) with a focal length of 25mm and is calibrated with respect to the robot workspace. The camera is mounted on a tripod and views the robot and workcell from a 2m distance.

3.4   Alignment and Tracking

The objective of this experiment was to remain constant relative pose between the target and the robot hand. A typical need for such an application is during the grasp of a moving object or if the robot is performing a grasping task while moving. Position based visual servoing is used to minimize the error in pose between the hand coordinate system an a reference point defined in the object coordinate frame. When a stable tracking is achieved grasping may be performed.

Fig. 10. From an arbitrary starting position in figure a), the end–effector is precisely servoed into a predefined reference position with respect to the target object, figure b). If the object starts moving, the visual system tracks the motion (pose) of the object and the robot is controlled keeping the constant relative pose between the hand and the object.

3.5   Grasping

Here, pose estimation is used to perform a set of "learned" grasps. For each object we generate a set of reference points from which the grasp sequence should be initialized. These reference points depend on the pose of the object relative to the camera, i.e. the object might be placed horizontally, vertically, etc. First, we estimate the current pose of the object. After this, the manipulator is guided to the reference point that is either on a side (for a horizontal placement of the object) or above the object (if the object is placed vertically on the table). When the positioning is achieved, the "learned" grasp is performed.

3.6   Model Based Servoing

Here, the robot starts to move from an arbitrary position holding the grasped object. The task is to bring the object to some final or "thought" position where "teach by showing" approach was used [39].

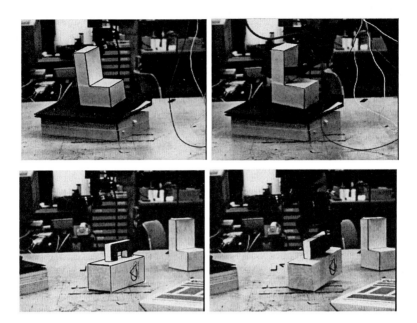

Fig. 11. Two examples of grasps: left) the vision system determines the pose of the object within the robot workspace and right) the completed grasp.

The whole sequence consists of the following steps:

- The "teach by showing" approach is initiated in the following manner: the object is placed at some arbitrary position and its pose with respect to the camera coordinate system is estimated by matching points between the object and geometrical model using the approach presented in Section 3.1. Image coordinates of corner points are used to build a vector of desired image positions. After this, the object is placed to a new, initial position and tracking sequence is initiated. The vector of desired positions may also be known *a–priori*.
- In each frame, the current image positions of corner features are estimated using the technique presented in Section 3.2.
- The image Jacobian [39] is estimated and the control law obtained using a simple position controller.

4   Discussion

Vision is a highly flexible sensory modality for estimation of the state of the environment. This functionality can be used for construction of facilities for interaction with humans and objects for a variety of applications. In this paper it has been described how vision may be used for interpretation of gestures and how similar methods also may be used for recognition and servoing to facilitate

a   b

c   d

e

Fig. 12. a) Start pose, b) Destination pose, c), d), e) Intermediate images of visual servo sequence. This particular test was performed mainly in order to test the visual tracking system. The object undergoes a rotational motion around camera's y–axis. However, the tracking is remained during the whole servoing loop.

grasping and manipulation. The presented methods have been evaluated in the context of an service robot application that operated in a domestic setting. A robust performance has been reported for the particular applications considered. The methods described provide basic facilities for interaction in a natural environment.

For interaction in natural environments it is necessary for the user to be able to provide fairly general commands and the robot must be able engage in a dialogue where turn-taking allow resolution of ambiguities and learning from past experience. So far the basic facilities for interaction has been provided but future research will need to consider how these methods can be integrated in a service robot application that can be used by regular people. To provide such an artifact there is a need for research on instruction by demonstration, which involves methods for dialogue design, task and behavioural modelling, and learning for generalisation etc. In addition the current methods allow recognition and interaction with prior defined objects. For operation in a general setting it is also necessary to provide methods that allow automatic acquisition of object models and grasping strategies. While the present paper has described a basic set of functionalities for a service robot there is still fundamental research to be carried out before service robots are ready to enter our homes.

Acknowledgment

This research has been sponsored by the Swedish Foundation for Strategic Research through its Center for Autonomous Systems. The support is gratefully acknowledged.

The research presented in this paper was carried out in the context of the Intelligent Service Robot project and as such some of the results are based on prior work by Patric Jensfelt, Anders Orebck, Olle Wijk, David Austin, Lars Petersson and Magnus Andersson. Without their assistance this work would not have been possible. Finally the work has benefited from discussions with Jan-Olof Eklundh and Peter Allan.

References

1. P. Wallace, *AgeQuake: Riding the Demographic Rollercoaster Shaking BUsiness, Finance and our World.* London, UK: Nicholas Brealey Publishing Ltd., 1999. ISBN 1–85788–192–3.
2. S. Thrun, "Probabilistic algorithms in robotics," *AI Magazine*, vol. 21, pp. 93–110, Winter 2000.
3. P. Jensfelt, D. Austin, and H. I. Christensen, "Towards task oriented localisation," in *Intelligent Autonomous Systems – 6* (E. Pagelle, F. Groen, T. Aria, R. Dillmann, and A. Stenz, eds.), (Venice, IT), pp. 612–619, IAS, IOS Press, July 2000.
4. M. Andersson, A. Orebäck, M. Lindström, and H. Christensen, *Intelligent Sensor Based Robots*, vol. 1724 of *Lecture Notes in Artificial Intelligence*, ch. ISR: An Intelligent Service Robot, pp. 291–314. Heidelberg: Springer Verlag, October 1999.
5. G. Fink, N. Jungclaus, F. Kummert, H. Ritter, and G. Sagerer, "A distributed system for integrated speech and image understanding," in *Intl. Symp on Artificial Intelligence*, (Cancun, Mexico), pp. 117–126, 1996.
6. G. A. Fink, C. Schillo, F. Kummert, and G. Sagerer, "Incremental speech recognition for multi-modal interfaces," in *IEEE 24th Conf. on Industrial Electronics*, (Aachen), pp. 2012–2017, September 1998.

7. J. Cassell, "A framework for gesture generation and interpretation," in *Computer Vision for machine interaction* (R. Cipolla and A. Pentland, eds.), pp. 191–215, Cambridge University Press, 1998.
8. T. Sterner, J. Weawer, and A. Pentland, "Real-time american sign language recognition using desk and wearable computer based video," *IEEE-PAMI*, vol. 20, pp. 1371–1375, Dec. 1998.
9. R. Cipolla, P. Hadfield, and N. Hollinghurst, "Uncalibrated stereo vision with pointing for a man-machine interface," in *IAPR workshop on machine vision application*, (Tokyo), December 1994.
10. R. Cipolla, N. Hollinghurst, A. Gee, and R. Dowland, "Computer vision in interactive robotics," *Assembly Automation*, vol. 16, no. 1, 1996.
11. M. Soriano, B. Martinkauppi, S. Huovinen, and M. Laassonen, "Skin colour modelling under varying illumination conditions using skin locus for selecting training pixels," in *Real-Time Image Sequence Analysis – RISA-2000* (O. Silven and J. Heikkilæ, eds.), (Oulu, Finland), pp. 43–49, Infotech, Oulu University, August 2000.
12. Y. Bar-Shalom and T. Fortmann, *Tracking and Data Association*. New York, NY.: Academic Press, 1987.
13. L. R. Rabiner, "A tutorial on Hidden Markov Models and selected applications in speech," *IEEE Proceedings*, vol. 77, pp. 257–286, February 1989.
14. S. Young, *The HTK Book*. Cambridge University, UK, 1995.
15. L. Petersson, D. Austin, D. Kragić, and H. Christensen, "Towards an intelligent robot system," in *Proceedings of the Intelligent Autonomous Systems 6, IAS-6*, (Venice), pp. 704–709, July 2000.
16. K. Tarabanis, P. Allen, and R. Tsai, "A survey of sensor planning in computer vision," *IEEE Transactions on Robotics and Automation*, vol. 11, no. 1, pp. 86–104, 1995.
17. S. Edelman, ed., *Representation and recognition in vision*. Cambridge, MA: The MIT Press, 1999.
18. D. Roobaert, "Improving the generalisation of Linear Support Vector Machines: an application to 3D object recognition with cluttered background," in *Proceedings of the International Joint Conference on Artificial Intelligence, IJCAI'99, Workshop on Support Vector Machines*, pp. 29–33, 1999.
19. G. Hager, W. Chang, and A. Morse, "Robot feedback control based on stereo vision: Towards calibration-free hand-eye coordination," *IEEE Control Systems Magazine*, vol. 15, no. 1, pp. 30–39, 1995.
20. W. Wilson, C. W. Hulls, and G. Bell, "Relative end-effector control using cartesian position based visual servoing," *IEEE Transactions on Robotics and Automation*, vol. 12, no. 5, pp. 684–696, 1996.
21. E. Malis, F. Chaumette, and S. Boudet, "Positioning a coarse–calibrated camera with respect to an unknown planar object by 2 1/2D visual servoing," in *5th IFAC Symposium on Robot Control (SYROCO'97)*, vol. 2, (Nantes, France), pp. 517–523, September 1997.
22. A. Bicchi and V. Kumar, "Robotic grasping and contact: A review," in *Proc. of the IEEE Int. Conf. on Robotics and Automation*, pp. 348–353, 2000.
23. K. B. Shimoga, "Robot grasp synthesis algorithms: A survey," *International Journal of Robotics Research*, vol. 15, pp. 230–266, June 1996.
24. D. Lowe, "Fitting parameterized three–dimensional models to images," *IEEE Transactions on Pattern Analysis and Machine Intelligence*, vol. 13, no. 5, pp. 441–450, 1991.

25. D. Koller, K. Daniilidis, and H. Nagel, "Model-based object tracking in monocular image sequences of road traffic scenes," *International Journal of Computer Vision*, vol. 10, no. 3, pp. 257–281, 1993.

26. E. Marchand, P. Bouthemy, and F. Chaumette, "A 2D–3D Model–Based Approach to Real–Time Visual Tracking," Technical report ISSN 0249-6399, ISRN INRIA/RR-3920, Unité de recherche INRIA Rennes, IRISA, Campus universitaire de Beaulieu, 35042 Rennes Cedex, France, March 2000.

27. P. Wunsch and G. Hirzinger, "Real-time visual tracking of 3D objects with dynamic handling of occlusion," in *Proceedings of the IEEE International Conference on Robotics and Automation, ICRA'97*, vol. 2, pp. 2868–2873, 1997.

28. A. Miller and P. Allen, "Examples of 3D grasp quality computations," in *Proc. of the IEEE International Conference on Robotics and Automation*, vol. 2, pp. 1240–1246, 1999.

29. L. Roberts, "Machine perception of three–dimensional solids," *Optical and Electrooptical Information Processing*, 1965.

30. M. Fischler and R. Bolles, "Random sample concensus: A paradigm for model fitting with applications to image analysis and automated cartography," *Comm. ACM*, vol. 24, pp. 381–395, 1981.

31. M. Abidi and T. Chandra, "A new efficient and direct solution for pose estimation using quadrangular targets: algorithm and evaluation," *IEEE Transactions on Pattern Analysis and Machine Intelligence*, vol. 17, pp. 534–538, May 1995.

32. D. DeMenthon and L. Davis, "New exact and approximate solutions of the three–point perspective problem," *IEEE Transactions on Pattern Analysis and Machine Intelligence*, vol. 14, pp. 1100–1105, November 1992.

33. R. Horaud, B. Conio, and O. Leboulleux, "An analytical solution for the perspective–4–point problem," *Computer Vision, Graphics and Image Processing*, vol. 47, pp. 33–44, 1989.

34. R. Tsai, "A versatile camera calibration technique for high–accuracy 3D machine vision metrology using off-the–shelf TV cameras and lenses," *IEEE Journal of Robotics and Automation*, vol. 3, pp. 323–344, 1987.

35. J. Yuan, "A general photogrammetric method for determining object position and orientation," *IEEE Transactions on Robotics and Automation*, vol. 5, pp. 129–142, April 1989.

36. D. DeMenthon and L. Davis, "Model-based object pose in 25 lines of code," *International Journal of Computer Vision*, vol. 15, pp. 123–141, 1995.

37. R. C. H. Araujo and C. Brown, "A fully projective formulation for Lowe's tracking algorithm," Technical report 641, The University of Rochester, CS Department, Rochester, NY, November, 1996.

38. J. Foley, A. van Dam, S. Feiner, and J. Hughes, eds., *Computer graphics - principles and practice*. Addison-Wesley Publishing Company, 1990.

39. S. Hutchinson, G. D. Hager, and P. I. Corke, "A tutorial on visual servo control," *IEEE Transactions on Robotics and Automation*, vol. 12, no. 5, pp. 651–670, 1996.

Vision and Touch for Grasping

Rolf P. Würtz

Institute for Neurocomputing, Ruhr-University Bochum, Germany
Rolf.Wuertz@neuroinformatik.ruhr-uni-bochum.de
http://www.neuroinformatik.ruhr-uni-bochum.de/PEOPLE/rolf/

Abstract. This paper introduces our one-armed stationary humanoid robot GripSee together with research projects carried out on this platform. The major goal is to have it analyze a table scene and manipulate the objects found. Gesture-guided pick-and-place This has already been implemented for simple cases without clutter. New objects can be learned under user assistance, and first work on the imitation of grip trajectories has been completed.

Object and gesture recognition are correspondence-based and use elastic graph matching. The extension to bunch graph matching has been very fruitful for face and gesture recognition, and a similar memory organization for aspects of objects is a subject of current research.

In order to overcome visual inaccuracies during grasping we have built our own type of dynamic tactile sensor. So far they are used for dynamics that try to optimize the symmetry of the contact distribution across the gripper. With the help of those dynamics the arm can be guided on an arbitrary trajectory with negligible force.

1 Introduction

In order to study models for perception and manipulation of objects we have set up the robot platform *GripSee*. The long-term goal of this research project is to enable the robot to analyze a table scene and to interact with the objects encountered in a reasonable way. This includes the recognition of known objects and their precise position and pose as well as the acquisition of knowledge about unknown objects. A further requirement is an intuitive mode of interaction to enable a human operator to give orders to the robot without using extra hardware and requiring only minimal training. As the most natural way of transferring information about object locations is pointing, we have chosen to implement a hand gesture interface for user interaction

This article reviews some of the steps we have taken towards that goal and first achievements. The whole project has only been possible through the cooperation of a large group, and I am indebted to Mark Becker, Efthimia Kefalea, Hartmut Loos, Eric Maël, Christoph von der Malsburg, Abdelkader Mechrouki, Mike Pagel, Gabi Peters, Peer Schmidt, Jochen Triesch, Jan Vorbrüggen, Jan Wieghardt, Laurenz Wiskott, and Stefan Zadel for their contributions.

G.D. Hager et al. (Eds.): Sensor Based Intelligent Robots, LNCS 2238, pp. 74–86, 2002.
© Springer-Verlag Berlin Heidelberg 2002

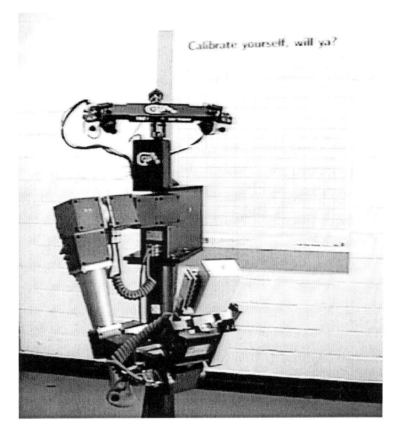

Fig. 1. GripSee's 3 DoF stereo camera head and 7 DoF manipulator.

2   Overall System

The robot hardware is shown in figure 1 and consists of the following components:

- A modular robot arm with seven degrees of freedom (DoF), kinematics similar to a human arm, and a parallel jaw gripper;
- a dual stereo camera head with three DoF (pan, tilt, and vergence) and a stereo basis of 30 cm for two camera pairs with different fields of view (horizontally 56° with color and 90° monochrome, respectively);
- a computer network composed of two Pentium PCs under QNX and a Sun UltraSPARC II workstation under Solaris.

Image acquisition is done by two color frame grabber boards controlled by one of the PCs, which also controls the camera head and performs real-time image processing for, e.g., hand tracking. The second PC controls the robot arm. Since image data has to be transferred between the processors, they are networked with FastEthernet to achieve sufficient throughput and low latencies.

Our software is based on the C++-library *FLAVOR* ("Flexible Library for Active Vision and Object Recognition") developed at our institute. FLAVOR comprises functionality for the administration of arbitrary images and other data types, libraries for image processing, object representation, graph matching, image segmentation, robotics, and interprocess communication [16].

3   Correspondence-Based Recognition

Recognition from visual data is required at two points in the desired functionality, namely the recognition of hand gestures and the recognition of objects.

In contrast to methods employing invariants or signatures our recognition methods rely on establishing a *correspondence map* between a stored aspect and the actual camera image. A big advantage of this type of recognition is that precise position information is recovered in addition to the object identity.

This position information in image coordinates can be used to measure 3-D position required for grasping by a combination of fixation and triangulation. For this to work the cameras must be calibrated relative to the position of the end effector. We have developed our own neural network-based method to do this, which is described in [10, 11].

3.1   Features

The difficulty of the correspondence problem depends on the choice of features. If grey values of pixels are taken as local features, there is a lot of ambiguity, i.e., many points from very different locations share the same pixel value without being correspondent. A possible remedy to that consists in combining local patches of pixels, which of course reduces this ambiguity. If this is done too extensively, i.e., if local features are influenced by a large area, the ambiguities disappear if identical images are used, but the features become more and more sensitive to distortions and changes in background.

As a compromise vectors of features at various scales (and orientations) can be used. One possibility is a wavelet transform based on complex-valued Gabor functions [6, 30, 31], with wavelets parameterized by their (two-dimensional) center frequency. At each image point, all wavelet components can be arranged into one feature vector, which is also referred to as a "jet". These features have turned out to be very useful for face and gesture recognition.

Depending on the application other feature vectors are more appropriate. The combination of different features has been systematically studied in [20]. For the object recognition described here, a combination of Mallat's multiscale edges [8] and object color has yielded good results. For the gesture recognition Gabor wavelets have been used together with color relative to a standard skin color. Generally, feature vectors seem to be better candidates for establishing correspondence maps than scalar features.

3.2 Elastic Graph Matching

Solving the correspondence problem from feature similarity alone is not possible, because the same feature may occur in an image more than once. Therefore, the relative position of features has to be used as well. One standard method for this is elastic graph matching: the stored views are represented by sparse graphs, which are vertex labeled with the feature vectors and edge labeled with the distance vector of the connected vertices. Matching is done by first optimizing the similarity of an undistorted copy of the graph in the input image and then optimizing the individual locations of the vertices. This results in a distorted graph whose vertices are at corresponding locations to the ones in the model graph. This procedure is described in full detail in [6, 28, 29].

3.3 Attention Control

Conceptually, the solution of the correspondence problem includes segmentation of the scene – when a correct correspondence map is available, the background is eliminated as well. Practically, the matching algorithms become time consuming and error prone when the area covered by the object proper becomes smaller relative to the image size. Therefore, our recognition system usually employs some attention control to select areas that are more likely than others to contain an object. It is important that this be fast enough to allow a reaction by the system. Therefore, only simple cues such as color or motion are used.

A second step to simplify the recognition tasks is *fixation*, i.e. centering both cameras on an object point. This also has the advantage to minimize perspective distortions.

3.4 Bunch Graphs

The major drawback of most correspondence-based recognition systems is that the computationally expensive procedure of creating a correspondence map must be done for *each* of the stored models. In the case of face recognition, this has been overcome by the concept of *bunch graphs* [28, 29]. The idea is that the database of models is arranged in such a way that corresponding graph nodes are already located at corresponding object points, e.g., a certain node lies on the left eye of all models. For large databases, this reduces the recognition time by orders of magnitudes. The resulting face recognition system has performed very well in the FERET test, an evaluation by an independent agency. It was one of two systems that underwent that test without requiring additional hand-crafted information such as the position of the eyes in the images to be analyzed. The performance on the difficult cases, where the images of persons were taken at widely different times was clearly better than the competitors' [15]. As a matter of fact, only one competitor underwent the test without requiring further information regarding the position of the eyes. This again illustrates how difficult the correspondence problem really is.

Object

> – List of stable **Poses**
> – (Mass)

Pose

> – List of **Views**
> – List of **Grips**

View

> – 3-D Orientation in object coordinates
> – Elastic **Graph**
> – Position of a fixation point relative to the graph

Graph

> – List of **Vertices** labeled with **Feature vectors**
> – List of **Graph Edges**

Grip

> – 3D orientation in object coordinates
> – 3D position in object coordinates
> – Gripper opening
> – (Max. pressure)
> – (Surface properties, etc.)

Fig. 2. Schema of our object representation. Items in parentheses are important object parameters which currently cannot be extracted from sensor data.

The bunch graph method exploits the fact that all faces are structurally very similar and correspondences inside the database make sense at least for a subset of salient points. For general object recognition this is much more difficult. Consequently, there is currently no suitable data format for storing many aspects of many objects. The organization of the database of all aspects of *one* object is subject of current research in our group [26, 14, 27, 13]. Another successful application lies in the recognition of hand gestures, where is has been used to make the recognition background invariant [24, 21, 23].

3.5   Object Representation

An object representation suited for visually guided grasping has to integrate 2D visual features and 3D grip information about a known object, in order to apply a known grip when the situation requires it. Autonomous learning of the representation is highly desirable, therefore complicated constructs like CAD models are not considered. Rather, we adopt the view that visual recognition and application of a grip is mainly a recollection of what has been seen or done before, with the necessary slight modifications to adapt to the situation at hand.

The visual representation is view-based, i.e., for each different orientation of the object a set of visual features is stored, which are extracted from the left and right stereo images and grouped into a *model graph*, which preserves the

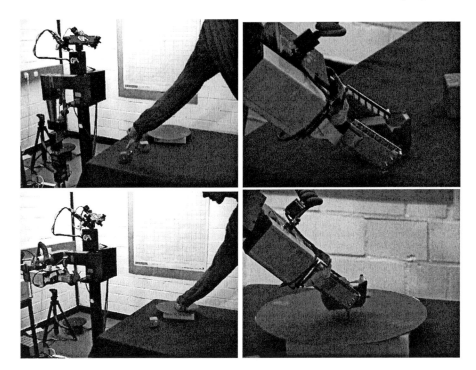

Fig. 3. Pick- and place behavior. The user points at one of the objects with a hand gesture (top left). The hand gesture is recognized and the closest object is picked up at the angle indicated by the gesture (top right). The robot waits for a new gesture (bottom left) and places the object at the indicated position (bottom right).

topological relationships between the features. These model graphs are stored in a library for different objects with different orientations and are used to recognize known objects with a graph matching process, which is invariant under translation and scale. The grips are then associated with the views. A schema of our object representation can be found in figure 2.

4 Pick and Place Behavior

The visual recognition techniques outlined above can be applied to everyday objects as well as to hand gestures. Like all correspondence based methods they have the advantage to yield detailed position information in addition to object identity. This makes them well suited as subsystems for a humanoid robot that can analyze and manipulate a table scene under human control. Thus, we have successfully implemented a behavior module, which allows a user to point to one of the objects on the table and have the robot pick it up and place it at a desired position. So far it is assumed that the robot knows all the objects from their visual appearance and how to grasp them.

The system starts in a state, in which the robot stands in front of a table with various known objects and waits for a human hand to transmit instructions. The interaction with the human is initiated by moving a hand in the field of view of the robot. This movement is tracked using a fusion of skin color and motion cues. In order to point to an object the hand must stop, which is detected by the robot using the sudden vanishing of the motion cue during tracking.

Now a rough position of the hand is known, which is fixated by the camera head. After fixation, the type of hand gesture posture is determined by graph matching. This matching process also yields a refined estimate of the position of the hand's center, which is fixated. The actual identity of the hand gesture (fist, flat, etc.) can be assigned arbitrary additional information to be conveyed to the robot. We currently use it to code for the grasping angle relative to the table (steep grip vs. low grip, etc.).

It is expected that the hand now points at an object, thus the fixation point is lowered by a fixed amount of about 10 cm. Now, a simple localization algorithm detects the object, which is then fixated. Now the object recognition module determines the type of object, its exact position and size in both images and its orientation on the table.

Like in the posture recognition the cameras fixate on a central point of the recognized object (which is a feature known for each each view of each object). This yields the most precise object position that can be expected. Obviously, this precision is crucial for reliable grasping.

Now a grip suitable for the object and according to what was demanded by the hand gesture is selected from the set of grips attached to the object and transformed into world coordinates. A trajectory is planned, arm and gripper are moved towards the object, and finally the object is grasped. Another trajectory transfers it to a default position conveniently located in front of the cameras for possible further inspection.

Then the field of view is again surveyed for the operator's hand to reappear. In this state, it should be pointing with a single defined gesture to the three-dimensional spot where the object is to be put down. The 3D position is again measured by the already described combination of fixation movements and gesture recognition. Then the arm moves the the object to the desired location and the gripper releases it.

5   User-Assisted Object Learning

The construction of the object database is a time consuming process, which includes taking images of the object from all possible orientations and coding useful grips from different angles. A good part of our work is dedicated to automate this construction.

Learning of a new object (or more precisely, one stable pose of a new object) is initiated by putting the object onto the table and having the robot create the various views by moving the object around. This procedure has two serious difficulties. First, a good grip must already be known for the robot to manipulate

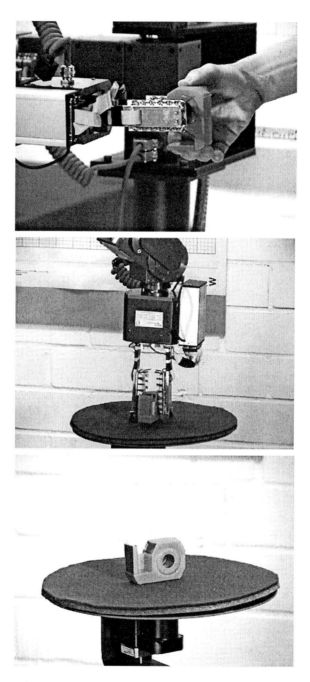

Fig. 4. Learning the representation of a new object. The object is placed into the gripper such that a stable grip is known (top). The robot puts the object onto a turntable (middle) and images of the various views are recorded without interference from the gripper (bottom).

the object in a predictable manner. Second, the actual orientation of the object should be known with good precision, because the error is likely to accumulate over the various views. The problem of learning new grips from scratch can only be solved by relying on tactile information. We have currently constructed and implemented tactile sensors on our gripper (see section 6), but the extraction of detailed object information is subject of future research.

In this situation, we have decided to solve both problems by what we call *user-assisted learning*. The general idea is that the acquisition of knowledge is as autonomous as possible, but a human operator still makes decisions about what is important and thus guides the process. Concretely, in the current case our learning procedure is as follows. The operator presents the novel object by putting it into the gripper (which has a defined position and orientation at that moment), in a position and orientation that are ideal for grasping. They thus define both a default grip and the object coordinate system. The robot closes the gripper, puts the object onto the center of a turntable, *fixates* on the grasping center, and takes a stereo image pair of the first object view. Then, the turntable is rotated by a specified increment and a second view is taken. This is repeated until a full circle of object views is acquired. After acquisition, all images are rotated around their center to compensate for the rotation associated with a combination of tilt and vergence.

After the images of the views are taken, they are converted into a collection of *labeled graphs*. In this learning step, it is assumed that the background is uniform (which is the case for the surface of the turntable) in order to avoid the necessity of complicated segmentation methods and to assure as clean graphs as possible in the representation (see [4] for details).

Two alternatives to using the turntable may be considered, namely having the robot place the object onto the table in all necessary orientations or holding it in the gripper while storing the visual information. Both have not been pursued so far in order to minimize the mechanical strain on the hardware. The second possibility, poses the additional problem of segmenting and subtracting the gripper itself from the image of the object. The quality of the object recognition depends rather critically on the quality of the model graphs. Our system therefore contains the possibility to modify the node positions by hand, but under good illumination the automatically created graphs are usually good enough.

6   Tactile Sensors

In the course of the project it has become clear that visual information in a humanoid setup is not accurate enough for precise grasping. It should be enhanced with tactile information for fine tuning during the application of a grip. As none of the currently available sensors met our requirements for a robust system that was very sensitive for dynamic touch, a new type of sensor has been developed [18, 19]. In combination it complements two static elements (for x-

and y- direction, respectively) with 16 dynamic sensor elements on each jaw of the gripper.

As a first approach to behavior guided by tactile sensing we have implemented a method for guiding the robot arm manually on a desired trajectory. The signal from each sensor is binarized with a suitable threshold and triggers a predefined reflex movement, which is represented by a translational and a rotational movement vector relative to the gripper. The translation vector moves the contact position towards a target position, e.g., the center between both gripper jaws. This results in a radial vector field for the translation, where each direction is given by a vector pointing from the target position towards the sensor position. The rotation vector turns the gripper such as to maximize the number of sensor contacts, i.e., it attempts a parallel orientation of the gripper jaw to the hand's surface. The opening width of the gripper is controlled to keep loose contact with the hand. In case of contact with both gripper jaws it receives a small opening signal, and in case of one-sided contact a small closing signal.

This control can be used to guide the robot arm on a desired trajectory by putting a hand between the jaws. If adjusted suitably, the movement components of each sensor element add up to a motion which minimizes the asymmetry of the contact around the hand and thus follows the hand's movement through the configuration space. Given a simple static object with parallel grasping surfaces, the same strategy will eventually converge to a situation with a symmetric contact distribution around the desired grasping position on each gripper jaw. For grasping, this can be followed by pressure-controlled closing of the gripper. Turning this method into a good grasping strategy with minimal requirements on object shape and developing a systematic way to learn optimal vector fields are subjects of current research. Full details about the dynamics and the overall robot control can be found in [7].

7   Related Work

A comprehensive study of the state of the art in robotic grasping is clearly beyond the scope of this paper. To put the work into perspective, I classify the methods used in terms of the possibilities provided in [3]. The control structure is hierarchical in the sense that the vision system estimates a 6D grasping position, and then a grasping trajectory is calculated. Possible grips are associated with object identities and part of the objects' description. The 3D position and the object's orientation on the table are derived from the recognition procedure, the remaining DoF of the grip orientation is derived from the user interaction. Different from the servoing systems in, e.g. [2], our system is currently "endpoint-open-loop", as no attempt is made to follow the end effector visually. A complete "endpoint-closed-loop" system like the one described in [5] would be desirable, but we will have to develop a suitable visual model of the end effector first that could accommodate the quite complicated occlusions arising at the critical moment of actual grasping. Tracking the manipulator with algorithms like in [9] is not enough in such situations. In order to circumvent these problems, we are currently planning to correct the control errors resulting from visual inaccuracies using the tactile sensors.

Our system is designed to resemble the situation in human perception as closely as possible, which distinguishes it from the work done on industrial manipulators. The latter are usually non-redundant, and the cameras can be positioned according to the needs of the task at hand. Robot systems similar to our approach include [25] and [12]. The attention mechanisms and active depth estimate bear resemblance to the system described in [1].

8   Outlook

In the following I give an outlook on projects that are currently underway in order to enhance the robot's capabilities.

8.1   Imitation Learning

In addition to learning the visual appearance of objects the learning of grips is important. As it is hard to define what constitutes a good grip an attractive method is to have the robot imitate trajectories performed by a human. We have measured the trajectories of a pair of thumb and index finger from a Gabor transform with specially adapted parameters [22] with encouraging results. Another project tries to estimate the arm position in whole-body gestures. First results have been published in [17].

8.2   Further Problems

Among the problems to be tackled next is the interpretation of the output of the tactile sensors and the classification of situations such as a sliding object, a perfect grip, and an unstable grip from the time series displayed by the tactile sensors during grasping. A further task will be to learn a representation of 3D space in a way that is suitable for manipulation. Concretely, the robot body (and possibly other parts) must be represented so that collisions can be reliably avoided. The relative calibration of vision, touch and proprioception must be improved, and it is highly desirable to have it updated during normal operation. The optimization of grips from sensor data has already been mentioned. On the visual side the organization of the object database to make rapid recognition possible is the most pressing issue. Finally, the integration of the sensor information to a coherent percept of the environment and a good organization of the overall behavior must be pursued.

Acknowledgments

Financial support from the projects NEUROS (01 IN 504 E 9) and LOKI (01 IN 504 E 9) by the German Minister for Education and Research and from the RTN network MUHCI by the EU is gratefully acknowledged.

References

1. Sven J. Dickinson, Henrik I. Christensen, John K. Tsotsos, and Göran Olofsson. Active object recognition integrating attention and viewpoint control. *Computer Vision and Image Understanding*, 67:239–260, 1997.
2. G.D. Hager. A modular system for robust positioning using feedback from stereo vision. *IEEE Trans. Robotics and Automation*, 13(4):582–595, 1997.
3. Seth Hutchinson, Gregory D. Hager, and Peter I. Corke. A tutorial on visual servo control. *IEEE Transactions on Robotics and Automation*, 12(5):651–670, 1996.
4. Efthimia Kefalea. *Flexible Object Recognition for a Grasping Robot*. PhD thesis, Computer Science, Univ. of Bonn, Germany, March 1999.
5. Danica Kragić and Henrik I. Christensen. Cue integration for visual servoing. *IEEE Transactions on Robotics and Automation*, 17(1):18–27, 2001.
6. Martin Lades, Jan C. Vorbrüggen, Joachim Buhmann, Jörg Lange, Christoph von der Malsburg, Rolf P. Würtz, and Wolfgang Konen. Distortion invariant object recognition in the dynamic link architecture. *IEEE Transactions on Computers*, 42(3):300–311, 1993.
7. Eric Maël. *Adaptive and Flexible Robotics for Visual and Tactile Grasping*. PhD thesis, Physics Dept., Univ. of Bochum, Germany, December 2000.
8. S. Mallat and S. Zhong. Characterization of signals from multiscale edges. *IEEE Transactions on Pattern Analysis and Machine Intelligence*, 14:710–732, 1992.
9. Kevin Nickels and Seth Hutchinson. Model-based tracking of complex articulated objects. *IEEE Transactions on Robotics and Automation*, 1997.
10. Mike Pagel, Eric Maël, and Christoph von der Malsburg. Self calibration of the fixation movement of a stereo camera head. *Autonomous Robots*, 5:355–367, 1998.
11. Mike Pagel, Eric Maël, and Christoph von der Malsburg. Self calibration of the fixation movement of a stereo camera head. *Machine Learning*, 31:169–186, 1998.
12. Josef Pauli. Learning to recognize and grasp objects. *Autonomous Robots*, 5(3/4):407–420, 1998.
13. Gabriele Peters. *Representation of 3D-Objects by 2D-Views*. PhD thesis, Technical Faculty, University of Bielefeld, Germany, 2001. In preparation.
14. Gabriele Peters and Christoph von der Malsburg. View Reconstruction by Linear Combinations of Sample Views. In submitted to: *International Conference on Knowledge Based Computer Systems (KBCS'2000), Mumbai, December 17 - 19, 2000*, 2000.
15. P. Jonathan Philips, Hyeonjoon Moon, Syed A. Rizvi, and Patrick J. Rauss. The FERET evaluation methodology for face-recognition algorithms. *IEEE Transactions on Pattern Analysis and Machine Intelligence*, 22(10):1090–1104, 2000.
16. Michael Rinne, Michael Pötzsch, Christian Eckes, and Christoph von der Malsburg. Designing Objects for Computer Vision: The Backbone of the Library FLAVOR. Internal Report IRINI 99-08, Institut für Neuroinformatik, Ruhr-Universität Bochum, D-44780 Bochum, Germany, December 1999.
17. Achim Schäfer. Visuelle Erkennung von menschlichen Armbewegungen mit Unterstützung eines dynamischen Modells. Master's thesis, Physics Dept., Univ. of Bochum, Germany, February 2000.
18. Peer Schmidt. Aufbau taktiler Sensoren für eine Roboterhand. Master's thesis, Physics Dept., Univ. Bochum, Germany, December 1998.
19. Peer Schmidt, Eric Maël, and Rolf P. Würtz. A novel sensor for dynamic tactile information. *Robotics and Autonomous Systems*, 2000. In revision.

20. J. Triesch and C. Eckes. Object recognition with multiple feature types. In L. Niklasson, M. Bodén, and T. Ziemke, editors, *Proceedings of the ICANN'98, International Conference on Artificial Neural Networks*, pages 233–238. Springer, 1998.

21. J. Triesch and C. von der Malsburg. A gesture interface for robotics. In *FG'98, the Third International Conference on Automatic Face and Gesture Recognition*, pages 546–551. IEEE Computer Society Press, 1998.

22. J. Triesch, J. Wieghardt, C. v.d. Malsburg, and E. Maël. Towards imitation learning of grasping movements by an autonomous robot. In A. Braffort, R. Gherbi, S. Gibet, Richardson, and D. J., Teil, editors, *Gesture-Based Communication in Human-Computer Interaction, International Gesture Workshop, GW'99, Gif-sur-Yvette, France, March 17-19, 1999 Proceedings*, volume 17 of *Lecture Notes in Computer Science*. Springer-Verlag, 1999. ISBN 3-540-66935-3.

23. Jochen Triesch. *Vision-Based Robotic Gesture Recognition*. PhD thesis, Physics Dept., Univ. of Bochum, Germany, 1999.

24. Jochen Triesch and Christoph von der Malsburg. Robust classification of hand postures against complex backgrounds. In *Proceedings of the Second International Conference on Automatic Face and Gesture Recognition*, pages 170–175. IEEE Computer Society Press, 1996.

25. J.K. Tsotsos, G. Verghese, S. Dickinson, M. Jenkin, A. Jepson, E. Milios, F. Nuflot, S. Stevenson, M. Black, D. Metaxas, S. Culhane, Y. Ye, and R. Mann. PLAYBOT: A visually-guided robot to assist physically disabled children in play. *Image and Vision Computing*, 16:275–292, 1998.

26. J. Wieghardt and C. von der Malsburg. Pose-independent object representation by 2-d views. In *IEEE International Workshop on Biologically Motivated Computer Vision, May 15-17, Seoul*, 2000.

27. Jan Wieghardt. *Learning the Topology of Views: From Images to Objects*. PhD thesis, Physics Dept., Univ. of Bochum, Germany, July 2001.

28. Laurenz Wiskott. *Labeled Graphs and Dynamic Link Matching for Face Recognition and Scene Analysis*. Reihe Physik. Verlag Harri Deutsch, Thun, Frankfurt am Main, 1996.

29. Laurenz Wiskott, Jean-Marc Fellous, Norbert Krüger, and Christoph von der Malsburg. Face recognition by elastic bunch graph matching. *IEEE Transactions on Pattern Analysis and Machine Intelligence*, 19(7):775–779, 1997.

30. Rolf P. Würtz. *Multilayer Dynamic Link Networks for Establishing Image Point Correspondences and Visual Object Recognition*. Verlag Harri Deutsch, Thun, Frankfurt am Main, 1995.

31. Rolf P. Würtz. Object recognition robust under translations, deformations and changes in background. *IEEE Transactions on Pattern Analysis and Machine Intelligence*, 19(7):769–775, 1997.

A New Technique for the Extraction and Tracking of Surfaces in Range Image Sequences

X. Jiang[1], S. Hofer[1], T. Stahs[2], I. Ahrns[2], and H. Bunke[1]

[1] Department of Computer Science, University of Bern, Switzerland
[2] DaimlerChrysler AG, Research and Technology Center, Ulm, Germany

Abstract. Traditionally, feature extraction and correspondence determination are handled separately in motion analysis of (range) image sequences. The correspondence determination methods have typically an exponential computational complexity. In the present paper we introduce a novel framework of motion analysis that unifies feature extraction and correspondence determination in a single process. Under the basic assumption of a small relative motion between the camera and the scene, feature extraction is solved by refining the segmentation result of the previous frame. This way correspondence information becomes directly available as a by-product of the feature extraction process. Due to the coupled processing of frames we also enforce some degree of segmentation stability. First results on real range image sequences have demonstrated the potential of our approach.

1 Introduction

The ability of dynamic scene analysis is essential to computer vision as well as human vision. In dealing with dynamic scenes, the relative motion between the camera and the imaged scene gives rise to the apparent motion of the objects in a sequence of images. Although considerable work has been done in this domain, the main emphasis of research has been on the intensity image modality.

Range images provide the three-dimensional shape of the sensed object surfaces in the field of view. The explicitness of this additional geometric information has turned out to be very useful for a variety of vision tasks [19]. Today, range cameras are becoming available that are able to acquire high-resolution range images in (quasi) real-time [5, 7, 29]. With these new developments in 3D imaging techniques, it is now realistic to address the problem of dynamic scene analysis in a sequence of range images.

A popular paradigm of motion analysis in a (range) image sequence consists of three stages:

- segmentation of each image of the sequence and extraction of key features,
- establishment of correspondences of the key features between frames,
- computation of motion using the feature correspondences.

Alternatively, motion can be analyzed by means of range flow computation [10, 16, 28, 30]. Here there is no need to solve the feature correspondence problem. As

G.D. Hager et al. (Eds.): Sensor Based Intelligent Robots, LNCS 2238, pp. 87–100, 2002.

a disadvantage, the motion information concerns merely the individual pixels; motion associated with higher-level structures such as surfaces and objects has to be determined separately. Overview of motion analysis in general and optical flow computation in particular can be found in [1, 2, 14, 18] and [6], respectively.

In the present work the correspondence determination paradigm is adopted. However, we propose a framework of motion analysis, in which the feature extraction and the correspondence determination are unified in a single stage. This novel approach has a number of advantages, including stable feature extraction and efficient correspondence determination. First results on real range image sequences have demonstrated the potential of our method.

The remainder of the paper is structured as follows. In the next section we motivate our work by a brief discussion of related literature. Then, some basic results fundamental to the new algorithm are given, followed by a description of the new algorithm itself in Section 4. In Section 5 experimental results are presented. Finally, some discussions conclude the paper. Note that the actual computation of motion parameters using known feature correspondences is not part of this work and the readers are referred to the survey paper [26].

2   Motivation

Various features can be used for motion analysis and their properties directly affect the stability, reliability, and robustness with respect to noise and distortions in input data. Thus far, researchers have investigated the use of 3D points [11, 12, 23]. Such local features are highly sensitive to noise and can easily be completely occluded. On the other hand, global features such as line segments [13] and surfaces [3, 24, 27] are more robust with respect to noise and occlusion.

In the correspondence determination paradigm for motion analysis, the features are independently extracted from the frames of an image sequence. Given the features of two successive frames, the correspondences are typically found by means of methods like graph matching [12], tree search [13, 27], and maximal clique detection [24], that have provably an exponential computational complexity in the worst case.

Segmentation and extraction of surfaces in range images turns out to be a difficult task and cannot be considered as solved [15, 22, 25]. In particular, there is generally no guarantee of a stable segmentation. Even for the case of small changes from frame I_{t-1} to frame I_t, their segmentation results may be different, for instance, due to an over-segmentation of some surface of I_{t-1} in I_t. This instability causes another potential problem to the correspondence determination paradigm.

In this work we assume a small relative motion between the camera and the scene. Considering the high-speed range cameras available today, this assumption is reasonable. Given the segmentation result of frame I_{t-1}, the next frame I_t is not segmented from scratch, but based on the segmentation of I_{t-1}. In this manner the segmentation result of I_t is strongly coupled with that of I_{t-1}, thus forcing some degree of segmentation stability. More importantly, the feature

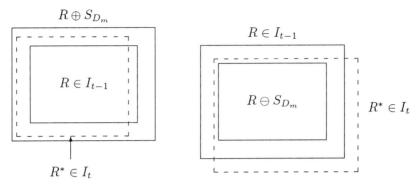

Fig. 1. Basic results: R^* is bounded by the contour of $R \oplus S_{D_m}$ (left); $R \ominus S_{D_m}$ is guaranteed to be within R^* (right).

correspondences become directly available as a by-product of the segmentation process at no extra cost. These properties make the proposed approach to motion analysis an attractive alternative to the popular correspondence determination paradigm.

From a fundamental point of view our approach possesses some similarity to the numerous works dealing with tracking and 3D reconstruction of contour deformation by means of active contours [8, 9]. While the interest there is in the tracking of, typically, a small number of contours, we fully cover all surfaces of the imaged scene.

3   Basic Results

A movement of a surface in the scene from time $t-1$ to t causes its corresponding image region R in frame I_{t-1} to become some R^* in frame I_t. Such a movement may be any combination of rotations and translations, and has six degrees of freedom. The caused change in the images varies accordingly: R^* corresponds to a rotated and translated version of R when the spatial translation is parallel to the image plane and the spatial rotation is around an axis perpendicular to the image plane. If the camera comes closer/further to the scene, then R^* results from a scaling of R. In general, the deformation of R to R^* may become more complex than these basic image transformations and even their combinations.

We use D_m to represent the maximal displacement of an image pixel from R to R^* due to the relative motion between the camera and the scene. Let the contour of R and R^* be denoted by C and C^*, respectively. Then, any deformation of R to R^* can be modeled by a continuous function f:

$$f: \quad C \rightarrow C^*, \ \forall p \in C(|p - f(p)| \leq D_m) \tag{1}$$

that maps the contour C to C^*.

Let R_v be R translated by a directional vector v of length D_m. Obviously, R^* is contained in the union of R_v for all v. In the terminology of mathematical morphology, this fact is expressed by:

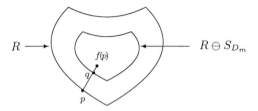

Fig. 2. $f(p)$ cannot be inside $R \ominus S_{D_m}$.

$$R^* \subseteq R \oplus S_{D_m}$$

where $\oplus$ denotes the dilation operation and S_{D_m} represents a circular structural element of radius D_m, see Figure 1 for a graphical illustration. Therefore, we have:

Property 1. The relationship $R^* \subseteq R \oplus S_{D_m}$ holds for any mapping function f of form (1).

Now we establish another result that is fundamental to the new feature extraction algorithm presented in the next section.

Property 2. The relationship $R \ominus S_{D_m} \subseteq R^*$ holds for any mapping function f of form (1). Here $\ominus$ represents the erosion operation.

Proof A graphical illustration of this property is given in Figure 1. Under the mapping function f, each point p on the contour of R, C, is mapped to $f(p)$ such that $|p - f(p)| \leq D_m$. Note that $f(p)$ cannot be inside $R \ominus S_{D_m}$. Otherwise, let q be the intersection point of the straight line segment $\overline{pf(p)}$ with the contour of $R \ominus S_{D_m}$, see Figure 2. According to the definition of erosion, q is at least D_m apart from the contour of R, thus $|p - q| \geq D_m$. Therefore, we get:

$$|p - f(p)| = |p - q| + |q - f(p)| > D_m$$

which is a contradiction. Since $f(p)$ on the contour of R^* is not inside $R \ominus S_{D_m}$, $R \ominus S_{D_m} \subseteq R^*$ holds.   QED

The discussion above can be summarized as follows. The region R^* in frame I_t resulting from R from frame I_{t-1} is bounded by the contour of $R \oplus S_{D_m}$. Although R^* is still undetermined, we know that $R \ominus S_{D_m}$ definitely lies within R^*. This latter property provides the basis for our feature extraction algorithm.

4   Feature Extraction Algorithm

We adopt the region-growing approach for surface extraction. In contrast to the correspondence determination paradigm, however, the frames are not processed independently. Instead, the segmentation result of frame I_{t-1} is taken over and refined to produce the segmentation result of frame I_t.

input: RM_{t-1} (region map at time $t-1$); I_t (range image at time t)
output: RM_t (region map at time t)

$RM_t = RM_{t-1}$;
/* old regions */
for each region $R \in RM_t$ **do** generate seed region R^s from R;
/* new regions */
generate seed regions R^s in unlabeled areas in RM_t;
/* region expansion */
for each seed region $R^s \in RM_t$ **do** expand R^s using data from I_t;

Fig. 3. Algorithm outline.

Under the basic assumption of a small relative motion between the camera and the scene, the term D_m and therefore the deformation of region $R \in I_{t-1}$ to $R^* \in I_t$ is small. From the last section we know that $R \ominus S_{D_m}$ is part of the unknown region R^*. Consequently, we can make use of $R \ominus S_{D_m}$ as a seed region for the region-growing process to generate R^*. For each region $R \in I_{t-1}$, a seed region can be created for I_t this way. With the time, new surfaces may come up in image frames. For these incoming regions no previous information is available. Accordingly, their seed regions have to be generated in a different manner. After the generation of all seed regions, the region-growing process is started. An outline of the feature extraction algorithm is given in Figure 3. Note that we still need to segment the first image of a sequence. This can be done by any segmentation known from the literature. From the second frame on our feature extraction algorithm becomes applicable.

4.1   Handling of Old Surfaces

The seed region R^s for an unknown region $R^* \in I_t$ is generated by $R \ominus S_{D_m}$. To be reliable, R^s should be of some minimum size. Otherwise, it is excluded from the region-growing process. On the one hand, a disappearing surface causes a too small seed region. On the other hand, a constantly visible surface of very small size may be discarded as well. This latter point will be further discussed in Section 4.4

We are faced with one problem when using the strategy of seed region generation above. After erosions, $R \ominus S_{D_m}$ may become disconnected. Figure 4 illustrates such a situation where $R \ominus S_{D_m}$ consists of two parts. There are two potential solutions for this problem:

(a) allow disconnected (seed) regions;
(b) consider the disconnected parts as different seed regions.

In (a) no special handling is needed. After region-growing the parts may become a single connected region or remain disconnected (the number of disconnected

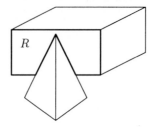

Fig. 4. Seed region $R \ominus S_{D_m}$ is disconnected.

parts may be reduced). In the latter case the disconnectedness of the region is propagated to the succeeding frames, indicating their common origin. This way we also preserve the chance to reunify them in the future when the occlusion disappears. In contrast, the disconnected parts result in different region in solution (b) even when they become connected after region-growing or later in the image sequence. In this case an explicit merge operation is needed. In our current implementation we have used solution (a).

4.2   Handling of Incoming Surfaces

Incoming surfaces manifest themselves as image areas that cannot be merged to the existing regions. We may consider these unlabeled areas after growing the existing regions. Again, seed regions are generated and expanded. In this approach two passes are needed to handle the existing and the incoming surfaces separately. Alternatively, we may also postpone handling of unlabeled areas to the next frame. There seed regions are generated at the same time as for the existing regions. Then, a single pass of region-growing deals with all seed regions. In our current implementation we have used the second approach, which is also the version given in the algorithm outline in Figure 3.

There is no previous information for the incoming surfaces. Accordingly, their seed regions have to be generated in a different manner. In an unlabeled area we simply put a square seed region at a certain distance apart from the area boundary. Potentially, more than one seed region can be set as long as they are distant enough from each other.

4.3   Region-Growing

The region-growing process is guided by surface approximations. Let the approximating surface function be $g(\boldsymbol{a}, x, y, z) = 0$ defined by a parameter vector $\boldsymbol{a}$. The choice of g is dependent on the applications. Examples are planes, bivariable polynomial functions, and quadrics. For each seed region R^s an initial surface function $g(\boldsymbol{a}_s, x, y, z) = 0$ is computed by the least-squares method. A pixel (x_0, y_0, z_0) adjacent to R^s is merged to R^s if the fit error test with the maximal tolerable fit error T_f:

$$e(g(\boldsymbol{a}_s, x, y, z), x_0, y_0, z_0) \leq T_f$$

survives successfully. Here $e()$ represents a fit error function. Typically, this may be the function approximation error, the orthogonal distance to surface, or the algebraic approximation error (for quadrics). The region-growing process is repeated until no more merge can be made. Finally, the surface function is updated using the merged pixels as well to yield a more reliable surface representation. At this place a cycle of region-growing and surface approximation can be started until the region becomes stable.

4.4   Correspondence Determination

In our framework of feature extraction the correspondence determination is trivial. A region $R^* \in I_t$ either evolves from a region $R \in I_{t-1}$ or is newly generated. In the latter case a new surface emerges and will be tracked from now on.

Using this simple strategy, we lose the correspondence information sometimes. The reason is that the computed seed region $R \ominus S_{D_m}$ may be too small to be really useful. Then, it is excluded from the region-growing process. This particular unlabeled area will be considered as a potential incoming surface. Note that this problem is associated with very small regions only. As far as such regions become larger later in the image sequence, their tracking will become stable.

5   Experimental Results

The feature extraction algorithm described in the last section has been implemented in C language under both UNIX and MS Windows. For the initial segmentation of the first image of a sequence, an edge-based approach was used, in which edge points resulting from an edge detection operator [20] are grouped to produce a complete region map [21]. The current program version is limited to planar surface patches. Accordingly, the surface function for region-growing is defined by:

$$g(\boldsymbol{a}, x, y, z) \;=\; Ax + By + Cz + D \;=\; 0.$$

The principal component analysis technique was applied to compute a plane function by minimizing the sum of squared Euclidean (orthogonal) distances. The fit error function is computed by the orthogonal distance of a point (x_0, y_0, z_0) to the plane:

$$e(g, x_0, y_0, z_0) \;=\; \frac{|Ax_0 + By_0 + Cz_0 + D|}{\sqrt{A^2 + B^2 + C^2}}.$$

We have used an active range camera (see Figure 5) developed by DaimlerChrysler Aerospace (DASA Space Infrastructure) to acquire range image sequences. It uses diode laser floodlight in pulsed operation and a CCD-camera with an ultra fast shutter. The scene is illuminated with a variable pulsed laser light (10-100ns) passing through a micro-lens arrays creating a homogeneous beam. The light reflected from the scene is filtered to reduce the surrounding

Fig. 5. Range camera used in the experiments.

lights and the ultra fast electronic shutter of the CCD camera is used to transform the received optical energy and the time delay into range information. Using no scanner, the range information is obtained by time of flight measurement simultaneously for all pixels. This range camera has no moving parts (mirrors) and provides dense range images of 640×480 pixels together with corresponding intensity images at a frame rate of 7Hz. The measured field extends from 0.5m to 15m and the accuracy is $\pm$ 2% of the measured field.

In the following results on four image sequences are discussed. The first sequence shown in Figure 6 consists of thirteen images, in which a box sitting on a cabinet undergoes a clockwise rotation. All other surfaces belonging to this cabinet, the floor, and the two cabinets in the background remain constant. In the initial segmentation one surface of the box was not segmented well due to image noise caused by its large angle to the camera. This surface disappears and a new one emerges from the right side, as can be seen in the segmentation result of the subsequent images. The moving surfaces of the box have been correctly tracked. Figure 7 shows the initial segmentation (left) and the segmentation result of the last frame (right) of the second sequence of twelve images, in which two chairs move towards the center of the field of view. The third sequence of twenty-one images shown in Figure 8 illustrates the gradual closing of a drawer of the cabinet in the foreground. The fourth sequence shown in Figure 9 contains an office scene and has forty-eight frames. Here the camera is smoothly rotated from left to right. In Figure 9 the initial segmentation and the results of three frames by our algorithm are presented. The greylevel images of the first and the last frame are given for a better understanding of the contents of the scene. In addition a rendering of the first and the last frame is shown as well, which illustrates a problem of the range camera in use. It gets confused by very dark areas and cannot provide reliable data in this case, see the various calibration patterns on the wall. In this image sequence the left surface of the large cabinet has disappeared in frame 10. Some time later the right surface of the small cabinet becomes completely visible in frame 37. Finally, part of a second chair emerges in the last frame. Note that corresponding regions are not necessarily given the same greylevel in Figure 9.

The segmentation results reveal some over-segmentations. Basically, this occurs in two situations. Over-segmented regions in the initial segmentation will

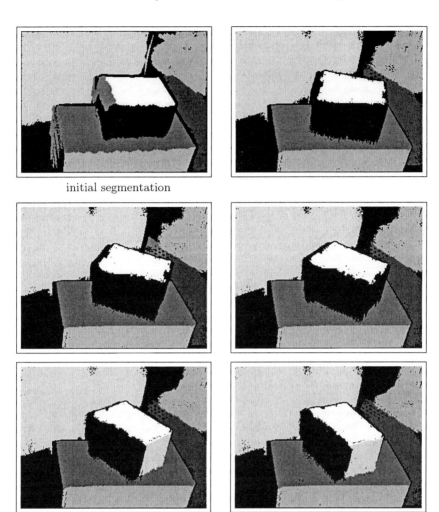

initial segmentation

Fig. 6. Image sequence: box.

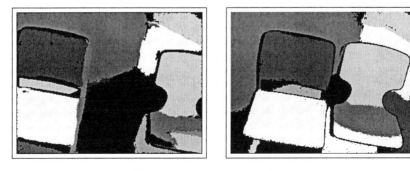

Fig. 7. Image sequence: chairs.

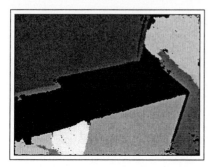

Fig. 8. Image sequence: drawer.

be propagated into the succeeding frames since we do not perform any merge operation. An example is the supporting plane of the large cabinet in Figure 9. Also, if part of a surface cannot be included in the corresponding region (due to noise in the current frame), there will be a new seed region generated and it may be successfully expanded. Most of the over-segmentation instances encountered in the experiments were caused this way.

For the reduced image resolution 320×240, the average computation time for the four image sequences amounts to 2.7s, 3.2s, 3.2s, and 2.3s, respectively, per frame on a SUN Ultra5 workstation. The (highly optimized) edge-based segmentation program [21, 20] used for the first frame of a sequence is typically slightly faster and requires about 70–100% of the segmentation time per frame. Currently, the code of the new algorithm has not been optimized and there is some potential of speed-up. Further speed-up can be achieved in various ways. Important in this context is the fact that the performance of our feature extraction algorithm depends on the number of necessary erosions and the related number of expansions determined by the inter-frame changes, which in turn is a function of object motion speed and camera frame rate. The DaimlerChrysler range camera has a frame rate of 7Hz. If this frame rate is increased, say to a video rate, then the computation time can be significantly reduced. As a matter of fact, the DaimlerChrysler range camera provides a higher frame rate at lower image resolutions and this feature has not been utilized in the experiments thus far. Currently, the control parameter for erosions, D_m, is manually set and an erosion by S_{D_m} basically covers all possible motion directions to the largest extent. Some form of dynamic setting of D_m, say according to informations about the geometry of segmented regions and camera parameters, and/or by means of a Kalman filter, will make more predicted motion hypotheses and therefore enhance the performance. Under these considerations we expect that the computational demand of our feature extraction algorithm is not higher than that of the individual segmentation of frames in general. Finally, the correspondence information is directly available at no extra cost in our case. Often, earlier works on correspondence computation [24, 27] have only shown experimental results on sequences in which all objects undergo a uniform motion. Considering the

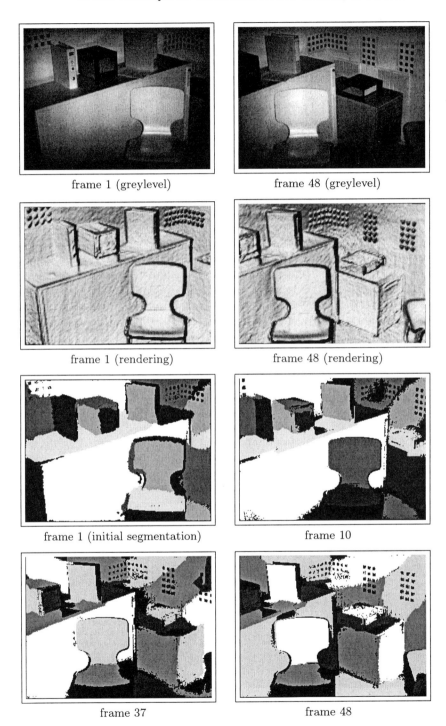

frame 1 (greylevel)   frame 48 (greylevel)

frame 1 (rendering)   frame 48 (rendering)

frame 1 (initial segmentation)   frame 10

frame 37   frame 48

Fig. 9. Image sequence: office.

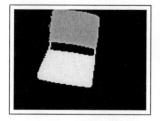

Fig. 10. Tracking of a chair.

difficulty and the computational burden that is expected in dealing with general motion, even an occasionally slightly higher computation time of our approach compared to the individual segmentation of frames seems justified.

Our unified framework of feature extraction and correspondence determination enables the tracking of all surfaces of the scene. If desired, the same framework can be applied to only track a subset of surfaces. For instance, a particular object may be recognized in some frame and tracked in the following frames. Figure 10 shows such a scenario where a chair is recognized (manually marked) first and its movement from left to right is then tracked. In this case we perform a focused tracking. As a consequence, the processing is done faster than in the case of a full tracking. Roughly speaking, the reduction in computation time depends on the total size of the tracked surfaces.

6   Conclusion

Traditionally, motion analysis is based on a correspondence determination using features independently extracted from each image of a sequence. The solutions suggested in the literature for the correspondence determination problem have typically an exponential computational complexity. In the present paper we have introduced a novel framework of motion analysis that unifies feature extraction and correspondence determination in a single process. Under the basic assumption of a small relative motion between the camera and the scene, feature extraction is solved by refining the segmentation result of the previous frame. This way correspondence information becomes directly available as a by-product of the feature extraction process. Due to the coupled processing of frames we also enforce some degree of segmentation stability. First results on real range image sequences have demonstrated the potential of our approach.

The proposed framework is not limited to planar surfaces. In future the current implementation will be extended to work with curved surfaces. As discussed in the last section, some form of dynamic setting of the erosion parameter based on more predicated motion hypotheses is able to further enhance the performance of our feature extraction algorithm.

In addition to dealing with dynamic scenes for robotic applications, motion analysis in the way presented in this paper is potentially useful in other contexts. One example is model generation by registration of multiple views. Usually, different views of an object are registered by means of the iterative closest point algorithm whose convergence in global optimum crucially depends on the initial inter-view transformation estimation [17]. Here we can take a series of range images from one view to another. After processing of the sequence using the method presented in this paper, the correspondences between the first frame (one view) and the last frame (another view) are able to provide a reliable estimation of the inter-view transformation. The same idea may be useful to analyzing articulated objects [4] as well.

References

1. J.K. Aggarwal and N. Nandhakumar, On the computation of motion from sequences of images – A review, Proceedings of IEEE, 76(8): 917-935, 1988.
2. J.K. Aggarwal, Q. Cai, W. Liao, and B. Sabata, Nonrigid motion analysis: Articulated and elastic motion, Computer Vision and Image Understanding, 70(2): 142–156, 1998.
3. M. Asada, M. Kimura, Y. Taniguchi, and Y. Shirai, Dynamic integration of height maps into a 3D world representation from range image sequences, International Journal of Computer Vision, 9(1): 31–53, 1992.
4. A.P. Ashbrook, R.B. Fisher, C. Robertson, and N. Werghi, Segmentation of range data into rigid subsets using surface patches, Proc. of ICCV, 201–206, 1998.
5. C.M. Bastuscheck, Techniques for real-time generation of range images, Proc. of CVPR, 262–268, 1989.
6. S.S. Beauchemin and J.L. Barron, The computation of optical flow, ACM Computing Surveys, 27(3): 433–467, 1995.
7. J.A. Beraldin, M. Rioux, F. Blais, L. Cournoyer, and J. Domey, Registered intensity and range imaging at 10 mega-samples per second, Optical Engineering, 31(1): 88–94, 1992.
8. A. Blake and A. Yuille (Eds.), Active vision, The MIT Press, 1992.
9. A. Blake and M. Isard, Active contours, Springer-Verlag, 1998.
10. K. Chaudhury, R. Mehrotra, C. Srinivasan, Detecting 3-D motion field from range image sequences, IEEE Transactions on Systems, Man, and Cybernetics, Part B: Cybernetics, 29(2): 308–312, 1999.
11. H.H. Chen and T.S. Huang, Maximal matching of two three-dimensional point sets, Proc. of ICPR, 1048–1050, 1986.
12. J.-C. Cheng and H.-S. Don, A graph matching approach to 3-D point correspondences, International Journal of Pattern Recognition and Artificial Intelligence, 5(3): 399–412, 1991.
13. A. Escobar, D. Laurendeau, J. Cote, and P. Hebert, Tracking moving objects using range data, Proc. of 30th Int. Symposium on Automotive Technology & Automation: Robotics, Motion and Machine Vision in the Automotive Industries, 77–84, 1997.
14. H. Haussecker and H. Spies, Motion, in: Handbook of Computer Vision and Applications (B. Jähne, H. Haussecker, and P. Geissler, Eds.), Academic Press, 1999.
15. A. Hoover, G. Jean-Baptiste, X. Jiang, P.J. Flynn, H. Bunke, D. Goldgof, K. Bowyer, D. Eggert, A. Fitzgibbon, and R. Fisher, An experimental comparison of range image segmentation algorithms, IEEE Transactions on Pattern Analysis and Machine Intelligence, 18(7): 673–689, 1996.

16. B.K.P. Horn and J.G.Harris, Rigid body motion from range image sequences, CVGIP: Image Understanding, 53(1): 1–13, 1991.
17. H. Hügli and C. Schütz, Geometric matching of 3D objects: Assessing the range of successful initial configuration, Proc. of Int. Conf. on Recent Advances in 3-D Digital Imaging and Modeling, 101–106, 1997.
18. T.S. Huang and A.N. Netravali, Motion and structure from feature correspondences: A review, Proceedings of IEEE, 82(2): 252-268, 1994.
19. R.C. Jain and A.K. Jain (Eds.), Analysis and interpretation of range images, Springer-Verlag, 1990.
20. X. Jiang and H. Bunke, Edge detection in range images based on scan line approximation, Computer Vision and Image Understanding, 73(2): 183–199, 1999.
21. X. Jiang, An adaptive contour closure algorithm and its experimental evaluation, IEEE Transactions on Pattern Analysis and Machine Intelligence, 22(11): 1252–1265, 2000.
22. X. Jiang, K. Bowyer, Y. Morioka, S. Hiura, K. Sato, S. Inokuchi, M. Bock, C. Guerra, R.E. Loke, and J.M.H. du Buf, Some further results of experimental comparison of range image segmentation algorithms, Proc. of 15th Int. Conf. on Pattern Recognition, Vol. 4, 877–881, Barcelona, 2000.
23. A. Joshi and C.-H. Lee, On the problem of correspondence in range data and some inelastic uses for elastic nets, IEEE Transactions on Neural Networks, 6(3): 716–723, 1995.
24. N. Kehtarnavaz and S. Mohan, A framework for estimation of motion parameters from range images, Computer Vision, Graphics, and Image Processing, 45(1): 88–105, 1989.
25. M.W. Powell, K.W. Bowyer, X. Jiang, and H. Bunke, Comparing curved-surface range image segmenters, Proc. of ICCV, 286–291, 1998.
26. B. Sabata and J.K. Aggarwal, Estimation of motion from a pair of range images: A review, CVGIP: Image Understanding, 54(3): 309–324, 1991.
27. B. Sabata and J.K. Aggarwal, Surface correspondence and motion computation from a pair of range images, Computer Vision and Image Understanding, 63(2): 232–250, 1996.
28. H. Spies, B. Jähne, and J.L. Barron, Regularised range flow, Proc. of ECCV, 785–799, 2000.
29. J. Tajima and M. Iwakawa, 3-D data acquisition by rainbow range finder, Proc. of ICPR, 309-313, 1990.
30. M. Yamamoto, P. Boulanger, J.A. Beraldin, and M. Rioux, Direct estimation of range flow on deformable shape from a video range camera, IEEE Transactions on Pattern Analysis and Machine Intelligence, 15(1): 82–89, 1993.

Dynamic Aspects of Visual Servoing and a Framework for Real-Time 3D Vision for Robotics

Markus Vincze, Minu Ayromlou, Stefan Chroust, Michael Zillich,
Wolfgang Ponweiser, and Dietmar Legenstein

Institute of Flexible Automation, Vienna University of Technology
1040 Vienna, Austria
{vm,ma,mz,wp,dl}@infa.tuwien.ac.at
http://www.infa.tuwien.ac.at

Abstract. Vision-based control needs fast and robust tracking. The conditions for fast tracking are derived from studying the dynamics of the visual servoing loop. The result indicates how to build the vision system to obtain high dynamic performance of tracking. Maximum tracking velocity is obtained when running image acquisition and processing in parallel and using appropriately sized tracking windows. To achieve the second criteria, robust tracking, a model-based tracking approach is enhanced with a method of Edge Projected Integration of Cues (EPIC). EPIC uses object knowledge to select the correct feature in real-time. The object pose is calculated from the features at every tracking cycle. The components of the tracking system have been implemented in a framework called Vision for Robotics (V4R). V4R has been used within the EU-funded project RobVision to navigate a robot into a ship section using the model data from the CAD-design. The experiments show the performance of tracking in different parts of the ship mock-up.

1 Introduction

The goal of a vision-based control system is to track the motion of an object. Examples are the end-effector of a manipulator to grasp a part or a person with an active vision head. The cameras can be either fixed in the work space or mounted on a robot or active head.

This paper presents the results of the analysis of the dynamics of a vision-based control system (or a visual tracking system) and the implementation of a CAD-based tracking system. The analysis of the dynamics is used to derive the architecture of the vision system (parallel image acquisition and image processing) and parameters (window size, sampling). The implementation uses this principle and presents the tool "Vision for Robotics" (V4R).

The dynamic performance of vision-based control of motion is specified by the dynamic properties that can be reached. Focusing on the goal to track an object, significant properties are the velocity and the acceleration the target can make without being lost. The analysis of the visual servoing loop started with

G.D. Hager et al. (Eds.): Sensor Based Intelligent Robots, LNCS 2238, pp. 101–121, 2002.

the work in [3]. Several visual servoing systems have been built and operate at different cycle rates and latencies. The work in [30] shows that the tracking velocity can be optimised. The derivations are summarised below.

Using these results a vision tool (V4R) can be designed. The system handles the image processing to track features and the estimation of the object pose from these features. As a consequence of the dynamics analysis, the vision processing is window-based and including the pose estimation it operates at frame rate. Each feature is tracked in one window. Features can be lines or arcs or region-based trackers. Feature tracking has been made more robust than classical edge-based schemes with a method of Edge Projected Integration of Cues (EPIC). It allows to track edges independent of background or motion.

After a systems overview (Section 2) the results of the dynamics analysis are presented in Section 3. Section 4 present V4R and Section 5 gives experiments.

1.1   Related Work

A large number of systems for *visual servoing* (the control of a motion using input of visual sensors) have been built in the last decades (for an extensive review see [17] and the workshop notes of IROS'97, ICRA'98, and ECCV'98). However, the number of commercial applications is small.

When analysing the present state-of-the-art, two major roadblocks can be identified. (1) The *integration of system components* is often neglected, although integration of mechanism, control, and visual sensing and processing is essential to achieve good performance of the system. And (2), the *robustness of the visual input* limits performance and applications. This is supported by the observation that successful applications are commonly restricted to special environments (results for autonomous car driving are impressive [6], however visual processing is tailored to the specific task) or prepared objects (markers, high contrast to background, manual initialization). It also relates to automatic operation, a typical problem for many systems that require manual initialization.

The perspective of the control loop to integrate the components of a visual servoing system is reviewed in relation to the goal of optimising the system performance. The analysis of system performance is motivated by the wide range of systems that have been built. In particular the cycle times and processing times of existing systems vary greatly. In robotics processing time often exceeds acquisition time and several pictures are skipped [11, 26]. Systems for juggling [24], to catch satellites [33], and for vehicle steering [6] use frame rate in a fully parallel system. A system to grasp an object with a robot operates at field rate [31]. Only very specific tasks can use on-the-fly image acquisition and processing [3]. Visual servoing systems for active vision, in particular fixation control, run with highly different cycle rates and processing times. A small processing time ($3ms$ [14], $8ms$ [1]) reduces latency. Latency is then governed by image acquisition at field rate. Most systems need processing times of several times the frame rate [5], therefore increasing latency substantially. Some systems run acquisition and processing at frame rate [22]. Few more work are referenced in Section 3.

From the perspective of building an integrating framework, there have been several attempts for vision-based control. A set of basic tasks for image-based visual servoing has been defined in [7] and extended to a software environment for eye-in-hand visual servoing [20]. It has been used to follow pipe systems using one camera. The visual tracking tool XVision can be used for stereo visual servoing [12]. Servoing uses points and lines on the robot gripper and on the object for alignment. Both approaches need manual initialization of the visual features.

The "Vision as Process" project developed a software system to integrate vision and an active head and the interactions between them [4]. Integration united 2-D image with 3-D structure data to control the head motions. The experiments with these systems indicate the lack of robust vision methods. Objects have different gray values on each surface (Chap. 9 & 10 in [4]), use good contrast background [13, 20], or white markers and objects on black ground [15].

Robustness has been approached in recent years by adding redundancy to the vision system (e.g., using multiple cameras, resolutions, features, or temporal constraints [6, 10]). Filtering and prediction are the common control methods to improve robustness [31]. However, if the underlying visual process is not reliable, quality of prediction is limited. One method of improving robustness of 3D object tracking is to use a CAD-model [19, 26]. The main problem is changing background and high computational demands. For example, in a space application with an object of different characteristics from surface to surface, edge detection is reliable but requires dedicated hardware to run at frame rate [33].

In humans the integration of cues has been found as a likely source of the excellency to cope with varying conditions. Active vision research was the first to utilize cue integration, for example using voting of simple cues [23]. Regarding systems that detect objects and features, present state of the art in integration handles two cues, such as edge and motion [25] or colour and image position [32]. Flexibility is added by introducing a framework that can switch between single cue extraction methods [27] but it does not integrate cues to obtain better, combined robustness or accuracy.

2 System Overview

Controlling the motion of a robot requires to find and track an object in a sequence of images and to extract pose information of this object. Industrial grasping tasks have been solved by using a ground plane to render the problem 2D and by imposing proper lighting conditions (for example, the vision tools of ABB or Adept). For most navigation tasks the pose is also 2D and can be resolved using other sensors than vision. However, practically all manipulation tasks in more realistic scenarios for service or personal robots require full 3D information. The proposed V4R system specifically sets out to solve 3D vision tasks and considers the 2D task as a sub task of the 3D requirement. Using 2D image information alone, V4R controls the motion of the vision head.

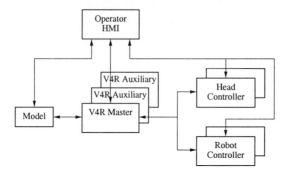

Fig. 1. Components of a visual servoing system.

The task of the vision process consists of finding and tracking the object in the image(s). To define the task a description of the target object (e.g., in form of a model) is needed. V4R takes the model and uses the data of one or more cameras (or views) to initialize tracking and pose determination. Fig. 1 shows the components of the visual servoing system. An operator selects the object relevant for the task and starts the robot motion. In case of multiple vision systems and heads, the vision master fuses the data and sends only one control command to the robot.

V4R is designed to allow the use of any model and image processing technique. This duality of image description to model description is typical of existing approaches. An example is aspect based tracking using Eigen-images [21]. The model is the representation obtained of sampling the object at different aspects. The feature is the region used for Eigen-tracking. Another example is image-based servoing. The model provides image locations for features and also specifies the type of tracker to use [7, 12].

The approach presently implemented is based on 3D object models typically available from CAD-systems. The key idea is that the features (line, vertex, arc, face) and attributes (color, texture, machined) of the CAD-system are exploited in the image processing algorithms. Furthermore, CAD models are commonly available in industry and require no exploration or learning procedures. If needed, the models could be built up using these techniques. Section 4 will described how the model is utilized to render tracking more robust. A further advantage is the explicit 3D representation that allows to link the visual process to grasp or assembly planning techniques [16].

3   Dynamics Analysis

Fig. 2 shows the basic visual servoing loop for the control of either the head or the robot in Fig. 1. The blocks marked by $V(z), C(z)$, and $R(z)$ constitute the discrete transfer functions of the vision system, the feedback controller, and the mechanism (robot or pan/tilt unit), respectively. The vision system V(z) provides

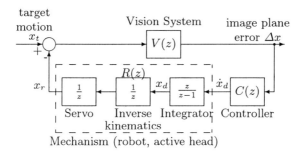

Fig. 2. Block diagram of visual servoing.

a image based pixel error ΔX, which is transformed to a velocity command by the controller $C(z)$. ΔX is the controlled variable.

In his work [3] Corke showed that the above loop is the appropriate description. As the target motion x_t is not directly measurable it has to be treated as a non-measurable disturbance input.

It is important to note that this diagram is valid for active vision systems as well as a fixed or moving camera controlling the motion of a robot. The difference is the controlled variable and the way of how to obtain the control signal from the image data.

$R(z)$ contains an integrator to generate the position set point x_d for the control of an axis of the mechanism, one optional latency $1/z$ for times needed for data transfer or the calculation of the inverse kinematics, and the servo, which is ideally treated as a unit delay. This is justified, since the underlying position loop has a higher sampling rate than the vision system and therefore an interpolation of the position set point is possible. This has been analysed and confirmed in [3]. An extension of treating the mechanism as a PT2 element is presented in [30].

3.1 The Latency of the Vision System

The vision system executes image acquisition and image processing. Image acquisition time t_{ac} depends on camera and system technology. CMOS, RAM and other direct-access cameras allow arbitrary pixel access and transfer, therefore $t_{ac} = t_{ac}(n_p)$, where n_p is the number of pixels within window W. More common are CCD cameras (video standards RS170 or CCIR), where t_{ac} is the image frame time $t_f = 33.\dot{3}$ or $40ms$, respectively. When operating with field shuttering, $t_{ac} = t_f/2$. High-speed cameras reduce frame time down to $t_f = 1ms$ (a 16×16 pixel sensor [18]). For all these sensors $t_{ac} = const = t_f$ (t_f is used subsequently for all systems). Shuttering and frame time impose a technological constraint on cycle time, the *frame-time constraint* $T \geq t_f$. As CMOS and high-speed cameras are rare, the basic case of a constant image acquisition time is investigated. Note, however, that smaller image acquisition times do not change the basic considerations.

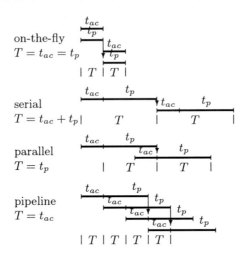

Fig. 3. Different architectures to acquire and process images.

Visual servoing can operate at different sampling times T of the discrete control system of Fig. 2. The four main configurations are on-the-fly, serial, parallel, and pipeline processing. Note that every layer of the pipeline has the same computing power as used in the other cases. Fig. 3 shows the four cases pointing out image acquisition time t_{ac} (field or frame time for CCD cameras), image processing time t_p, and the corresponding sampling time T of the discrete control system. The arrows mark the time instant when the processed data is available for further use, e.g., to place the tracking window and to move the mechanism.

In summary, the vision system $V(z)$ introduces n unit delays, one for acquiring the image and the other n-1 for processing the data. The transfer function is $V(z) = \frac{k_v}{z^n}$, where k_v is the gain which relates the target pose to an image plane displacement describing the projection from 3D-space onto the image plane.

The resulting closed-loop transfer function of the image plane error as a reaction to the target motion can be written as

$$F_w(z) = \frac{\Delta X(z)}{x_t(z)} = \frac{k_v z(z-1)}{z^{n+1}(z-1) + k_v C(z)} \tag{1}$$

where $zn + 1$ is the latency of the vision system and the controller and the inverse kinematics is neglected (but could be added). The transfer function is the basis for any controller design in Section 3.3.

3.2   Visual Tracking

An important parameter is window size as it determines processing time and thus latency of the vision system. A valid estimate of processing time t_p of common tracking techniques (see discussion in [28]) states that t_p is proportional to the

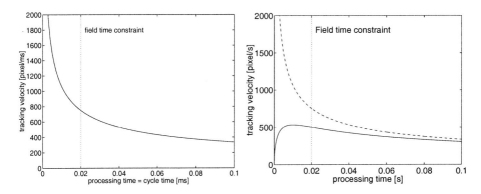

Fig. 4. Tracking velocity over processing time for a colour blob tracker $D_{pix} = 2.22e-5$. *Left:* Without constant times. *Right:* with a constant time of $0.01s$.

number of pixels within the window. For a window size with the side length $2r$, processing time t_p is given by

$$t_p = 4D_{pix}r^2 \qquad (2)$$

where D_{pix} is the time necessary to evaluate one pixel. This relationship holds for a wide class of algorithms, since the search region for the target in the window is a 2D area. However, the later derivations are also valid, if there are additional constant or linear terms in eq. (2) (compare Section 3.6).

To illustrate the consequence of this relationship on tracking, consider the task of following a coloured object in a fixed camera. The object can move in the image plane at each cycle the maximum given by the window size, that is, $\pm r$, which is in general smaller than the size of the image or the sensor size. The velocity v of the target in the image is then $v = r/T$. In the simplest case $T = t_p$ and $r = (t_p/(4D_{pix}))^{1/2}$. Both r and T depend on the processing time t_p. A very short image processing time results in a small window and a small cycle time T. For a larger and larger image the radius r increases linearly while the image processing time and therefore T increases quadratically. This relationship is displayed in Fig. 4 *Left* for the case of parallel image acquisition and processing. In this case the cycle time of the system is identical to the processing time.

The diagram demonstrates that shorter and shorter cycle times result in an increasingly higher tracking velocity. The field time constraint is introduced by commercial CCD-cameras. Lower processing times are possible but then tracking needs to wait for acquiring the next image and tracking velocity does not increase. However, using high-speed cameras shorter acquisition times are possible. In this case also processing time is reduced to be always as long as acquisition time and the tracking velocity increases as shown.

Fig. 4 *Right* gives the case of a constant time t_c in the tracking cycle, which is the common case due to communication times or constant processing overhead. The cycle time $T = t_c + t_p$. Higher cycle rates still increase the velocity but

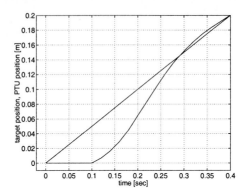

Fig. 5. Tracking response to a ramp input. The latency of the system delays the first response. The control method is responsible for any additional increase of the tracking error.

the property of increasing the velocity for lower cycle times is lost and a clear optimum is found.

3.3   Controller Design

The performance of the visual servoing loop depends on the controller design and has been studied, for example, in [3, 2]. Corke studied the behaviour of the basic control loop, designed an optimized PID controller and investigated a feed-forward controller. The summary in [17] shows that a feed-forward controller performs better than PID controllers. The rationale is the predictive character. Similar approaches use a Kalman filter [31] or a Generalized Predictive Controller [9] to compensate for the delay introduced in the control loop.

   To study the effect of latency on the control, assume that the task is to follow the motion of an object. In order to provide a ramp following behaviour without any steady-state error, a double integrator in the open-loop transfer function is necessary. This can be achieved by a classical PID-controller. Starting out from Corke's work [3] on controller design, an improved (25%) PID controller has been presented in [30]. Fig. 5 shows the response to a ramp input. Due to the latency in the loop, the error in the image increases to a maximum before the control compensates for it.

   The steepness of the ramp indicates the maximum average velocity that can be tracked. The higher the velocity, the higher will be the maximum deviation in the image plane. In order to keep the target within the tracking window, the critical measure is the maximum deviation from the centre of the search window. Therefore, an optimal controller needs to minimize the *image plane pixel error*.

   Besides this requirement, the controller should also avoid oscillations when following the target. Therefore it is appropriate to minimize the absolute value of the pixel error within a given time interval. The mean deviation over an interval or a least square optimisation are investigated in [30]. The result of using either

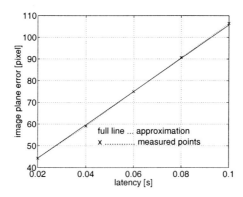

Fig. 6. Measured and calculated image plane error.

metric is a measure C for the quality of the control method. C relates the maximum pixel error to the latency of the vision system. It can be interpreted as a quality measure of the control algorithm with the ideal value of $C = 1$.

The behaviour shown in Fig. 5 shows a slight increase of the tracking error. The optimised PID controller uses a single dominant pole and shows better performance as the controller in [3], however it does not perfectly make up for the ramp and $C = 9/8$. This experiment can be made for several latencies in the system. Fig. 6 shows that the tracking error in the image increases linearly with the latency.

Studying the dynamic performance of a visual servoing system the tracking properties (Fig. 4) must be related with the control responses to latency.

3.4 Dynamic Performance

The dynamic performance of visual servoing is specified by the dynamic properties that can be reached. Focusing on the goal to track an object, significant properties are the velocity and the acceleration the target can make without being lost. (Recovery from loss requires search for the target, and this again follows the same laws, since the object is moving.) A basic property of dynamic performance is the *maximum average velocity* (MAV) of the target. The ramp input of Fig. 5 simulates this behaviour.

Given a method that solves the visual processing task, the dynamic performance of a visual servoing system is influenced by four factors: the computing hardware, the control method, the processing architecture, and processing parameters such as window size and sub-sampling. The effects of communication times between components can be added to the processing time. Two of the factors, hardware and control method, show a simple scaling effect on the dynamic performance.

More complex is the effect of the processing architecture used and the relation to the most significant processing parameter of tracking, the window size.

The differences in the processing architecture result from the configuration of acquiring and processing the images: on-the-fly, serial, parallel, and pipeline.

3.5 Image Plane Error versus Latency

To be able to optimize the performance it is necessary to find a general relation between MAV (that is, image plane error) and latency. For several architectures resp. latencies the measured image plane error has been given in Fig. 6. The relationship is expressed with

$$\Delta X = k \cdot k_v \cdot T(lat_{Mech} + C \cdot lat_{Vis}) = k \cdot k_v \cdot T \cdot C_I \qquad (3)$$

where lat_{Mech} and lat_{Vis} are the latencies of the mechanism and the vision system, respectively, denoted as multiples of the sampling time T.

3.6 Optimal System Architecture

In the last section the image plane pixel error was calculated for a given latency and the rising factor of the ramp, that is the velocity of the target. The objective is to relate MAV to a given latency in the visual control loop. Together with the image acquisition time t_{ac}, mainly t_p determines the latency of the vision system. MAV is the limited by the image plane error. Rewriting eq. (3) gives,

$$k \cdot k_v = \frac{\Delta X}{T \cdot C_I} \qquad (4)$$

which shows tracking velocity $k \cdot k_v$ in units of pixels per second as a function of both the maximum image plane error ΔX and the latency of the visual feedback system denoted by C_I. With identity (2), substituting r for ΔX in eq. (4), and substituting the values of C_I for the corresponding system architecture we obtain tracking velocity as a function of processing time t_p. The following equations show the results for the on-the-fly, pipeline, parallel, and serial architectures.

$$v_{otf} = a/t_{ac} \qquad (5)$$
$$v_{pipe} = a/(lat_{Mech}t_{ac} + C(t_{ac} + t_p)) \qquad (6)$$
$$v_{par} = a/(lat_{Mech}max(t_p, t_{ac}) + C(t_{ac} + t_p)) \qquad (7)$$
$$v_{ser} = a/((lat_{Mech} + C)(t_{ac} + t_p)) \qquad (8)$$

where $a = \frac{\sqrt{t_p}}{2\sqrt{D_{pix}}}$.

In actual implementations, processing is executed at the very instant of execution and $t_p = t_{ac}$. This t_p is used to compare the architectures. Window size is computed from eq. (2) for all settings according to the given processing time.

Fig. 7 shows tracking velocity as a function of processing time in all four cases. The image acquisition time is $20ms$ and the latency of the mechanism is assumed to be one unit delay (Section 3).

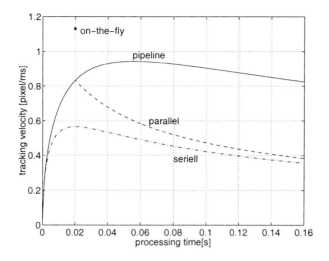

Fig. 7. Tracking performance vs. processing time ($t_{ac} = 20ms$, $lat_{Mech}=1$; $D_{pix}=$ 2.22-5 s).

The Figure shows that on-the-fly processing gives best performance. However, for most vision methods it is unobtainable. Therefore parallel and pipeline processing should be used. Differentiating eq. (5) with respect to t_p gives the maximum tracking velocity, the optimal processing time, and the optimal number of processing steps in the pipeline [30]. Note, that in the parallel case, $t_p = 20ms$ is optimal, which means that the processing time t_p is equal to the acquisition time t_{ac} and the total latency is 40ms. Also note that the location of the optimum is therefore independent of D_{pix}. The parallel architecture always has this maximum. The optimal velocity of course depends on D_{pix}, because D_{pix} is used to determine the window size from the condition $t_p = 4D_{pix}r^2 = 20ms$.

3.7 Summary and Guidelines

The dynamic relations of visual servoing have been studied. The investigation of optimal dynamic performance indicate the following guidelines when building a visual servoing system.

– On-the-fly processing gives best performance. However, most image processing algorithms cannot be executed only on pixels, even with future computing power.
– Hence, given a processing power, parallel processing gives best performance. The best dynamic performance is reached when processing time is equal to acquisition time. The size of the image window is calculated from the processing time and the cost to process one pixel.
– If processing power of the parallel system is multiplied in a pipeline configuration, pipeline processing can give better performance. If using a pipeline, a

certain number of steps in the pipeline gives the best dynamic performance. However, building a pipeline is far inferior to using the same computing power in a parallel system.

- Optimal system architecture is fixed: adding computer power for image processing increases the performance without changing the optimal architecture. More computing power allows to increase the window size that can be processed in a given time. Therefore the performance improves, but the architecture remains the same.
- The optimum for a parallel system is independent of the time to process one pixel. This indicates that the parallel architecture should be used independent of the vision method used to extract the target.

The procedure of the performance evaluation proposed is independent of the controller design. The dynamic performance is scalable with the quality of the control. A similar argument shows that any other performance metric would scale the overall performance, the relation between the architectures remains the same. The superiority of the on-the-fly and the parallel architecture for a vision system can be always observed.

The work in [28] also showed that a *space-variant* image tessellation can have superior dynamic performance for visual tracking in a fixed camera. This advantage is gained at the loss of imaging resolution.

The *consequence* of the dynamics analysis is to build a window-based tracking system that operates at the highest possible rate. The consequence is also that the the motion of the robot should be restricted to the maximum allowable tracking velocity to avoid loss of features. Another option is to reduce the resolution of the image to gain speed at the cost of accuracy. The next section will now introduce the window-based framework implemented.

4   Vision for Robotics (V4R)

The results above indicate that image processing must operate at the highest rate possible to obtain best dynamic performance. Therefore field/frame rate is optimal when using conventional CCD cameras and processing should only operate on windows of appropriate size. The work also showed that image processing time is linear with the number of pixels in the window, which holds true for all tracking methods of the papers cited above and the method introduced below.

Consequently, the Vision for Robotics (V4R) framework has been implemented. Fig. 8 shows the main components. The scene contains objects that are described with the CAD or wire-frame model. A view contains view dependent data such as camera pose and the image representation of the object seen from this view.

A critical step that is often neglected in visual servoing approaches is the need for an automated initialization. V4R progresses as follows. If a pose estimate of an object is available, the object model is used to predict the location of the object and its features. Otherwise, the features and their attributes can be used to instantiate search functions as implemented in [29]. A recognition step can

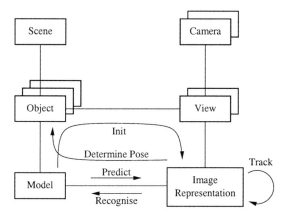

Fig. 8. The main classes and methods of V4R.

be used to confirm that the features found constitute an aspect (a view) of the model.

Tracking is a core component of vision-based control of motion. Every view sees a subset of the full wire-frame model of the object. Based on image features the wire-frame is tracked (see Section 4.1). A fast and robust method for feature tracking has been devised (Section 4.2). All features that are tracked is used to extract the 3D object information utilising the relation between image features and model features.

As both feature tracking and feature search (driven by the visual search functions and the need of pose determination) operate in parallel, feature dependencies and new search results are constantly incorporated into the model while targets are in motion.

In summary, V4R provides the following features.

- Tracking using live images, image sequences or mpeg video streams.
- Class structure for tracking with implementations of line and arc (ellipse) trackers using EPIC (Section 4.2), junctions and blob (region) trackers (more coming continuously). The topology of features (junctions, parallel lines, closed polygons) is used to select among ambiguous tracking candidates (for an example see Section 5).
- Interface (TCP/IP protocol) to a CAD-modelling tool to use generic object models.
- Initialisation can be performed manually, from a projection of the model into the image, or from finding potential object regions using colour adapted to the ambient lighting situation.
- Pose estimation using any line, 2D point, 3D point, and surface normal vector information. The inputs can be weighted and outliers are detected and re-initialised for tracking.

In the following section the basic tracking capability is outlined in more detail. The second prototype of the V4R software is available from the authors.

114   Markus Vincze et al.

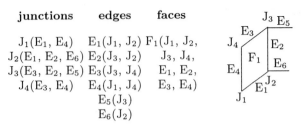

junctions	edges	faces
$J_1(E_1, E_4)$	$E_1(J_1, J_2)$	$F_1(J_1, J_2,$
$J_2(E_1, E_2, E_6)$	$E_2(J_3, J_2)$	$J_3, J_4,$
$J_3(E_3, E_2, E_5)$	$E_3(J_3, J_4)$	$E_1, E_2,$
$J_4(E_3, E_4)$	$E_4(J_1, J_4)$	$E_3, E_4)$
	$E_5(J_3)$	
	$E_6(J_2)$	

Fig. 9. Storing edges E, junctions J, and faces F (regions, surfaces) in a wire-frame model.

4.1 Tracking a Wire-Frame Structure

In each view a partial wire-frame model is built up from the features seen. The search functions use this graph-based model as database, and for adding or deleting edges (line, ellipse arc) and junctions found in the image. Fig. 9 gives an example of a simple object partially seen in the image.

Tracking the wire-frame model found in the image means the updating of all features and feature relations at each cycle. In detail, tracking proceeds as follows. The wire-frame contains a set of N edges $\mathcal{E}$ and a set of M junctions $\mathcal{J}$ (presently regions are not used for tracking). An edge e_i is denoted by its states in the image, for example, the location x, y, orientation o and length len: $e_i(x, y, o, len) \in \mathcal{E}$ for $i = \{1, \cdots, N\}$. Similarly, a junction is given by $j_j(x, y) \in \mathcal{J}$ with $j = \{1, \cdots, M\}$. The relations between edges and junctions in a wire-frame model are then given by

$$f_1 : \mathcal{E} \mapsto \mathcal{J}^m \qquad m = \{0, \cdots, 2\} \qquad (9)$$
$$f_2 : \mathcal{J} \mapsto \mathcal{E}^n \qquad n = \{2, \cdots, \infty\} \qquad (10)$$

where m indicates that an edge can exist without any or up to two junctions and n indicates the number of edges intersecting at a junction. A wire-frame $\mathcal{WF}$ is the union of edges and junctions, that is $\mathcal{WF} = \mathcal{E} \cup \mathcal{J}$. (If faces are also tracked, they can be added quiet easily to the following scheme.)

Updating the wire-frame at each tracking cycle using the data from the image **I** is then described by the following procedure:

```
update Wire-Frame WF:
    update Edge E: ∀eᵢ ∈ E    f₃ : L × I ↦ E
           if   f₃(E,I) = eᵢ      // edge found
           else f₃(E,I) = ∅       // mistrack
    update Junction J: ∀jⱼ ∈ J    f₄ : f₁(J) ↦ J
```

where f_3 is the result (edge found or not found) of the tracking function outlined in section 4.2 and f_4 is the function of finding the intersection between all edges intersecting at one junction. Tracking the features using the wire-frame update procedure is executed at each cycle before further search functions can be called.

4.2   Fast and Robust Edge Tracking

A basic need of a vision-based control system that shall operate in a common indoor environment is robustness of visual image processing. Robustness indicates the ability to resist variations in the appearance of the object, the lighting, and the environment.

Tracking an edge proceeds in three steps: (1) warping an image window along an edge and edge detection, (2) EPIC - integration of cue values to find a list of salient edgels, and (3) a probabilistic (RANSAC [8]) scheme to vote for the most likely geometry, for example a line or an ellipse. These steps are outlined in the next three sections for a line tracker and constitute f_3 from the update wire-frame procedure. Ellipse tracking is analogues.

The resulting tracker shows stable behaviour when lighting conditions vary (as edgels will still form a line/ellipse although cue values change) and when background comes close to foreground intensity or colour (as the discontinuity still yields edgels though of low significance, but again a series of edgels is obtained along what seems to be the edge).

Warping Image and Edge Detection. Every edge e_i is tracked by warping a part of the image along the edge. For lines this is a warped rectangular image, ellipses use several tracker lines around the circumference. As the object moves, the edge must be re-found in the new window. Each tracker holds an associated state vector comprising the basic edge parameters (x, y, o, len) and the cue values found at the last tracking cycle. These values are extracted from the image $\mathbf{I}(x)$ from search lines orthogonal to the edge. The indices $left$ and $right$ denoted the sides of the edgel with respect to the orientation of the edge.

In a first step edgels are found by computing the first derivative of the intensity, $grad\ \mathbf{I}(x)$, using a Prewitt filter of size 8×1 (other common filters give similar results, the Prewitt filter can be implemented effectively as it uses only ones). The positions of the k local maxima x_{Mk} define the edgels and the intervals along the search line. For each interval the cue values can now be used to discriminate the edgels into the object edge and background disturbances using EPIC.

EPIC: Edge-Projected Integration of Cues. The basic idea of EPIC is to project the values of cues such as intensity, colour, texture, or optical flow, shortly called *cues*, to the nearest edgel. Using the likelihood for these edgels renders subsequent line (or ellipse) detection more robust and effective.

The new cue value c_{side} for each interval is calculated using a histogram technique[1]. The index *side* refers to the two possible sides of the edgel, *left* and *right*. Knowledge of the object is used to select only the side that belongs to the target object while the background side is not regarded. This knowledge is found

---

[1] Experiments with median values gave results of similar robustness, however the complexity of computing the median is not linear.

from the object model and used to set the weights w_{side}. If the model indicates an object, the respective weight w_{side} is 1, otherwise it is 0.

The likelihood l_k that an edgel k is a "good" edgel along the line to be tracked is evaluated to

$$l_k = \frac{1}{W} \sum_{i=1}^{n} w_{left} C_{left,i} + w_{right} C_{right,i} \quad \text{with } i = 1, .., \text{ number of cues} \quad (11)$$

No parameters need to be used to weight the cues differently to each other, since for each cue the values $C_{side,i}$ are calculated using a Gaussian distance measure. Each reference cue value from the last tracking cycle $t - 1$ is stored as a pair of a mean value μ^{t-1} and a standard deviation value σ^{t-1}. The cue value of the present cycle c^t_{side} (the value found from histogramming) is used to find the values $C_{side,i}$ as follows

$$C_{side,i} = \exp\left(-\frac{(\mu^{t-1} - c^t_{side})^2}{2\sigma^{t-1}}\right) \quad (12)$$

Once the final line is found, new values μ^t and σ^t are calculated from the c^t_{side} values of the good edgels.

The scheme is very effective because the features contain attributes that give indications to the cues (intensity, colour, texture, ...) of the object. If the edge is a contour edge, the object can be only on one side and the respective weight is set to one. In case the edge is not a contour edge, e.g., the edge between two visible surfaces of the object, then both weights are set to one.

The advantage of EPIC is that, based on the localisation of edgels, cues can be easily integrated and the list of cues given above can be enhanced with any other type of cue. The principle idea is to use these cues to choose only "good" edgels. For example, trials to incorporating colour added robustness to distinguish the correct edges from shadows and highlights. This renders the next step very efficient.

Voting for Most Likely Line. In the final step, the edge geometry (line, ellipse) is found using a probabilistic [8] verification step, which adds the geometric constraint to obtain further robustness. The investigation of the likelihood to find the correct geometry is the best justification for the combination of eq. (11) with the RANSAC scheme. A good line is found if two good edgels are selected to define the line. If the likelihood that an edgel is "good" is given by g, a good line is found with likelihood g^n with $n = 2$ (for an ellipse $n = 5$). Repeating the random selection of n edgels to define the edge geometry k times gives the likelihood of finding the correct edge e to

$$e = 1 - (1 - g^n)^k. \quad (13)$$

This relationship depends strongly on g. Therefore limiting the number of "good" edgels using EPIC is extremely effective to reduce k (exponentially for the same

Fig. 10. Tracking a ladle in front of an office scene. The two top image show miss-tracking when using only gradient information. The bottom line uses EPIC for robust finding of edgels.

e). The result is the capability to find lines more robustly and at real-time ($2ms$ are needed for a line in a 40×40 window on a 266 MHz Pentium PC). The same method can be used to robustly track an ellipse in $10ms$ (Fig. 10).

5   Experiments

The feasibility and robustness of the approach is demonstrated with a walking robot walking into a ship section. This application is part of the RobVision project funded by the EU (see also robvision.infa.tuwien.ac.at). The goal is to move an eight legged pneumatically driven walker into the ship for welding and inspection tasks. The model (see Fig. 1) of the specific features in a given camera view is provided automatically from the CAD data of the ship. The user only specifies the path of the robot. In RobVision the head of DIST (University of Genova, head: Giulio Sandini) provide the capability to look at different directions.

Fig. 11 shows snapshots from a typical motion of the robot in the ship cell. For this motion the head is held constant. Angular rotations are compensated using gyros in a direct feedback loop. However, the vertical displacement cannot be measured and must be tracked.

The window-based approach of tracking has, in general, problems to handle closely spaced features, for example, the parallel lines in the bottom front. Local windows cannot discriminate between these features. Therefore a simple ordering scheme for two neighbouring lines has been implemented, which also proved feasible to discriminate between more than two close parallel lines. Another

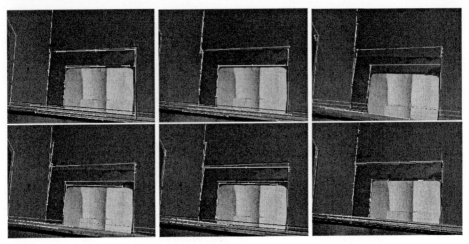

Fig. 11. A sequence of images taken every 120 ms from the walking robot. Observe that tracking is regained quickly after the sudden downward motion of the pneumatically driven robot.

Table 1. Summary of mean and standard deviation of position and orientation accuracy obtained. The reference of the roll (α) angle could not be measured with sufficient accuracy.

	position *mm*			orientation *deg*		
	x	y	z	α	β	γ
mean	5.49	3.29	8.48	0.0	0.92	0.35
std	4.88	8.62	51.25	0.0	0.62	0.13

local validation method is the use the junctions to indicate the correct line intersections.

Fig. 12 plots the pose values (the pose of the robot with respect to the ship model) in metres for position and radians for orientation. The robot motion is relatively slow but the high jerk of the pneumatic actuators introduce high deviations from the path, which can be tracked as given in the Figure before.

In another experiment the robot was walked eight times into the mock-up and its final pose was measured. Fig. 13 shows the measured position values. The reference values have been found using conventional measurement technique, which is also responsible for the constant offset introduced in the z direction due to the bending bottom plate of the mock-up when the weight of the robot stands on it. Table 1 summarises the accuracy that could be obtained in the experiments walking the robot in the mock-up.

In summary, the robot could be positioned within the mock-up with an accuracy of about 1 cm per meter distance to the mock-up. The maximum dimensions of the mock-up is three meters. This is sufficient to fulfill the task of placing a robot near the welding seems. The final correction will be executed using a laser seam following sensor.

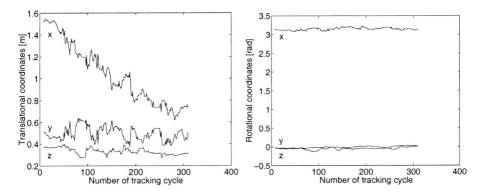

Fig. 12. The six components of pose during the motion of the walking robot.

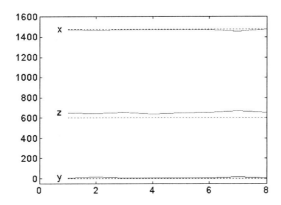

Fig. 13. The position of the robot determined by the RobVision system for eight trials. The position is measured in mm, the dashed line is the reference.

6   Conclusion and Future Work

A software system as tool for vision-based control of motion systems has been presented. The system tries to solve three typical deficiencies of existing approaches. First, a generic software structure for vision-based control of motion is presented. It is general in terms of enabling the use of any model and image processing methods and an example using edge features is presented that operates at field rate. Second, the image processing architecture is chosen such that optimal dynamic performance is obtained using parallel image acquisition and image processing and a windowing scheme for tracking. And third, special emphasis is given to robust feature (line, junction, ellipse) detection using Edge Projected Integration of Cues (EPIC).

While tracking is fast and shows good robustness, the maximum allowed object motion is still restricted. The use of inertial sensors proved effective to compensate fast angular motions. This aspect will be investigated more closely by integrating a full 6D inertial sensor suit with the optical tracking methods.

Another open issue is the automatic initialisation. Projection is a good start if a first pose estimate exists. However, a generic initialisation method need to incorporate results from object recognition into the framework, which has been initially tested but needs substantial work to operate robustly.

Acknowledgements

This work is partly supported by the Austrian Science Foundation (FWF) under grant P13167-MAT and Project GRD1-1999-10693 FlexPaint.

References

1. Asaad, S., Bishay, M., et.al., "A Low-Cost, DSP-Based, Intelligent Vision System for Robotic Applications," *IEEE ICRA*, pp. 1656-1661, 1996.
2. Barreto, J., Peixoto, P., Batista, J., Araujo, H., "Improving 3D Active Visual Tracking," Int. Conf. on Vision Systems ICVS'99, pp. 412-431, 1999.
3. Corke, Peter I.: *Visual Control of Robots: High Performance Visual Servoing*, Research Studies Press (John Wiley), 1996.
4. Crowley, J.L., Christensen, H.I.: *Vision as Process;* Springer Verlag, 1995.
5. Dias, J., Paredes, C., Fonseca, I., Araujo, H., Batista, J., Almeida, A., "Simulating pursuit with machine experiments with robots and artificial vision," *IEEE Trans. RA*, Vol. 14(1), pp. 1-18, 1998.
6. Dickmanns, D.E., "The 4D-Approach to Dynamic Machine Vision," *Conf. on Decision and Control*, pp. 3770-3775, 1994.
7. Espiau, B., Chaumette, F., Rives, P., "A New Approach to visual Servoing in robotics," *IEEE Trans. RA* 8, pp. 313-326, 1992.
8. Fischler, M.A., Bolles, R.C.: *Random Sample Consensus: A Paradigm for Model Fitting;* Communications of the ACM Vol.24(6), pp.381-395, 1981.
9. Gangloff, J.A., de Mathelin, M., Abba, G.: *6 DOF High Speed Dynamic Visual Servoing using GPC Controllers;* IEEE ICRA, pp.2008-2013, 1998.
10. Gee, A., Cipolla, R.: *Fast Visual Tracking by Temproal Consensus;* Image and Vision Processing 14, pp. 105-114, 1996.
11. Grosso, E., Metta, G., Oddera, A., Sandini, G.: "Robust Visual Servoing in 3D Reaching Tasks," *IEEE Trans. RA* Vol.12(5), pp. 732-741, 1996.
12. Hager, G.D.: *A Modular System for Robust Positioning Using Feedback from Stereo Vision;* IEEE RA Vol.13(4), pp.582-595, 1997.
13. Hager, G.D., Toyama, K.: "The XVision-System: A Portable Substrate for Real-Time Vision Applications," *Computer Vision and Image Understanding* 69(1), pp. 23-37, 1998.
14. Heuring, J.J., Murray, D.W., "Visual Head Tracking and Slaving for Visual Telepresence," *IEEE Int. Conf. Robotics & Automation*, pp. 2908-2914, 1996.
15. Horaud, R., Dornaika, F., Espiau, B.: *Visually Guided Object Grasping;* IEEE RA Vol.14(4), pp.525-532, 1998.
16. Hu, Y., Eagleson, R., Goodale, M.A.: *Human Visual Servoing for Reaching and Grasping: The Role of 3-D Geometric Features;* IEEE ICRA, pp.3209-3216, 1999.

17. Hutchinson, S., Hager, G.D., Corke, P.: "Visual Servoing: A Tutorial," *IEEE Trans. RA* Vol.12(5), 1996.
18. Ishii, I., Nakabo, Y., Ishikawa, M.: "Target Tracking Algorithm for 1ms Visual Feedback System Using Massively Parallel Processing", *IEEE ICRA*, pp. 2309-2314, 1996.
19. Kosaka, A., Nakazawa, G.: *Vision-Based Motion Tracking of Rigid Objects Using Prediction of Uncertainties;* ICRA, pp.2637-2644, 1995.
20. Marchand, E., "ViSP: A Software Environment for Eye-in-Hand Visual Servoing," IEEE ICRA, pp.3224-3229, 1999.
21. Nayar, S.K., Nene, S.A., Murase H., "Subspace methods for robot vision", *IEEE Robotics and Automation* 12(5), 750-758, 1996.
22. Nelson, B.J., Papanikolopoulos, N.P., Khosla, P.K.: *Robotic Visual Servoing and Robotic Assembly Tasks;* IEEE RA Magazine, pp. 23-31, 1996.
23. Pirjanian, P., Christensen, H.I., Fayman, J.A., "Application of voting to fusion of purposive modules: An experimental investigation," Robotics and Autonomous Systems 23(4), pp.253-266, 1998.
24. Rizzi, A.A., Koditschek, D.E.: "An Active Visual Estimator for Dexterous Manipulation," *IEEE Trans. RA* Vol.12(5), pp. 697-713, 1996.
25. Y. Shirai, Y. Mae, S. Yamamoto: *Object Tracking by Using Optical Flows and Edges;* 7th Int. Symp. on Robotics Research, pp. 440-447, 1995.
26. Tonko, M., Schäfer, K., Heimes, F., Nagel, H.H.: " Towards Visually Servoed Manipulation of Car Engine Parts," *IEEE ICRA*, pp. 3166-3171, 1997.
27. Toyama, Kentaro; Hager, Gregory D.: *Incremental focus of attention for robust vision-based tracking,* International Journal of Computer Vision 35(1), Pages 45-63, 1999.
28. Vincze, M., Weiman, C.: "On Optimising Window Size for Visual Servoing," *IEEE ICRA*, pp. 2856-2861, April 22-24, 1997.
29. Vincze, M., Ayromlou, M., Kubinger, W., "An Integrating Framework for Robust Real-Time 3D Object Tracking,' ' *Int. Conf. on Vision Systems*, Gran Canaria, pp. 135-150, 1999.
30. Vincze, M.: Real-time Vision, Tracking and Control -Dynamics of Visual Servoing; ICRA '00 IEEE Int. Conf. on Robotics and Automation, San Francisco, pp. 644-649, April 24-28, 2000.
31. Wilson, W.J., Williams Hulls, C.C., Bell, G.S.: "Relative End-Effector Control Using Cartesian Position Based Visual Servoing," *IEEE Trans. RA* 12(5), pp. 684-696, 1996.
32. Wren, C.R., Azarbayejani, A., Darrell, T., Pentland, A.P.: *Pfinder: Real-Time Tracking of the Human Body;* IEEE Transactions on Pattern Analysis and Machine Intelligence Vol.19(7), pp.780-785, 1997.
33. Wunsch, P., Hirzinger, G.: " Real-Time Visual Tracking of 3D-Objects with Dynamic Handling of Occlusion," *IEEE ICRA*, 1997.

Partitioned Image-Based Visual Servo Control: Some New Results

Peter Corke[1], Seth Hutchinson[2], and Nicholas R. Gans[2]

[1] CSIRO Manufacturing Science & Technology
Pinjarra Hills
Australia 4069
pic@cat.csiro.au
[2] Beckman Institute for Advanced Technology
University of Illinois at Urbana-Champaign
Urbana, Illinois, USA 61801
{seth,ngans}@uiuc.edu

Abstract. In image-based visual servo control, since control is effected with respect to the image, there is no direct control over the Cartesian velocities of the robot end effector. As a result, trajectories that the robot executes, while pleasing in the image plane, can be quite contorted in the Cartesian space. In this paper we describe a partitioned approach to visual servo control that overcomes this problem. In particular, we decouple the z-axis rotational and translational components of the control from the remaining degrees of freedom. Then, to guarantee that all features remain in the image throughout the entire trajectory, we incorporate a potential function that repels feature points from the boundary of the image plane. We illustrate the new control scheme with a variety of simulations and laboratory experiments.

1   Introduction

Visual servoing is a maturing approach to the control of robots in which tasks are defined visually, rather than in terms of previously taught Cartesian coordinates. In visual servo systems, information obtained from the vision system is used to control the motion of the robot in real-time, as opposed to older systems that used vision only to determine the initial state of the world, prior to task execution. There are numerous advantages to a visual servoing approach. Part position tolerance can be relaxed, as can the open-loop accuracy specification of the robot. The ability to deal with parts that are moving comes almost for free. Comprehensive overviews of the basic approaches, current applications, and open research issues can be found in a number of recent sources, including [18, 16, 5, 13].

Typical modern visual servo systems have a control structure that is hierarchical, with information from the vision system being used to provide setpoints to the low-level joint controllers. The robot's joint-level controller bears the burden of tracking velocity commands from the visual servo control level. Stability

G.D. Hager et al. (Eds.): Sensor Based Intelligent Robots, LNCS 2238, pp. 122–140, 2002.

of visual servo systems has been addressed by several researchers, including [7, 5, 23, 20]. Stability analyses typically focus on the image-based control law (described below in Section 2), and ignore robot dynamics (although there are a few exceptions to this). In hierarchical systems, ignoring robot dynamics does not pose a real problem, since these latter stability issues are treated in the design of the joint-level controller. Even though many of the stability issues have been dealt with effectively, there are a number of open problems regarding the performance of visual servo systems. In this paper we describe some of these performance issues, and propose a new visual servo scheme to improve performance for several types of visual servoing tasks.

Broadly speaking, there are two basic approaches to visual servo control: Image-Based Visual Servo (IBVS), and, Position-Based Visual Servo (PBVS). In IBVS, which will be briefly reviewed in Section 2, an error signal is measured in the image, and is mapped directly to actuator commands (see, e.g., [26, 10]). In PBVS systems, features are extracted from an image, and subsequently used to compute a (partial) 3D reconstruction of the environment or of the motion of a target object in the environment [30, 11, 25]. An error is then computed in the task space, and it is this error that is used by the control system. Thus, the actual control problem confronted by a PBVS system is the classical robotics problem of tracking a Cartesian trajectory.

IBVS approaches have seen increasing popularity, largely due to the shortcomings of PBVS systems. With PBVS, any errors in calibration of the vision system will lead to errors in the 3D reconstruction, and subsequently to errors during task execution. In addition, since the control law for PBVS is defined in terms of the 3D workspace, there is no mechanism by which the image is directly regulated. Thus it is possible that objects of interest (including features that are being used by the visual servo system) can exit the camera's field of view.

There are, however, also problems associated with IBVS systems. For an IBVS system the control law involves the mapping between image space velocities and velocities in the robot's workspace. This mapping is encoded in the image Jacobian (which will be briefly reviewed below). As one would expect, singularities in this Jacobian (which occur as a function of the relative position and motion of the camera and the object under observation) lead to control problems. This is, perhaps, the most persistent problem arising in IBVS systems. Second, since control is effected with respect to the image, there is no direct control over the Cartesian velocities of the robot end effector. Thus, trajectories that the robot executes, while producing images that are visually appealing, can appear quite unintuitive and far from optimal in the Cartesian space. In [3], Chaumette introduced a specific example that seems to produce both of these problems in a synergistic way: the camera moves in an extremely suboptimal trajectory with respect to the Cartesian path, while being driven toward a singularity in the image Jacobian. In Section 3, we will describe these performance issues in more detail, and provide a geometric explanation for them.

These performance problems with IBVS systems have led to the recent introduction of several hybrid methods [20, 21, 6]. Hybrid methods use IBVS to control certain degrees of freedom while using other techniques to control the remaining degrees of freedom. In Section 4 we describe a number of these hybrid approaches, and how they address specific performance issues.

In Sections 5 and 6, we present a new partitioned visual servo control scheme that overcomes a number of the performance problems faced by previous systems. The basic idea is to decouple the z-axis motions (including both the translational component and rotational component) from the other degrees of freedom, and to derive separate controllers for these z-axis motions. Our new approach improves performance, particularly for tasks that require large Z-axis rotation including the example of Chaumette[3]. We then incorporate techniques borrowed from the robot motion planning literature to guarantee that all features remain within the field of view.

Throughout the paper, we illustrate various concepts and methods with simulation results. We note here that in all simulations, the image features are the coordinates of the vertices of a unit square (1 m side length) in the XY-plane intersecting the Z-axis at $z = 8$ m. The camera uses a central projection model, with focal length $\lambda = 8$ mm, square pixels of side length $10\,\mu$m, and the principal point is (256, 256).

2   Traditional IBVS

In this section we present a very brief review of Image-Based Visual Servo control. Let $r = (x, y, z)^T$ represent coordinates of the end-effector, and $\dot{r} = (T_x, T_y, T_z, \omega_x, \omega_y, \omega_z)^T$ represent the corresponding end-effector velocity, composed of a linear velocity $\mathbf{v} = (T_x, T_y, T_z)^T$ and angular velocity $\omega = (\omega_x, \omega_y, \omega_z)^T$. Let $f = (u, v)^T$ be the image-plane coordinates of a point in the image and $\dot{f} = (\dot{u}, \dot{v})^T$ the corresponding velocities. The image Jacobian relationship is given by

$$\dot{f} = J(r)\dot{r}, \tag{1}$$

with

$$J = \begin{bmatrix} \dfrac{\lambda}{z} & 0 & \dfrac{-u}{z} & \dfrac{-uv}{\lambda} & \dfrac{\lambda^2 + u^2}{\lambda} & -v \\[2mm] 0 & \dfrac{\lambda}{z} & \dfrac{-v}{z} & -\dfrac{\lambda^2 - v^2}{\lambda} & \dfrac{uv}{\lambda} & u \end{bmatrix} \tag{2}$$

in which λ is the focal length for the camera. Derivations of this can be found in a number of references including [16, 1, 12].

The image Jacobian was first introduced by Weiss et al.[26], who referred to it as the *feature sensitivity matrix*. It is also referred to as the *interaction matrix* [7] and the **B** matrix [23]. The most common image Jacobian is based on the motion of points in the image (e.g., [10, 7, 14, 23, 27, 28]), but other image features have been used in visual servo schemes, including the distance between two points in the image plane and the orientation of the line connecting those two points [10], perceived edge length [29], the area of a projected surface and the relative areas of two projected surfaces [29], the centroid and higher order moments of a projected surface [2, 19, 29, 32], the parameters of lines in the image plane [4, 7] and the parameters of an ellipse in the image plane [7]. Of course, each different image feature requires its own specific image Jacobian, and these can be found in the references listed above.

Equation (1) can be decomposed, and written as

$$\dot{f} = J_v(u, v, z)\mathbf{v} + J_\omega(u, v)\omega, \tag{3}$$

in which $J_v(u, v, z)$ contains the first three columns of the image Jacobian, and is a function of both the image coordinates of the point and its depth, while $J_\omega(u, v)$ contains the last three columns of the image Jacobian, and is a function of only the image coordinates of the point (i.e., it does not depend on depth). This decomposition is at the heart of the hybrid methods that we discuss below.

The simplest approach to IBVS is to merely use (1) to construct the control law

$$\mathbf{u} = \Gamma J^{-1}(r)\dot{f} \tag{4}$$

in which $\dot{f}$ is the desired feature motion on the image plane, Γ is a gain matrix, and $\mathbf{u} = \dot{r}$ is the control input, an end-effector velocity (this can be converted to joint velocities via the manipulator Jacobian). Of course this approach assumes that the image Jacobian is square and nonsingular, and when this is not the case, a generalized inverse, J^+, is used. Since (4) essentially represents a gradient descent on the feature error, when this control law is used, feature points move in straight lines to their goal positions. This can be seen in Figure 1(a).

More sophisticated control schemes can be found in a variety of sources, including [23] where state space design techniques are used, and [7] where the task function approach is used.

3   Performance Issues

A commonly mentioned criticism of IBVS is that the Cartesian paths often involve large camera motions, which are undesirable. Often the camera moves away from the target in a normal direction and then returns, a phenomenon we refer to as *camera retreat*. Such motion is not time optimal, requires large and possibly unachievable robot motion, and is a seemingly non-intuitive solution to the required image plane motion. Figure 1 illustrates the problem. In Figure 1(a), the feature points are seen to be driven on straight line trajectories to their goal positions, producing a large, and seemingly unnecessary, motion in the z-direction, seen in Figure 1(c).

In [3], Chaumette introduced an extreme version of this problem, which we refer to as the Chaumette Conundrum, illustrated in Figure 2. Here, the desired camera pose corresponds to a pure rotation about the optic axis by π rad, i.e., the image feature point with initial coordinates (u, v) has the desired coordinates $(-u, -v)$. Since control laws of the form given in (4) drive the feature points in straight lines even for the case of pure target rotation, in this case the feature

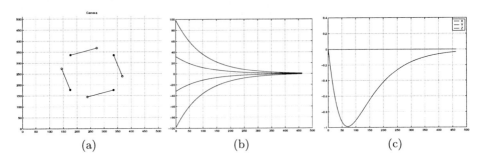

Fig. 1. IBVS for pure target rotation (0.3 rad). (a) Image-plane feature motion (initial location is ○, desired location is ●), (b) Feature error trajectory, (c) Cartesian translation trajectory.

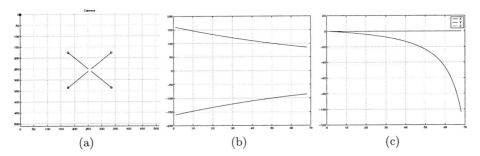

Fig. 2. Performance of classical IBVS with the Chaumette example. (a) Image-plane feature motion (initial location is ○, desired location is ●), (b) Feature error trajectory, (c) Cartesian translation trajectory.

points are driven toward the origin, which corresponds to a singularity in the image Jacobian. The singularity arises because the feature points will reach the origin when the camera retreats to a distance of infinity, at which all points collapse to the origin, and no motion can be observed. Thus, in the Chaumette Conundrum we observe two performance problems acting synergistically: (1) the controller is driven toward a singular configuration; (2) this singular configuration is approached asymptotically, and thus the system will servo forever without reaching the goal. We note that, as mentioned in [3], this problem cannot be detected by simply examining the image Jacobian, since the image Jacobian is well conditioned (at least initially). We use the term *IBVS failure* to refer to cases for which the system fails to achieve its goal.

At first it might seem that some rotational motion of the camera about its optic axis should be induced for the Chaumette Conundrum; however, this is not the case. The ω_z component of (4) is given by

$$\omega_z = (J^+)_6 \dot{f} \tag{5}$$

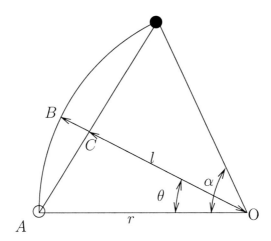

Fig. 3. Camera retreat model.

in which $(J^+)_6$ denotes the bottom row of the generalized inverse. In this particular case, even though $\dot{f} \neq 0$, the inner product is zero, i.e., the various contributions to rotational velocity cancel one another.

This camera retreat phenomenon can be explained in geometric terms, leading to a simple model that predicts the magnitude of the camera retreat motion. For the example of Figure 1, a pure rotational motion of the camera would cause the points to follow an arc from point A to point B, as shown in Figure 3. In order for the points to follow a straight line, as specified by (4), the scale must be changed so as to move the point from B to C. The required change in scale is given simply by the ratio of the distances OC and OB. The scale reduction attains its maximum value at $\theta = \alpha/2$ for which

$$\left(\frac{OC}{OB}\right)_{max} = \cos\frac{\alpha}{2}. \tag{6}$$

In the IBVS the reduction in scale is achieved by moving the camera away from the target. The reduction in the apparent length of the line segment is inversely proportional to the distance that the camera retreats, and therefore,

$$\frac{OC}{OB} = \frac{d_{targ}}{d} \tag{7}$$

in which d is the current distance to the target, and d_{targ} is the desired target distance, and assuming the camera is moving normal to the target. The maximum reduction is thus given by

$$\frac{d_{targ}}{d_{max}} = \cos\frac{\alpha}{2}. \tag{8}$$

For the Chaumette Conundrum, in which $\alpha = \pi$, the model accurately predicts infinite camera retreat.

At first it might seem that the introduction of line segment features would solve the problem, since the orientation of such a segment is unambiguous. Chaumette notes that such an approach is not guaranteed to solve the performance problems [3], and our own simulation results support this conclusion. Specifically, in simulations we added one extra row to the image Jacobian corresponding to a line segment angle feature [4]. Its effect was not significant. For the Chaumette Conundrum the addition of this feature does command *some* camera rotation, but this commanded rotational motion is nearly 3 orders of magnitude less than the Z-axis translation, even when scaling of feature magnitudes is taken into account.

There are a variety of possible solutions to this camera retreat problem. The requirement that points move in a straight line to their goal positions could be relaxed, giving rise to an image feature trajectory planning problem. The target depth, z, could be underestimated, causing the points to move in an arc instead of a straight line, reducing the magnitude of camera retreat (this is mentioned in [3]), but this will still fail for this example, in which no camera rotation occurs. Finally, the z-axis translational and rotational motions could be decoupled from the control law of (4), and separate controllers could be designed to enforce appropriate rotational and retreat motions. This latter approach leads to hybrid approaches that combine aspects of IBVS and PBVS systems. In Section 4 we describe several such approaches that have been recently introduced. Then, in Section 5 we introduce our new partitioned method.

4   Some Recent Hybrid Approaches

A number of authors [20, 21, 6] have recently addressed these problems by proposing hybrid control architectures. In each of these approaches, the decomposition of (3) is used. These methods rely on recent results in computing the epipolar geometry that relates a pair of images. In particular, the camera configurations that correspond to the initial and desired images are related by a homography matrix, which can be decomposed into the translational and rotational components of the motion between the two camera configurations. This homography matrix can be computed from a set of corresponding points in the initial and desired images [9].

For the special case of four coplanar points, the homography matrix that relates the current and desired images of these points is given by

$$\mathbf{H} = \mathbf{R} + \frac{\mathbf{t}}{d^*}\mathbf{n}^{*T},\qquad(9)$$

in which $\mathbf{R}$ is the rotation matrix describing the relative orientation of the current and desired camera coordinate frames, $\mathbf{t}$ is the displacement between the two frames, d^* is the distance from the camera frame to the plane containing the points, and $\mathbf{n}^*$ is the normal to the plane containing the four points. As has been shown in [8], this homography can be decomposed into a rotational component and a translational component. It is important to note that the translational component can be recovered only up to scale, and therefore, depth must be estimated if the translational component is to be used in a visual servo scheme (as is the case for [6]).

With 2.5-D visual servo [20], IBVS is used to control translational degrees of freedom, while the homography matrix (9) is used to estimate the desired rotational motion. A significant innovation in the 2.5-D visual servo method is their novel method for controlling the camera's rotational DOF. In [20], the rotation matrix is expressed as a rotation, θ, about an axis, $\mathbf{u}$. The resulting control is given by

$$\omega = \mathbf{u}\theta \tag{10}$$
$$\mathbf{v} = -\lambda J_v^{-1}\dot{f} + \lambda J_v^{-1}J_\omega \mathbf{u}\theta. \tag{11}$$

Thus, the rotational component of the control is computed directly from the computed desired rotation in 3D, and the translational component is computed by subtracting from the traditional IBVS control a term that accounts for the translational motion in the image induced by the rotation. Results in [3] and [20] show that this new method handles the problem of Figure 1 and eliminates camera retreat.

The only drawback to this approach seems to be that the commanded rotational motions may cause feature points to leave the image-plane. Note that with pure IBVS, the paths of the feature points are straight lines in the image plane, and this problem does not arise if the points are visible at the start and end of the motion (assuming that the viewable portion of the image plane is convex).

The problem of feature points leaving the image plane during 2.5-D visual servo motion motivated Morel et al. to propose a modified approach [21]. They use the same control as given by (10) and (11), but use a different feature vector (and, accordingly, an appropriate image Jacobian). Their modified feature vector includes the image coordinates (u_c, v_c) for a distinguished feature point (typically chosen near the center of the set of feature points), and

$$\sigma = \max_{(u_i,v_i)} \sqrt{(u_c - u_i)^2 + (v_c - v_i)^2}, \tag{12}$$

the maximal distance of any feature point to the distinguished feature point. They have shown results for which the distinguished feature point moves along a linear path straight to its desired image location, while the radius σ varies linearly with center displacement. This solution amounts to choosing a new image feature, σ, for depth control (i.e., T_z,), and the new features u_c, v_c to control translation parallel to the image plane (i.e., T_x, T_y). Thus, $\mathbf{v}$ is determined using a new type of IBVS, while ω is determined using PBVS.

Deguchi [6] takes the opposite approach from the 2.5-D scheme of Malis et al. In particular, he uses the decomposition of the homography matrix (9) to compute the translation velocity, leading to the control

$$\mathbf{v} = \hat{d}\left(\frac{t}{d^*}\right) \tag{13}$$

$$\omega = -\lambda J_\omega^{-1} \dot{f} + \lambda J_\omega^{-1} J_v \mathbf{v}. \tag{14}$$

Here, $\hat{d}$ is the estimated depth of the point in 3D and the ratio t/d^* is the scaled translation that is directly yielded by the decomposition of the homography matrix (9). Thus, the translational component of the control is computed directly from the estimated desired translation in 3D, and the rotational component is computed by subtracting from the traditional IBVS control a term that accounts for the motion in the image that is induced by the translation.

Deguchi also presents a second method, in which the essential matrix is used (instead of the homography matrix (9)) to compute the desired translational component. He uses the eight point algorithm to compute the essential matrix and its decomposition (see, e.g., [9]). This method yields essentially the same control as the first method, but the constraint that the four feature points be coplanar is removed.

All of the hybrid methods reported in [21, 20, 6] use the epipolar geometry to determine certain components of the camera motion while using an IBVS approach to determine the remaining component of the velocity. These methods rely on the online computation of the epipolar geometry [9], which amounts to computing a homography between two images, and require that the feature points on the object be coplanar. This homography is encapsulated in the fundamental matrix (for uncalibrated cameras) or essential matrix (for cameras with intrinsic parameters calibrated). The homography must then be decomposed to extract the rotational component and the problem of non-unique solutions must be dealt with.

We now describe our new approach, which does not exploit the epipolar geometry of the desired and initial images, and does not use any explicit 3D information.

5   A New Partitioned IBVS Scheme

Our approach is based on the observation that while IBVS works well for small motions, problems arise with large motions and particularly involving rotation about the z axis. Our proposed partitioned scheme singles out just Z-axis translation and rotation for special treatment, unlike the hybrid approaches mentioned above which treat all three rotational degrees of freedom specially. The motivation for the new partitioning is that camera retreat is a Z-axis translation phenomenon and IBVS failure is a Z-axis rotation phenomenon.

We partition the classical IBVS of (1) so that

$$\dot{f} = J_{xy}\dot{r}_{xy} + J_z\dot{r}_z \tag{15}$$

where $\dot{r}_{xy} = [Tx \; Ty \; \omega_x \; \omega_y]$, $\dot{r}_z = [T_z \; \omega_z]$, and J_{xy} and J_z are respectively columns $\{1, 2, 4, 5\}$ and $\{3, 6\}$ of J. Since $\dot{r}_z$ will be computed separately we can write (15) as

$$\dot{r}_{xy} = J_{xy}^+ \left\{ \dot{f} - J_z\dot{r}_z \right\} \tag{16}$$

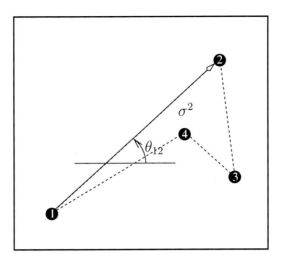

Fig. 4. Image features for new partitioned IBVS control.

where $\dot{f}$ is the feature point coordinate error as in the traditional IBVS scheme.

The Z-axis velocity, $\dot{r}_z$, is based directly on two new image features that are simple and computationally inexpensive to compute. The first image feature, $0 \le \theta_{ij} < 2\pi$, is the angle between the u-axis of the image plane and the directed line segment joining feature points i and j. This is illustrated in Figure 4. For numerical conditioning it is advantageous to select the longest line segment that can be constructed from the feature points, and allowing that this may change during the motion as the feature point configuration changes. The rotational rate is simply

$$\omega_z = \gamma_{\omega_z}(\theta^*_{ij} - \theta_{ij})$$

in which γ_{ω_z} is a scalar gain coefficient. This form allows explicit control over the direction of rotation, which may be important to avoid mechanical motion limits. For example if a hard stop exists at θ_s then

$$\omega_z = \gamma_{\omega_z}\mathrm{sgn}(\theta^*_{ij} - \theta_s)\mathrm{sgn}(\theta_{ij} - \theta_s)\left[\theta^*_{ij} - \theta_{ij}\right]$$

will avoid motion through that stop.

The second new image feature that we use is a function of the area of the regular polygon whose vertices are the image feature points. The advantages of this measure are that (1) it is a scalar; (2) it is rotation invariant thus decoupling camera rotation from Z-axis translation; (3) it can be cheaply computed. We compute this area using the the method of Wilf and Cunningham[31]. For n sequential boundary points labeled $1 \cdots n$ where point $P_0 \equiv P_n$, the p, q moments of the polygon are given by

$$m_{pq} = \frac{1}{p+q+2}\sum_{\ell=1}^{n}A_\ell\sum_{i=0}^{p}\sum_{j=0}^{q}\frac{(-1)^{i+j}}{i+j+1}\binom{p}{i}\binom{q}{j}x_\ell^{p-i}y_\ell^{q-j}\Delta x_\ell^i\Delta y_\ell^j \qquad (17)$$

where $A_\ell = x_\ell \Delta y_\ell - y_\ell \Delta x_\ell$, $\Delta x_\ell = x_\ell - x_{\ell-1}$ and $\Delta y_\ell = y_\ell - y_{\ell-1}$. The area of the polygon is just the $0, 0$ moment, m_{00}.

The feature that we choose to use is the square root of area

$$\sigma = \sqrt{m_{00}}$$

which has the dimension of length, giving this feature the units of pixels and thus a similar magnitude control gain as for the features $\dot{f}$. The camera z-axis translation rate is thus given by

$$T_z = \gamma_{T_z}(\sigma^* - \sigma). \tag{18}$$

The features discussed above for z-axis translation and rotation control are simple and inexpensive to compute, but work best when the target normal is within $\pm 40°$ of the camera's optical axis. When the target plane is not orthogonal to the optical axis its area will appear diminished, due to perspective, which causes the camera to initially approach the target. Perspective will also change the perceived angle of a line segment which can cause small, but unnecessary, z-axis rotational motion. Other image features can however be used within this partitioning framework. The ability to explicitly control z-axis translational motion is of particular benefit for controlling the field of view, as will be discussed in the next section. We also note that classical IBVS is very sensitive to depth estimation as the angle between the optical axis and the plane normal increases.

5.1   Experimental Results

Figure 5 shows a simulation of the performance of the proposed partitioned controller for the Chaumette Conundrum. The important features are that the camera does not retreat since σ is constant at $\sigma = 0$. The rotation θ monotonically decreases and the feature points move in a circle. The feature coordinate error is initially increasing, unlike the classical IBVS case in which feature error is monotonically decreasing.

A simulation that involves more complex translational and rotational motion is shown in Figure 6. The new features decrease monotonically, but the error in f does not decrease monotonically and the points follow complex curves on the image plane. Figure 7 compares the Cartesian camera motion for the two IBVS methods. The proposed partitioned method has eliminated the camera retreat and also exhibits better behavior for the X- and Y-axis motion. However the consequence is much more complex image plane feature motion that admits the possibility of the points leaving the field of view. This problem is discussed in the next Section.

We have also obtained experimental results using an eye-in-hand system in our lab. The system comprises a Puma 560 robot with a 480×640 grey-scale camera mounted on the end effector. The initial and goal images are shown in Figure 8. The images contain three integrated circuit chips on a black background. For this experiment, the initial image corresponds to a motion of the camera from the goal configuration of (15mm,10mm,-15mm) in the (x,y,z) directions and 15

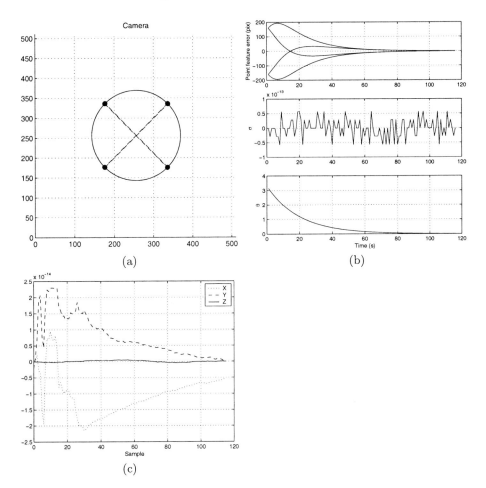

Fig. 5. Proposed partitioned IBVS for pure target rotation (π rad). (a) Image-plane feature motion (initial location is ○, desired location is ●), (b) Feature error trajectory, (c) Cartesian translation trajectory.

degrees about the z axis. The centroids of the three chips are used as the image features. In this experiment, the error decreased to zero after 113 iterations.

Figure 9 shows the error measurements of the locations of the three points, as well as for σ and θ. Figure 10 shows the error measurements for the same initial and goal images, but using the traditional IBVS approach.

6 Keeping Features in the Image Plane

In order to keep all feature points inside the viewable portion of the image plane at all times, we borrow collision avoidance techniques from the robot motion planning community. In particular, we establish a repulsive potential at the

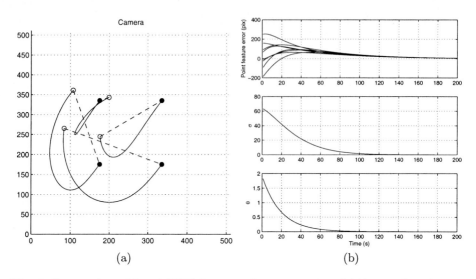

Fig. 6. Proposed partitioned IBVS for general target motion. (a) Image-plane feature motion (dashed line shows straight line motion for classical IBVS), (b) Feature error trajectory.

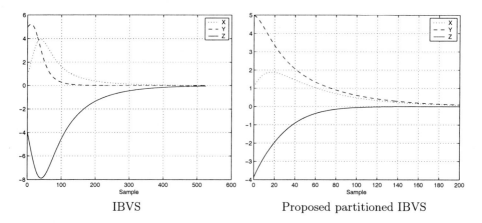

Fig. 7. Comparison of Cartesian camera motion for classic and new partitioned IBVS for general target motion.

boundary of the viewable portion of the image, and incorporate the gradient of this potential into the control law. We use the simple potential given by

$$
U_{rep}(u,v) = \begin{cases} \frac{1}{2}\eta \left(\frac{1}{\rho(u,v)} - \frac{1}{\rho_0} \right) & : \rho(u,v) \le \rho_0 \\ 0 & : \rho(u,v) > \rho_0 \end{cases} \tag{19}
$$

in which $\rho(u,v)$ is the shortest distance to the edge of the image plane from the image point with coordinates (u,v). The value ρ_0 specifies the zone of the image

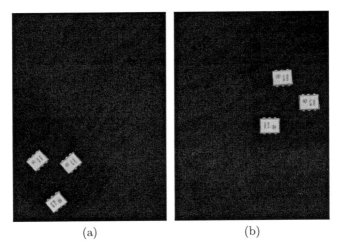

(a)   (b)

Fig. 8. The initial (a) and goal (b) images for the experiment.

in which U_{rep} affects the control; if the feature point is not with distance ρ_0 of the boundary, then the corresponding motion is not affected by U_{rep}. The value of η is a scalar gain coefficient.

For an $N_r \times N_c$ image, the value of ρ is easily computed as

$$\rho(u,v) = \min\{u, v, N_r - u, N_c - v\}. \tag{20}$$

If $\mathbf{n}$ is the unit vector directed from the nearest boundary to image feature point with coordinates (u, v), then $\nabla U_{rep} = F\mathbf{n}$, with F given by

$$F(u,v) = \begin{cases} \eta \left(\dfrac{1}{\rho(u,v)} - \dfrac{1}{\rho_0} \right) \dfrac{1}{\rho^2(u,v)} & : \rho(u,v) \le \rho_0 \\ 0 & : \rho(u,v) > \rho_0 \end{cases}. \tag{21}$$

Since a pure translation in the negative z-direction will cause feature points to move toward the center of the image, the value of F is mapped directly to the T_z component of the velocity command by combining it with the control given in (18). Because of chatter effects (where the feature points oscillate in and out of the potential field), we smooth and clip the resulting T_z, yielding the discrete-time controller

$$T'_z(k) = \mu T'_z(k-1) + (1-\mu)(\sigma^* - \sigma - F) \tag{22}$$

$$T_z = \min\{\max\{T'_z(k), T_{z_{min}}\}, T_{z_{max}}\}. \tag{23}$$

In simulation we found it advantageous to use asymmetric velocity clipping where $|T_{z_{max}}| < |T_{z_{min}}|$, that is, the camera can retreat faster than it can approach the target. This reduces the magnitude of the "bounces" off the boundaries of the image plane when points first enter the potential field. In practice this smoothing and clipping may not need to be explicitly implemented, since the real robot will have finite bandwidth and velocity capability.

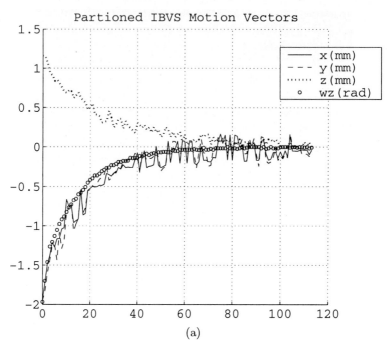

(a)

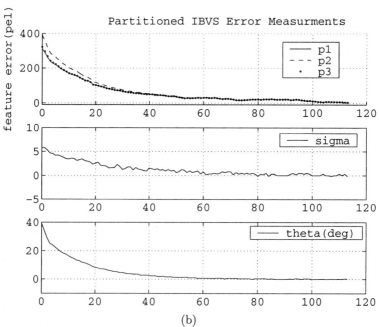

(b)

Fig. 9. (a) Camera motion and (b) feature error trajectory for the images shown in Figure 8 using the new controller.

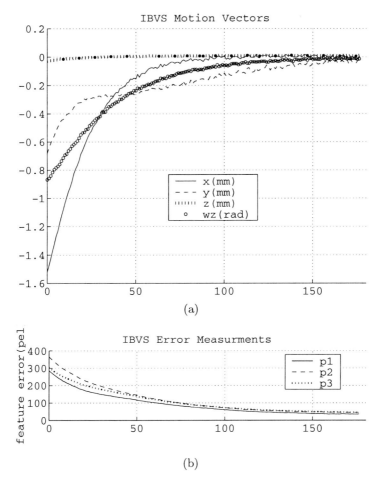

Fig. 10. (a) Camera motion and (b) feature error trajectory for the images shown in Figure 8 using a traditional IBVS approach.

The use of a potential field raises the issue of local minima in the field, but in our case, these issues do not arise, since the potential field is used merely to force a camera retreat, and since it will be possible for the system to achieve the goal when this retreat is effected (in this case we merely approach the performance of the classical IBVS system). Of course this assumes that no goal feature point locations lie within the influence of the potential field. Should this not be the case, then ρ_0 must be adjusted accordingly.

Results of the new partitioned IBVS with collision avoidance are shown in Figure 11. The target is larger than before, so as the camera rotates the feature points move into the potential field. The parameters used were $\eta = 5 \times 10^6$ and $\mu = 0.8$. It can be seen that as the points are rotated, they move into the potential field and then follow a path parallel to the edge, where the repulsion and scale demand are in equilibrium.

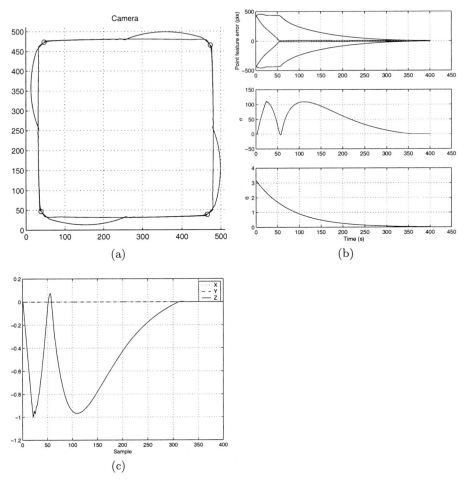

(a) (b) (c)

Fig. 11. Proposed partitioned IBVS with collision avoidance for pure target rotation (π rad). (a) Image-plane feature motion (initial location is ○, desired location is ●), (b) Feature error trajectory, (c) Cartesian translation trajectory.

For high rotational rates, the chatter phenomenon will occur, and at very high rates the points may pass through the potential field and become trapped *outside* the image plane. Rotational rate should properly be controlled by another loop, and this problem has strong similarities to that of controlling step size in numerical optimization procedures.

7   Conclusion

In this paper we have investigated some problems with classical image-based visual servoing and proposed a new partitioned visual servoing scheme that inexpensively overcomes these limitations. We have also provided simple geometric insight into the root cause of the undesirable camera retreat phenomenon, and the pathological case we have termed IBVS failure.

Other hybrid IBVS schemes have been recently proposed and are based on decoupling camera translational and rotational degrees of freedom. We have proposed a different decoupling and servo Z-axis rotation and translation using decoupled controllers based on two easily computed image features.

All hybrid schemes admit the possibility of points leaving the image plane, as does the approach that we described in 5. In this paper we consider this as a collision avoidance problem and employ potential field techniques to repel the feature points from the image plane boundary.

References

1. J. Aloimonos and D. P. Tsakiris. On the mathematics of visual tracking. *Image and Vision Computing*, 9(4):235–251, August 1991.
2. R. L. Andersson. *A Robot Ping-Pong Player. Experiment in Real-Time Intelligent Control.* MIT Press, Cambridge, MA, 1988.
3. F. Chaumette. Potential problems of stability and convergence in image-based and position-based visual servoing. In D. Kriegman, G. Hager, and S. Morse, editors, *The confluence of vision and control*, volume 237 of *Lecture Notes in Control and Information Sciences*, pages 66–78. Springer-Verlag, 1998.
4. Francois Chaumette. *La relation vision-commande: théorie et application à des tâches robotiques.* PhD thesis, L'Univesité de Rennes I, 1990.
5. Peter I. Corke. *Visual Control of Robots: High-Performance visual servoing.* Mechatronics. Research Studies Press (John Wiley), 1996.
6. K. Deguchi. Optimal motion control for image-based visual servoing by decoupling translation and rotation. In *Proc. Int. Conf. Intelligent Robots and Systems*, pages 705–711, October 1998.
7. B. Espiau, F. Chaumette, and P. Rives. A New Approach to Visual Servoing in Robotics. *IEEE Trans. on Robotics and Automation*, 8:313–326, 1992.
8. O. D. Faugeras and F. Lustman. Motion and Structure from Motion in a Piecewise Planar Environment. *Int'l Journal of Pattern Recognition and Artificial Intelligence*, 2(3):485–508, 1988.
9. O.D. Faugeras. *Three-Dimensional Computer Vision.* MIT Press, Cambridge, MA, 1993.
10. J.T. Feddema and O.R. Mitchell. Vision-guided servoing with feature-based trajectory generation. *IEEE Trans. Robot. Autom.*, 5(5):691–700, October 1989.
11. S. Ganapathy. Real-time motion tracking using a single camera. Technical Memorandum 11358-841105-21-TM, AT&T Bell Laboratories, November 1984.
12. R. M. Haralick and L. G. Shapiro. *Computer and Robot Vision.* Addison Wesley, 1993.
13. K. Hashimoto, editor. *Visual Servoing*, volume 7 of *Robotics and Automated Systems*. World Scientific, 1993.
14. K. Hashimoto, T. Kimoto, T. Ebine, and H. Kimura. Manipulator control with image-based visual servo. In *Proc. IEEE Int. Conf. Robotics and Automation*, pages 2267–2272, 1991.
15. Koh Hosoda and Minoru Asada. Versatile visual servoing without knowledge of true Jacobian. In *Proc. IROS*, September 1994.

16. S. Hutchinson, G. Hager, and P. Corke. A tutorial on visual servo control. *IEEE Trans. Robot. Autom.*, 12(5):651–670, October 1996.

17. Martin Jägersand, Olac Fuentes, and Randal Nelson. Experimental evaluation of uncalibrated visual servoing for precision manipulation. In *Proc. IEEE Int. Conf. Robotics and Automation*, 1996.

18. D. Kriegman, G. Hager, and S. Morse, editors. *The confluence of vision and control*, volume 237 of *Lecture Notes in Control and Information Sciences*. Springer-Verlag, 1998.

19. M. Lei and B. K. Ghosh. Visually-Guided Robotic Motion Tracking. In *Proc. Thirtieth Annual Allerton Conf. on Communication, Control, and Computing*, pages 712–721, 1992.

20. E. Malis, F. Chaumette, and S. Boudet. 2-1/2-d visual servoing. *IEEE Trans. Robot. Autom.*, 15(2):238–250, April 1999.

21. G. Morel, T. Liebezeit, J. Szewczyk, S. Boudet, and J. Pot. Explicit incorporation of 2d constraints in vision based control of robot manipulators. In Peter Corke and James Trevelyan, editors, *Experimental Robotics VI*, volume 250 of *Lecture Notes in Control and Information Sciences*, pages 99–108. Springer-Verlag, 2000. ISBN: 1 85233 210 7.

22. N. P. Papanikolopoulos and P. K. Khosla. Adaptive Robot Visual Tracking: Theory and Experiments. *IEEE Trans. on Automatic Control*, 38(3):429–445, 1993.

23. N. P. Papanikolopoulos, P. K. Khosla, and T. Kanade. Visual Tracking of a Moving Target by a Camera Mounted on a Robot: A Combination of Vision and Control. *IEEE Trans. on Robotics and Automation*, 9(1):14–35, 1993.

24. J.A. Piepmeier, G. McMurray, and H. Lipkin. A dynamic quasi-newton method for uncalibrated visual servoing. In *Proc. IEEE Int. Conf. Robotics and Automation*, pages 1595–1600, 1999.

25. A.A. Rizzi and D.E. Koditschek. Preliminary experiments in spatial robot juggling. In *Proc. 2nd International Symposium on Experimental Robotics*, Toulouse, France, June 1991.

26. A. C. Sanderson, L. E. Weiss, and C. P. Neuman. Dynamic sensor-based control of robots with visual feedback. *IEEE Trans. Robot. Autom.*, RA-3(5):404–417, October 1987.

27. R. Sharma, J-Y. Hervé, and P. Cucka. Dynamic Robot Manipulation Using Visual Tracking. In *Proc. of the 1992 IEEE International Conference on Robotics and Automation*, pages 1844–1849, May 1992.

28. S. B. Skaar, W. H. Brockman, and W. S. Jang. Three-Dimensional Camera Space Manipulation. *Int'l Journal of Robotics Research*, 9(4):22–39, 1990.

29. L. E. Weiss, A. C. Sanderson, and C. P. Neuman. Dynamic Sensor-Based Control of Robots with Visual Feedback. *IEEE Journal of Robotics and Automation*, 3:404–417, 1987.

30. D. B. Westmore and W. J. Wilson. Direct dynamic control of a robot using an end-point mounted camera and Kalman filter position estimation. In *Proc. IEEE Int. Conf. Robotics and Automation*, pages 2376–2384, 1991.

31. J.M. Wilf and R.T. Cunningham. Computing region moments from boundary representations. JPL 79-45, NASA JPL, November 1979.

32. B. Yoshimi and P. K. Allen. Active, uncalibrated visual servoing. In *Proc. IEEE Int'l Conf. on Robotics and Automation*, pages 156–161, San Diego, CA, May 1994.

Towards Robust Perception
and Model Integration

Bernt Schiele, Martin Spengler, and Hannes Kruppa

Perceptual Computing and Computer Vision Group
Computer Science Department
ETH Zurich, Switzerland
{schiele,spengler,kruppa}@inf.ethz.ch
http://www.vision.ethz.ch/pccv/

Abstract. Many of today's vision algorithms are very successful in controlled environments. Real-world environments, however, cannot be controlled and are most often dynamic with respect to illumination changes,
motion, occlusions, multiple people, etc. Since most computer vision algorithms are limited to a particular situation they lack robustness in the
context of dynamically changing environments. In this paper we argue
that the integration of information coming from different visual cues and
models is essential to increase robustness as well as generality of computer vision algorithms. Two examples are discussed where robustness
of simple models is leveraged by cue and model integration. In the first
example mutual information is used as a means to combine different
object models for face detection without prior learning. The second example discusses experimental results on multi-cue tracking of faces based
on the principles of self-organization of the integration mechanism and
self-adaptation of the cue models during tracking.

1 Introduction

Visual cues and object models have all their specific strengths and weaknesses
depending on the context and the dynamics of the environment. Therefore, combining different and complementary visual cues and object models promises to
increase the robustness and generality of today's computer vision algorithms. As
for recognition there exist a considerable number of highly successful algorithms.
All approaches, however, have their own inherent limitations. Similar observations hold for tracking and detection of objects. While there are many computer
vision algorithms using a single object model or visual cue, work on the *combination* of different cues and models is still in its infancy. Since no single model is
robust and general enough to cover all possible environmental conditions, their
combination promises to increase robustness and generality.

The goal of our research is to overcome limitations of individual models by
combining multiple models at different levels. In order to integrate and combine
complementary object models, the paper explores two different approaches. The
first approach uses a general framework based on mutual information which is

G.D. Hager et al. (Eds.): Sensor Based Intelligent Robots, LNCS 2238, pp. 141–158, 2002.

used to measure mutual agreement between different object models. Rather than learning a static integration scheme, the algorithm determines model parameters which maximize agreement among the employed models and the current image data. Since the image data is used directly, the integration mechanism dynamically adapts to environmental changes for example with respect to illumination conditions. The framework allows to combine different models dynamically and 'on-the-fly'. This makes the approach general and easily extendible to new object models.

The second approach explores the use of different visual cues for tracking. The quality of each visual cue is estimated and used as a measure of reliability. These reliabilities are used directly to combine visual cues. An important aspect of the approach is the adaptability of the object models as well as of the integration mechanism according to the actual context. These principles, called self-organization and self-adaptation have been proposed by Triesch and von der Malsburg [20]. The paper discusses their approach and evaluates the performance of different extensions of the system.

1.1   Related Work

Probably the most common approach to integrate information is to accumulate all responses (raw data or features) into a single vector. Taking a number of these vectors, standard machine learning techniques [6] can be used to learn a joint probabilistic model. Bayesian approaches can be taken to statistically model dependencies between the different data sources and sensors. However, the amount of training data required may be prohibitive [3]. [2] proposes a hierarchical mixture of experts [11] in which mixtures of appearance models are learned for the classification of objects. The common drawback of these approaches is that the integration is static in the sense that we cannot change the weighting of responses dynamically depending, for example, on their usefulness or the environmental conditions.

Combining different classifiers is a standard problem in pattern recognition (see for example [23, 12, 10]). Typically, different classifiers are trained individually and their parameters are fixed thereafter. It is the combiner's task to learn and choose the appropriate combination mechanisms depending on particular situations. In that sense only the combiner itself may be able to increase the robustness and generality of the individual classifiers. However, the results will most often be sup-optimal since the combiner does not have access to the classifiers themselves and cannot adapt their parameters. As before the combination scheme is fixed and static during runtime.

In the literature, several multi-cue tracking algorithms are reported. Even though they are called multi-cue most approaches use only one cue at a time by employing methods to select the "optimal" cue for the task at hand. For example Toyama and Hager [18] propose a layered hierarchy of vision based tracking algorithms. There declared goal is to enable robust, adaptive tracking in real-time. Depending on the actual conditions tracking algorithms are chosen: when the conditions are good, an accurate and precise tracking algorithm is chosen and when conditions deteriorate more robust but less accurate algorithms are chosen. This architecture is one of the rare examples where the declared goal is robustness which makes it a very interesting framework. In our belief, however, it is not only important to select an appropriate tracking algorithm but even more important to integrate the results of different algorithms and cues in order to achieve robust tracking. Crowley and Berard [5] propose to use three different tracking algorithms in a similar way as proposed by Toyama and Hager.

Eklundh et al. [22, 14] convincingly argue for a system's oriented approach to fixation and tracking in which multiple cues are used concurrently. The important visual cues are motion and disparity which enable to track and fixate single targets using a stereo camera head. It is not clear however how this approach may be extended to track multiple targets concurrently. We argue that simultaneous tracking of multiple targets and target hypotheses is important not only in the presence of multiple targets but also for recovery from tracking failure. Isard and Blake propose an algorithm called CONDENSATION [8] which is well suited to track multiple target hypotheses simultaneously. They extended their original approach [9] by a second cue (color in their case) which allows to recover from tracking failures. Even though not proposed by Isard and Blake their algorithm is well suited for integrating multiple cues.

Democratic Integration, an approach proposed by Triesch and Malsburg [20] implements *concurrent* cue integration: All visual cues contribute simultaneously to the overall result and none of the cues has an outstanding relevance compared to the others. Again, robustness and generality is a major motivation for the proposal. Triesch et al [19] convincingly argue for the need for adaptive multi-cue integration and support their claims with psychophysical experiments. Adaptivity is a key point in which democratic integration contrasts with other integration mechanisms. Following the classification scheme of Clark and Yuille [4], democratic integration implements weakly coupled cue integration. That is, the used cues are independent in a probabilistic sense[1]. Democratic integration and similar weakly coupled data fusion methods are also closely related to voting [17]. The weighted sum approach of democratic integration may be seen as weighted plurality voting as proposed in [13]. A more thorough analysis of the relations between sensor fusion and voting can be found in [1].

2   Using Mutual Information to Combine Object Models

This section presents a hierarchical method to combine an arbitrary number of different object models based on the concept of mutual information. A randomized algorithm is used to determine the model parameters which maximize mutual agreement between the models. To validate the effectiveness of the proposed method, experiments on human face detection by combining different face models are reported. The experimental results indicate that the performance of

---

[1] Due to the feedback loop of democratic integration its cues are not entirely independent anymore but indirectly coupled by the result they agreed upon.

simple visual object models can be leveraged significantly by combining them based on the proposed method.

The proposed method for the combination of multiple object models is based on the information-theoretic principle of *mutual information*. Mutual information of two discrete random variables W and Z is defined as:

$$I(W;Z) = \sum_i \sum_j p(W=i, Z=j) \log \frac{p(W=i, Z=j)}{p(W=i)p(Z=j)} \qquad (1)$$

In the equation, the summations (over i and j) are over the possible values of the random variables. In our case, these values are the different possible classification events according to the two object models. Since we treat each object model in the following as binary classifier there are two exclusive events per classifier. For the first model for example pixels may be labeled object ($W = 1$) or pixels may be labeled non-object ($W = 0$). The probabilities of these two events are estimated using the relative frequency of pixels labeled object and non-object:

$$p(W=i) = \frac{\#\{w_n = i\}}{N} \qquad (2)$$

where w_n corresponds to the classification of pixel n according to the first model and N is the total number of pixel in the image. Similarly the probabilities of the two events for the second model are estimated:

$$p(Z=i) = \frac{\#\{z_n = i\}}{N} \qquad (3)$$

where z_n corresponds to the classification of pixel n according to the second model. The combination of the two models results in four different events: pixels are labeled object by both models ($W = 1, Z = 1$), pixels are labeled non-object by both models ($W = 0, Z = 0$), or pixels are labeled object by one model and non-object by the other model ($W = 1, Z = 0$, and $W = 0, Z = 1$). Again, the probabilities of these events are estimated using the relative frequency of pixels classified as one of the four events:

$$p(W=i, Z=j) = \frac{\#\{w_n = i, z_n = j\}}{N} \qquad (4)$$

Since W and Z correspond to the outputs of different models, maximizing equation 1 corresponds to maximizing *cross-model mutual information*. Changing any of the parameters of these models obviously influences the classification of the pixels and therefore the cross-model mutual information. Let α_W be the set of parameters of the first model and β_Z the parameters of the second model. Using this notation, maximizing mutual information (or *MMI*) comes to:

$$I_{max}(W;Z) = max_{\alpha_W, \beta_Z} I(W;Z) \qquad (5)$$

Currently we use a randomized search algorithm to find the optimal model parameter α_W and β_Z. The actual value of $I_{max}(W;Z)$ is represented by a

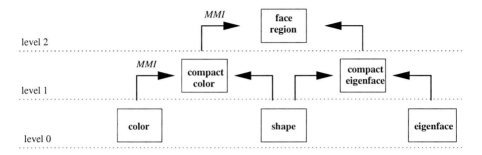

Fig. 1. Multi-level usage of MMI for combining object models.

fraction of a single bit of information. To undermine the relevance of mutual information in the context of object model combination, we briefly refer to the well-known Kullback-Leibler divergence. The KL-divergence between a probability mass function $p(W, Z)$ and a distinct probability mass function $q(W, Z)$ is defined as:

$$D(p(W,Z)||q(W,Z)) = \sum_{i,j} p(W = i, Z = j) log \frac{p(W = i, Z = j)}{q(W = i, Z = j)} \qquad (6)$$

The Kullback-Leibler divergence (also called relative entropy or information divergence) is often used as a distance measure between two distributions. By defining $q(W, Z) = p(W)p(Z)$ the mutual information can be written as the KL-divergence between $p(W, Z)$ and $p(W)p(Z)$:

$$I(W; Z) = D(p(W,Z)||p(W)p(Z)) \qquad (7)$$

Mutual information therefore measures the distance between the joint probability $p(W, Z)$ and the probability $q(W, Z) = p(W)p(Z)$, which is the joint probability under the assumption of independence. Conversely, it measures mutual dependency or the amount of information one object model contains about another. As a result mutual information can be used to measure *mutual agreement* between object models.

Note that while the mutual information between any number of discrete random variables could be evaluated directly by expanding equation 1 accordingly, we propose to combine models *pairwise*, essentially building up a hierarchy of combined responses. Figure 1 shows this principle of hierarchical pairwise combination of object models for face detection. At the bottom of this figure the models employed are shown, namely a skin color model, a facial shape model and a face template matcher. In the following we describe the individual face models and their hierarchical combination through MMI.

Level 0: Color, Eigenfaces and Shape: At the lowest level of the hierarchical combination scheme, three simple face detectors are employed. None of them is tuned particularly well to our dataset since we rather focus on the benefits of using mutual information for model combination than on tuning specific models.

We use a *skin color model* based on simple parametric model (see for example [16]) using a 3-dimensional Gaussian governed by the means of the r, g and b values and the associated covariance matrix. The parameter of the Gaussian are estimated using the standard maximum likelihood estimate using samples from several images of different people. Pixels are classified as face or non-face by thresholding the Mahalanobis distance, defined as $\Delta^2 = (x - \mu)^T \Sigma^{-1} (x - \mu)$.

Principal Component Analysis (PCA) is used as a generic way of dimensionality reduction preserving the essence of the data with respect to representation. PCA, as an approximation for face template matching, was introduced by [21] and [15], from which we adopt the terms *face space* and *eigenfaces*. We employ the eigenface method by iterating over each sub-image of a given input at the expected size of the face and by computing the distance to an averaged eigenface in the face space. The distance is then thresholded to classify each sub-image to contain a face or not.

While being independent of the actual visual input, we are treating *face shape* essentially as a third model. More specifically, we impose the constraint that the face region be elliptical. The parameters are therefore the position of the face region (given in 2D image coordinates) and the two main axes of the ellipse. One could extend the model by a fifth parameter, namely the rotation of the ellipse. This parameter has been omitted in the following experiments.

Level 1: Obtaining Compactness through MMI: At level 1 of the hierarchical combination scheme the color model is combined with the shape model and the eigenface model is combined with the shape model. In both cases the combination of the models results in a compact face region. As we will see in the experimental results described below this extraction of compact face regions allows to be robust with respect to a significant number of false positives.

Figure 2 illustrates the combination of color and shape using MMI. In this figure, the two images on level 0 depict the classification results of the color and shape model respectively using particular model parameters. In our implementation, the parameter space for shape is sampled to maximize mutual information while the parameters of the color model are held fixed. Three of the generated hypotheses at level 1 are shown in this figure, ranked by their mutual information from left to right. They represent the top three MMI maxima found by sampling the parameter space.

Level 2: MMI for Generating the Final Hypothesis: For the second and last level of the hierarchical combination scheme, MMI is used again to combine the intermediate outputs from level 1. More specifically, the mutual information of each of the 9 possible cross-model output pairs is computed. The global maximum from these computations amounts to the final hypothesis.

Figure 3 shows the top three maxima resulting from the combination of color and shape on the left side. Next to it are the top three maxima resulting from

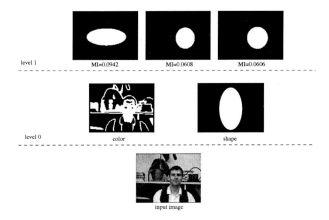

Fig. 2. Combination of Color and Shape based on MMI.

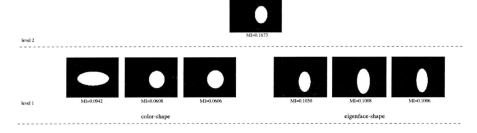

Fig. 3. MMI for generating the final hypothesis.

the combination of the eigenface model and the shape model. In this example, the combination of the third maxima from the color-shape combination and the second maximum from the eigenface-shape combination has the highest mutual information overall. These two are combined to generate the final hypothesis visualized by the image on level 2. As can be seen here, the unified hypothesis captures the computed face location as well as the size and shape of the face accurately. Note also that the mutual information value can be interpreted as a confidence measure for the final hypothesis.

2.1 Some Experimental Results

In order to understand the usefulness of MMI experiments have been conducted on a series of 252 test images each containing one face of seven different people. Each person in turn was recorded at three different poses (frontal, left and right), two different scales, two different lighting conditions and three different backgrounds. The first background is a white wall which was used as a reference background to train face models. The other backgrounds are cluttered scenes in order to evaluate the potentials and limitations of our approach. All images have

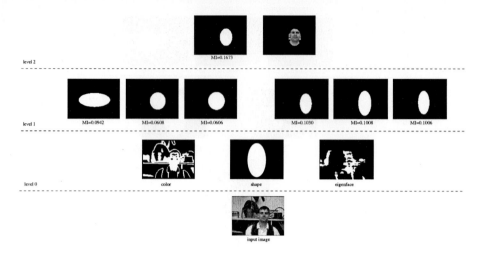

Fig. 4. MMI based model combination for face detection.

been smoothed and down-sampled to a resolution of 153x115 pixels. A third of the data (white background) has been used to train the face templates and the face color models described above. In the experiments about 0.2% of the total search space is grid-sampled to approximate the maximum mutual information values and associated model parameter.

Figure 4 shows the intermediate results from the pairwise combination of models and illustrates the final hypothesis. The individual images are arranged in analogy to figure 1. Starting on level 0, the left-most image depicts the outcome of applying the color model to the input image. In this case, pixels within a two-sigma interval are labeled as face color – shown as white pixels. Correspondingly, the picture to the right results from applying the eigenface model, while the image in the middle shows the elliptical face shape model at a specific location and scale.

The pictures on level 1 of figure 4, show the top-3 maxima of mutual information after combining color with shape (first three pictures from the left) and combining the eigenface model with shape, respectively. Again, face labeled pixels are shown as white. In this example, each individual maximum represents a good approximation of the actual face location. The final hypothesis shown on level 2 approximates the width of the face to within 3 pixels of accuracy, while the face height as well as the location of the face are approximated to within 1 pixel of accuracy.

In the next example, shown in figure 5, MMI has to deal with a situation in which one of the input models fails completely. The failure is caused by too many false positives on level 0 in the eigenface model and results in an explosion of false positives on level 1. Thus, the computed maxima of mutual information from combining eigenface with shape do not convey any information about location or size of the face. However, the combination of color and shape still

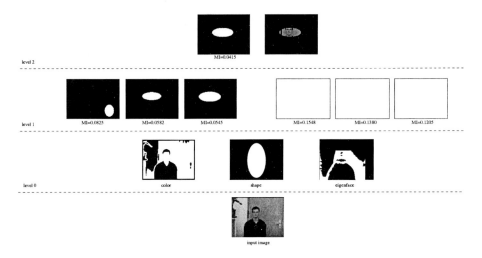

Fig. 5. Robustness to total failure of one model.

produces useful results. In the final stage the combination of the third maxima in both intermediate models yield the highest mutual information value. This demonstrates that MMI succeeds in propagating the available information through the hierarchy and produces a useful final result even when one of the three models fails completely. Note, that the precision of the final hypothesis is affected by loosing one model. In particular the width of the final hypothesis is not well estimated. Both the face location and the height of the face are still approximated within 5 pixels accuracy.

Finally, we are demonstrating a case in which MMI has to deal with partial failure in *two* models, shown in figure 6. As in the previous example, the test person is sitting in front of a wooden door. The color of the door makes detection of the face difficult because of its similarity to flesh tones. In consequence, both the color and eigenface model produce a large number of false positives on level 0. In this case, the combination with shape yields only one meaningful hypothesis in each of the two intermediate models shown on level 1. However, the final MMI computation finds the most meaningful hypothesis which gives the actual face location to within 6 pixels of accuracy. The approximated face height is still within a 8 pixels error while the face width can not be captured accurately here.

These three examples also justify the interpretation of mutual information as a confidence measure: The final hypothesis of the first example (figure 4) which had the greatest coherence among models also has the highest absolute mutual information. The two other examples (figures 5 and 6) which pose particular difficulties for face detection due to flesh-colored background have significantly lower mutual information values for their final hypotheses.

Discussion: The experiments show the validity and usefulness of mutual information for the combination of different object models. In particular the exper-

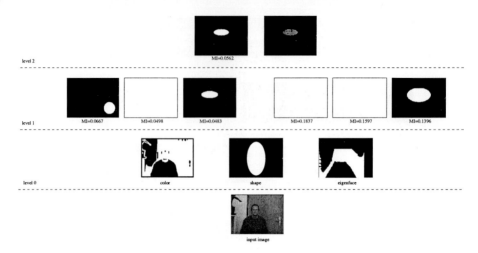

Fig. 6. Robustness to partial failure of two models.

iments show that the combination of object models provide significantly better results than any single model. The method works even when the individual model are very simple and prone to failure. Also, the calculated value of mutual information can be interpreted as a confidence measure.

3   Multi-cue Tracking of Faces

This section discusses the integration scheme depicted in figure 7. It is based on *Democratic Integration* introduced in [20]. Starting with an image sequence, saliency maps are computed for five different visual cues. Each cue has an associated weight reflecting the reliability of that cue in the current situation. Using these weights the integrator calculates a global saliency map. This superposition is used to estimate the target's most probable position. The weights of the different visual cues are estimated depending on their success in previous frames. More specifically, the quality of each visual cue is calculated with respect to the current position estimate. This quality measure is used to adapt the reliabilities i.e. the weights of each cue. Therefore, the integration process dynamically adapts to changes of the environmental conditions.

Due to the feedback loop of the proposed system maximal agreement between the involved cues can be reached. Two assumptions have to be fulfilled in order to obtain convergence and good overall performance: Firstly, at any given time a significant number of cues has to agree upon the target position and secondly, any environmental change may affect only a minority of the cues.

Self-Organized Sensor Integration. The method proposed by Triesch and von der Malsburg relies on two-dimensional saliency maps for each visual cue

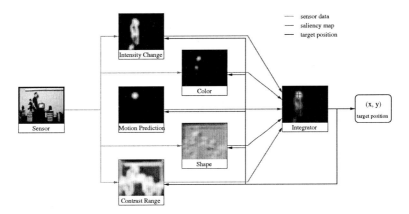

Fig. 7. Overall structure of multi-cue integration for tracking: five different visual cues are integrated. The maximum of the superposition is used to adapt the integration process as well as the visual cue models themselves.

involved. These saliency maps are denoted as $A_i(\mathbf{x}, t)$, where i indexes the different visual cues. We assume that $0 \le A_i(\mathbf{x}, t) \le 1$ where values close to 1 indicate high confidence about the target's presence at position $\mathbf{x}$ at time t. Typically, saliency maps are defined by a *similarity function* $S_i(P_i, F(\mathbf{x}, t))$ which measures the similarity of a prototype P_i and a feature vector $F(\mathbf{x}, t)$ extracted in image $I(\mathbf{x}, t)$ at position $\mathbf{x}$:

$$A_i(\mathbf{x}, t) = S_i(P_i, F(\mathbf{x}, t)), \quad 0 \le S_i(P_i, F(\mathbf{x}, t)) \le 1 \tag{8}$$

Here the prototypes P_i are held constant. Later an appropriate dynamics for prototype adaptation will be discussed. In order to integrate different visual cues, a global *reliability* $r_i(t)$ is estimated for each cue. It is assumed that $\sum_i r_i(t) = 1$. The saliency maps of the cues are combined to a *total estimation* $E(\mathbf{x}, t)$ by summation:

$$E(\mathbf{x}, t) = \sum_i r_i(t) A_i(\mathbf{x}, t) \tag{9}$$

So far, the integrating process does not consider any *a priori* knowledge about the target it ought to track. That is, the integration step is based entirely on single, spatially independent pixels. A simple but very useful way to deal with *a priori* knowledge like shape and size of expected targets is to convolve the superposition $E(\mathbf{x}, t)$ with a kernel T_{mask} which reflects some basic knowledge about the target:

$$E_{mask}(\mathbf{x}, t) = E(\mathbf{x}, t) * T_{mask} \tag{10}$$

As it turned out in the experiments this convolution is essential in order to obtain good overall performance of the system[2]. In the experiments described below we focus on head tracking which is why the applied kernel has elliptical shape approximating the shape of human heads.

---

[2] In the original method [20] no such *a priori* knowledge is used

The estimated target position $\hat{\mathbf{x}}(t)$ is defined as the global maximum of $E_{mask}(\mathbf{x}, t)$.

$$\hat{\mathbf{x}}(t) = \arg\max_{\mathbf{x}}\{E_{mask}(\mathbf{x}, t)\} \tag{11}$$

To achieve convergence towards robust tracking, the integrator adapts the reliabilities $r_i(t)$: The quality of the cues are measured against the system's overall performance, i.e. the superposition the cues agreed upon. This measurement is called *quality* $\tilde{q}_i(t)$ and has the following definition:

$$\tilde{q}_i(t) = \mathcal{R}(A_i(\hat{\mathbf{x}}(t)) - \langle A_i(\mathbf{x}, t)\rangle) \tag{12}$$

where $\langle \ldots \rangle$ denotes the average over all positions $\mathbf{x}$ and

$$\mathcal{R}(x) = \begin{cases} 0 : x \le 0 \\ x : x > 0 \end{cases} \tag{13}$$

is the *ramp* function. In words, quality $\tilde{q}_i(t)$ is defined as the distance between the cue's response at the estimated target position $\hat{\mathbf{x}}(t)$ and the cue's average response at time t, $\langle A_i(\mathbf{x}, t)\rangle$. Negative distances, i.e. cases where the response at the estimated position is lower than the average response, are truncated to 0.

It is worth noting that this definition is rather coarse. Other quality measurements may seem more natural and reasonable. For instance, one might define quality as the correlation between the cue's saliency map $A_i(\mathbf{x}, t)$ and the overall estimation $E_{mask}(\mathbf{x}, t)$. The Kullback-Leibler distance might be another appropriate quality measurement. The main advantage of the proposed quality definition is its low computational cost.

To change the cues' reliabilities, the relation between the reliability of a cue and its current quality is defined. This relation is expressed by the following *dynamics*:

$$\tau \dot{r}_i(t) = q_i(t) - r_i(t) \tag{14}$$

where τ is a time constant controlling the adaptation rate. Equation. (14) can be reformulated to express the adaptation step as Euler step:

$$r_i(t + \Delta t) = r_i(t) + \frac{\Delta t}{\tau}(q_i(t) - r_i(t)). \tag{15}$$

This equation couples a cue's reliability not directly to its quality but to its *normalized quality* $q_i(t)$ which is given by

$$q_i(t) = \frac{\tilde{q}_i(t)}{\sum_j \tilde{q}_j(t)}. \tag{16}$$

From equation. (14) or (15) one can derive that cues with quality measurements higher than their current reliability will tend to increase their reliability, whereas cues with quality lower than their current reliability will have decreased reliabilities. If a cue's quality remains zero, it will end up with reliability zero – the cue is suppressed entirely.

As mentioned before, the time constant τ influences the adaptation rate, i.e. it determines the rate at which the reliabilities are changed. This constant is considered to be large enough preventing the dynamics of $r_i(t)$ of being perturbed with high frequency-noise in the quality measurements $q_i(t)$.

Adaptive Visual Cues. Visual cue models extract certain features from the image sequence, e.g. localize areas in a given image which have a specific color or detect movement between two subsequent images. As a result, each visual cue model produces a two-dimensional *saliency map* $A_i(\mathbf{x}, t)$ which measures for each point $\mathbf{x} = (x, y)$ the expectation to find the target at the given position.

As before (equation 8) the saliency maps $A_i(\mathbf{x}, t)$ is calculated by the *similarity function* which compares a prototypical feature vector $P_i(t)$ with a feature vector $F_i(\mathbf{x}, t)$ extracted from image $I(\mathbf{x}, t)$ at position $\mathbf{x}$. The prototypes $P_i(t)$ are considered adaptive, in contrast to the non-adaptive prototypes P_i of the previous section:

$$A_i(\mathbf{x}, t) = S_i(P_i(t), F_i(\mathbf{x}, t)). \tag{17}$$

For adaptation, prototypes $P_i(t)$ and feature vectors $F_i(\mathbf{x}, t)$ at the estimated target position $\mathbf{x}$ are coupled by the following dynamics:

$$\tau_i \dot{P}_i(t) = F_i(\hat{\mathbf{x}}, t) - P_i(t) \implies P_i(t + \Delta t) = P_i(t) + \frac{\Delta t}{\tau_i}(F_i(\hat{\mathbf{x}}, t) - P_i(t)) \tag{18}$$

The time constants τ_i are not to be mistaken for the time constant τ describing the adaptation of the reliabilities (eq. 14). The prototype $P_i(t)$ converges towards the feature vector $\hat{F}_i(t)$ which is extracted from the current image. If the scene is stationary, i.e. $\hat{F}_i(t)$ is constant, $P_i(t)$ will converge to $\hat{F}_i(t)$ with time constant τ_i.

The prototype adaptation depends on the estimated target position in a second way: If no target is detected, i.e. the certainty $E(\hat{\mathbf{x}}, t)$ at position $\hat{\mathbf{x}}$ does not exceed a certain threshold Θ, the prototypes are adapted towards their default values. This behavior ensures that the system develops back to an initial state if the subject leaves the image and is no longer perceived as a potential target. If a valid estimation for the target's position is available then the adaptation converges the prototypes towards the current situation as described above.

Five Visual Cues for Facetracking. The *intensity change cue* detects motion in a grey-level image $\mathbf{s}(t)$ relative to its predecessor $\mathbf{s}(t-1)$. Motion is thus pixel-wise defined as the difference of intensity between two subsequent images. *Skin color detection* calculates for every pixel the probability of skin color. More specifically human skin color is modeled as a specific subspace of the HSI[3] color space. Depending only on the system's target position estimations $\hat{\mathbf{x}}(t)$, the *motion prediction cue* predicts the target's future motion to maintain motion continuity. In contrast to the original paper [20] we implemented a motion prediction unit using *Kalman filtering*. In order to determine potential head positions, the *shape template matching cue* correlates the grey-level input image $\mathbf{s}(t)$ and a head template for every position in the input image. The *contrast* cue extracts contrast pixel-wise from the input image $\mathbf{s}(t)$. Contrast, defined over the pixel's neighborhood, is compared to a adaptive model of contrast in order to detect pixels with salient contrast. All visual cues except the intensity change cue can be adaptive.

---

[3] Hue, Saturation, Intensity; see [7]

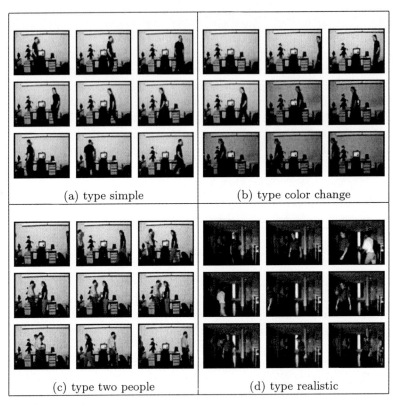

(a) type simple   (b) type color change

(c) type two people   (d) type realistic

Fig. 8. Samples of the four different types of images sequences tested.

3.1   Some Experimental Results

Figure 8 shows examples of different image sequences employed for testing of
the proposed integration scheme. In particular, four different types of image
sequences have been used. The first type (figure 8(a)) corresponds to the simple
situation of tracking a single person in an image sequence without environmental
changes. In the experiments below two such sequences are used, one with a person
moving from the right to the left and one with a person going back and forth.
The second type of images sequence (figure 8(b)) is more challenging since a
rather dramatic color change (from white to green illumination) occurs about
half through the images sequence. This sequence is a good test case for the
dynamic adaptivity of the integration scheme to environmental changes. The
third type of sequence (figure 8(c)) is interesting in two respects: firstly, more
than one person is present and secondly, the persons occlude each other. The
forth type of sequence (figure 8(d)) corresponds to a more realistic setting with
inhomogeneous and rather difficult lighting condition. Again we report results
on two different image sequences of this type.

In order to evaluate the effectiveness of the proposed integration scheme we
tested three different "levels" of the system. "Level 0" corresponds to the case

Table 1. Table summarizing the results of the different "levels" of the multi cue tracking system. Level 0 corresponds to fixed integration of the visual cues. Level 1 corresponds to the original proposition of democratic integration and level 2 corresponds to the use of visual cues enhanced with some a priori knowledge about the target

movie type	frames	level 0	level 1	level 2
(a) simple 1	78	97.4	75.6	87.2
(a) simple 2	151	82.1	48.3	88.7
(b) color change	70	61.4	11.4	97.1
(c) two people	114	31.6	9.6	80.7
(d) realistic 1	40	90.0	32.5	95.0
(d) realistic 2	299	36.5	18.4	82.6

where neither the visual cues nor the integration scheme are adaptive. This corresponds therefore to the "classical" case of using multiple visual cues in a predetermined and fixed way. The next stage of the system – "level 1" – corresponds to our implementation of the original system of Triesch and von der Malsburg [20]. Here, the visual cues as well as the integration mechanism are adaptive. Since the results of that level of the system were not robust enough we enhanced the visual cues by using a elliptical face template as introduced above (eq. 10). This stage of the system is called "level 2". This level of the system therefore uses some a priori knowledge about the target to be tracked.

Table 1 summarizes the results obtained with the different "levels" of the system. As one can see, the level 0 system obtains reasonable results for the two simple image sequences (a). In the more demanding case of a illumination color change (b) the performance of the system drops quite significantly. This is not surprising since no adaptation of the system to the new environmental condition is performed. In the case of two people (c), tracking performance is poor since the system has no means to distinguish between the two people. Which person is tracked at any time in the sequence is rather arbitrary explaining the poor performance. For the two image sequences taking under more realistic environmental conditions (d) the results are symptomatic: in the first sequence good tracking performance is obtained but in the second sequence the tracking performance is poor. Since the system cannot adapt to the situation tracking either succeeds or fails.

The level 1 system is expected to overcome the problems of the non-adaptivity of the level 0 system. Several sequences have been reported by Triesch and von der Malsburg where their system (corresponding to our level 1 system) succeeds to track in the presence of quite dramatic changes in illumination. For our sequences however, the results where rather disappointing. Basically for all cases the performance of the level 1 system was below the performance of the non-adaptive level 0 system. A more detailed analysis of the system revealed that the visual cues themselves where not well suited for face tracking. In order to make the cues more selective we therefore introduced some a priori knowledge about the shape of the target for the level 2 system.

Table 1 also shows the results for the level 2 system. First of all the results are significantly better than for the level 1 system. Compared to the level 0 system the level 2 fulfills the promises of the adaptivity of the system to environmental changes. In particular in the case of dramatic illumination color change (b) and the more realistic image sequences (d) tracking performance is always good. Also in the case of two people (c) the system successfully converges onto one target and tracks that person through the sequence. We'd like to emphasize that tracking performance of the level 2 system is always 80% and higher underlining the robustness of the system.

Discussion: Even though these experimental results are encouraging at least three observations have to be made: first of all, the proposed integration scheme is capable of tracking only one person at a time. Secondly, in the case of false positive tracking the system may lock onto a wrong target due to the adaptivity of the system. The third problem has to do with the significant number of parameters which are currently obtained heuristically and set manually. In order to obtain a truly robust and general tracking system more elaborate techniques such as learning have to be used for parameter setting.

In our believe, the problem of false positive tracking is the most fundamental problem of the system. Any tracking system has to be able to recover from failure. For the current system a bootstrapping mechanism might be added. We think however that is more promising to use a technique such as CONDENSATION [8] which explicitly allows tracking of multiple hypotheses simultaneously. We have started to extend the current system into that direction. Multiple hypotheses tracking may also help to overcome the first problem of single person tracking since it allows to track multiple targets such as people concurrently.

4   Conclusion and Discussion

In this paper we argue that the combination of different visual cues and models has the potential to increase the robustness and the generality of today's computer vision algorithms. Two examples are discussed where the robustness of simple models is leveraged significantly by means of cue and model integration.

The first example is based on an approach which combines different object models based on maximization of mutual information (MMI). The framework is generic and allows the individual cues to adopt different representations. Object models are combined into a single unified hypothesis *on-the-fly* without the need for specific training or reference data. The framework thus allows to easily add, reuse, remove and replace object models in a flexible manner. In a face detection experiment, three simple models for skin color, shape and eigenfaces are combined on-the-fly in an hierarchical fashion.

In the second example democratic integration [20] has been discussed, evaluated and extended. In that example five different visual cues are used for face tracking. Depending on the estimated quality of each visual cue their contribution to the integrated result can be adapted accordingly. This enables adaption of the integration scheme as well as the adaptation of the visual cues according to the environmental changes. Promising results have been obtained in the presence of dramatic illumination changes and realistic tracking environments.

References

1. C.G. Bräutigam. *A Model-Free Voting Approach to Cue Integration*. PhD thesis, Dept. of Numerical Analysis and Computing Science, KTH (Royal Institute of Technology), August 1998.
2. C. Bregler and J. Malik. Learning appearance based models: Mixtures of second moment experts. In *Advances in Neural Information Precessing Systems*, 1996.
3. T. Choudhury, B. Clarkson, T. Jebara, and A. Pentland. Multimodal person recognition using unconstrained audio and video. In *Proceedings of the 2nd International Conference on Audio-Visual Biometric Person Authentication*, 1998.
4. J. Clark and A. Yuille. *Data fusion for sensory information processing*. Kluwer Academic Publishers, Boston, Ma. – USA, 1994.
5. J.L. Crowley and F. Berard. Multi-modal tracking of faces for video communications. In *IEEE Conference on Computer Vision and Pattern Recognition*, 1997.
6. R. Duda and P. Hart. *Pattern Classification and Scene Analysis*. John Wiley & Sons, Inc., 1973.
7. Rafael C. Gonzalez and Richard E. Woods. *Digital Image Processing*. Addison-Wesley, 1993.
8. M. Isard and A. Blake. Condensation – conditional density propagation for visual tracking. *International Journal of Computer Vision*, 29(1):5–28, 1998.
9. M. Isard and A. Blake. Icondensation: Unifying low-level and high-level tracking in a stochastic framework. In *ECCV'98 Fifth European Conference on Computer Vision, Volume I*, pages 893–908, 1998.
10. A. Jain, P. Duin, and J. Mao. Statistical pattern recognition: A review. *IEEE Transactions on Pattern Analysis and Machine Intelligence*, 22(1):4–37, January 2000.
11. M.I. Jordan and R.A. Jacobs. Hierachical mixtures of experts and the EM algorithm. *Neural Computation*, 6(2), March 1994.
12. J. Kittler, M. Hatef, R. Duin, and J. Matas. On combining classifiers. *IEEE Transactions on Pattern Analysis and Machine Intelligence*, 20(3):226–239, March 1998.
13. D. Kragić and H. I. Christensen. Integration of visual cues for active tracking of an end-effector. In *IROS'99*, volume 1, pages 362–368, October 1999.
14. A. Maki, J.-O. Eklundh, and P. Nordlund. A computational model of depth-based attention. In *International Conference on Pattern Recogntion*, 1996.
15. B. Moghaddam and A. Pentland. Probabilistic visual learning for object representation. *IEEE Transactions on Pattern Analysis and Machine Intelligence*, 19(7):696–710, 1997.
16. N.Oliver, F.Berard, J.Coutaz, and A. Pentland. Lafter: Lips and face tracker. In *Proceedings IEEE Conf. Computer Vision and Pattern Recognition*, pages 100–110, 1997.
17. B. Parhami. Voting algorithms. *IEEE Transactions on Reliability*, 43(3):617–629, 1994.
18. K. Toyama and G. Hager. Incremental focus of attention. In *IEEE Conference on Computer Vision and Pattern Recognition*, 1996.

19. J. Triesch, D.H. Ballard, and R.A. Jacobs. Fast temproal dynamics of visual cue integration. Technical report, University of Rochester, Computer Science Department, September 2000.
20. J. Triesch and C. von der Malsburg. Self-organized integration of adaptive visual cues for face tracking. In *International Conference on Face and Gesture Recogintion*, pages 102–107, 2000.
21. M. Turk and A. Pentland. Eigenfaces for recognition. *Journal of Cognitive Neuroscience*, 3(1):71–86, 1991.
22. T. Uhlin, P. Nordlund, A. Maki, and J.-O. Eklundh. Towards an active visual observer. In *ICCV'95 Fifth International Conference on Computer Vision*, pages 679–686, 1995.
23. L. Xu, A. Krzyzak, and C. Suen. Methods of combining multiple classifiers and their applications to handwriting recognition. *IEEE Transactions on Systems, Man and Cybernetics*, 22(3):418–435, May/June 1992.

Large Consistent Geometric Landmark Maps

Wolfgang Rencken[1], Wendelin Feiten[2], and Martin Soika[2]

[1] Siemens Medical Solutions
Wolfgang.Rencken@med.siemens.de
[2] Siemens Corporate Technology, Information and Communications
Wendelin.Feiten,Martin.Soika@mchp.siemens.de

Abstract. The autonomous operation of an intelligent service robot in practical applications requires that the robot builds up a map of the environment by itself, even for large environments like supermarkets.
This paper presents a solution to the problem of building large consistent maps consisting of geometric landmarks. This solution consists of three basic steps:

- incremental extraction of geometric landmarks from range data
- recognition of previously mapped parts of the environment and identification of landmarks originating from the same structure and finally
- removing the inconsistencies by unifying those landmarks while retaining local relations between the other landmarks.

The recognition is based on comparing partial maps of geometric landmarks. This is done by enhancing an individual landmark with features derived from its environment. Care is taken that these features are invariant with respect to missing landmarks, rotation and translation of the map and varying landmark lengths. Based on this set of features, different landmarks originating from the same real world object can be identified.
For the purpose of correcting these inconsistencies the geometric relations between landmarks are modeled by links of variable length and variable angles between a link and the adjacent landmarks forming a flexible truss. Replacing two identified landmarks with their mean modifies length and angles of the related links, thus introducing energy into the truss. The overall energy in the truss is minimized by means of numerical optimization resulting in a consistent map.
Experience in the field with about 20 robots has shown that it is possible to build up maps of large environments robustly in real-time.

1 Introduction

A service robot, in order to perform useful tasks, needs to navigate within its working environment. One aspect of navigation deals with determining the position of the robot within its environment, usually known as the localisation problem. For this purpose, our robot uses a 2D map of geometric representations of physical landmarks (in our case, lines representing walls). The map can be acquired in several ways. It can be generated from CAD plans, measured by

G.D. Hager et al. (Eds.): Sensor Based Intelligent Robots, LNCS 2238, pp. 159–176, 2002.
© Springer-Verlag Berlin Heidelberg 2002

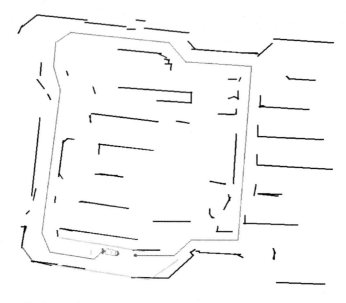

Fig. 1. The loop closing problem in a supermarket measuring $30mx40m$.

hand or built up automatically. Measuring maps by hand is very time consuming and CAD plans, if available at all, are usually incomplete and not accurate enough. Therefore the robot should build up the map by itself, based on sensor data acquired during either a guided teach-in run or autonomous exploration.

Consequently, a lot of work has been done on map building for autonomous mobile robots, mainly based on distance sensor readings and on vision. The sensor readings are in some way associated with positions and orientations of the robot and with models of objects in a global coordinate frame (where the *object model* could be as simple as one point on an object). This collection of data then serves as the map. The inherent bootstrap cycle in map building is, that in order to insert a measurement into the map, position and orientation of the robot have to be known, and that in order to estimate position and orientation of the robot, map information is required. This problem can be solved in comparatively small environments. These solutions for concurrent map building and localisation however do not scale to large environments. If the map is very large, small errors due to sensor inaccuracies (e.g. wheel slippage, range sensor noise) accumulate over time. The resulting errors can be corrected if the robot traverses an already mapped area. In large areas however the robot will sometimes re-enter a mapped area unguided by a map. In such a case, the accumulated error may be too large to be corrected locally. The map will become inconsistent and several instances of the same real world objects (see black and grey landmarks in Figure 1) are built up. This is known as the loop closing problem.

Clearly, such a map is not useful for robot navigation. If the robot uses contradictory landmarks for navigation, its position and orientation estimate will have serious errors prohibiting the reliable execution of path following and other tasks. The inconsistencies have to be removed, while correct map information has to be retained. Our method of building large scale consistent maps relies on three main ingredients:

- concurrent localisation and map building
- recognition of already mapped regions and identification of landmarks
- readjustment of position and orientation of landmarks

Since very much work has been done on concurrent localization and map building, this paper focuses on the aspect of recognizing previously visited areas and the aspect of correcting the mapping errors, after they have been identified. It summarizes work that has partially been covered in previous publications ([13], [14]), puts the pieces into the larger context and illustrates the overall procedure with new results.

The outline of the paper is as follows: In Section 2 the literature on recognizing already visited areas and on restoring consistency is reviewed. Then in Section 3 we give a brief outline and make some comments on the simultaneous localization and map building. The focus of this work is on recognizing previously visited areas in Section 4 and on restoring the map consistency in Section 5. The techniques developed in these sections are applied in real world experiments and the results are described in Section 6. Finally, in Section 7 some conclusions are drawn.

We apologise for not giving full details on parameters and algorithms due to the company's publication policy. More often than not, parameters are found experimentally.

2   Related Work

The task of building large consistent maps has been tackled by a number of authors, although it still remains a relatively unexplored area of research.

Most work on robot map building concentrates on extracting landmark data from sensor measurements and aggregating them over time and space (eg. [2] [3] [7] [9] [12]). More recently work has also been done on map recognition and map correction.

There are different ways of recognising that the robot has returned to a previously visited area.

One way of doing so, is to let the user tell the robot that it has returned to such an area [15]. This approach only allows for a reduced degree of autonomy of the map building process.

If the robot's map consists of an occupancy grid, the correlation of two grid maps, from different points in time, can tell the robot if it has returned [5]. This approach is very time consuming since a large correlation space has to be searched through.

On the other hand, if the robot's map consists of a set of stored sensor data (e.g. laser range scans), the recognition problem can be solved by directly matching the sensor data [10] or by correlating the angle and distance histograms deduced from the sensor data [17]. Both approaches are computationally very expensive and require a lot of memory, since a large set of scans has to be stored and correlated.

If the robot's map consists of geometric features which can be connected into star shaped polygons, the polygon of the current sensor data can be mapped to a set of precomputed polygons which are stored in the visibility graphs of the robot's environment [8]. This method is not very robust with respect to changes in the environment and sensor errors.

Chatila et al [3] have addressed the problem by assigning local object frames and a global map frame and by separating the mapping error into the error between an object and its local frame and the error between local frame and global frame. An initial local frame is assumed to be correct. As the map is built, objects are identified in new readings and the local frame of the new reading is corrected to allow for a good match of new data with existing objects. If there is a large correction, for example in the case of loop closing, these corrections are propagated back along the robot path with a decreasing strength. This strategy works for simple loops, but breaks down if the robot path contains points where many different loops intersect.

A probabilistic approach using grid based maps has been taken by Elfes[5]. He builds a large scale map by patching together local grid maps. Each local map is associated with a robot configuration. If the robot travels from one local map to the next, a transformation is stored corresponding to the displacement of the robot as measured by the odometry. This transformation is additionally corrected by matching overlapping parts of adjacent local maps. In the case of loop closing, the consistency of the rest of the map is ensured while correcting the current mismatch, by propagating the error back over all transformations made so far, taking into account the statistical properties of these former transformations. The problems with this approach are the computation time for the map matches and huge memory requirements for the grid maps.

Lu and Milios [10] also assign position and orientation to measurement frames, in this case laser range scans. Relations between scans are derived either from the robot's odometry as it travels from one to the other, or they are derived by matching different scans. This results in a network of pose estimates, based on relative coordinate transformations, with links between them. The deviation of the links from what they should have been is interpreted as energy in the system. The consistency of the map is ensured by minimizing this energy by means of a maximum likelihood estimation. The main drawbacks of this approach are the large memory requirements (all scans are stored) and the computational burden of inverting huge matrices. As an alternative they mention numerical optimization, which is the approach investigated in this paper.

Thrun et al [15] also use a grid based map. They assign a large enough memory for the entire environment. Then, the occupancy of the grid map and the robot position and orientation are estimated based on the measurements and the robot motion. This is done in two steps, where first the robot trajectory is estimated given a map and measurements, and then the map is estimated assuming

Fig. 2. Landmarks extracted from one laser scan

estimates of the robot trajectory. In the estimation of the map, each cell in a two-dimensional grid has to be updated. In the estimation of the robot position and orientation, each cell in a three-dimensional grid has to be updated. In spite of several approximations to the full solution, the memory and computation time requirements of this approach are prohibitive in our application.

In a subsequent paper [16], the authors overcome this drawback by using samples instead of complete grids and present a method that does mobile robot map building in real time.

Similar to Lu and Milios, Gutmann and Konolige [6] use registered sensor data (i.e. laser scans) as maps. Scan matching techniques are used to identify corresponding parts of scans and to adjust the poses of scans incrementally as the map is constructed. The loop closing problem is solved by map correlation, scan matching and by relaxation of the error using the techniques of Lu and Milios.

3   Landmark Extraction

Our application requires that the robot is capable of building up large scale maps by itself during normal operation, i.e. *on-line*. Furthermore limited amounts of memory are available for storing the mapping information. Therefore we have chosen to model the environment by means of landmarks. In this paper, a landmark will be considered as a finite, oriented line segment which is extracted from range sensor data originating from a laser scanner or sonar sensors. For information on how to extract planar landmarks from sonar data, please refer to [13] and [12].

In the last years it has become more usual to extract landmarks from laser scanner range data. For this, a great variety of algorithms can be used. In our system, line segments are extracted from laser scans by first segmenting the range data and then applying a least squares fit algorithm to the individual segments (see Figure 2). The landmarks extracted from individual scans are then aggregated over several scans using similar techniques as in the case of ultrasonic sensors. It should be noted that due to the extraction processes, the length of the line segments can vary considerably. Additional information such as colour or texture of the landmarks can be obtained from cameras.

4   Partial Map Match

To solve the loop closing problem, the robot has to be able to recognise that it has returned to an already visited area. This is done by means of partial map matching.

The map of the environment can be considered as a set of landmarks L. In the process of map building, newer landmarks are distinguished from older ones, and the new landmarks are compared to the old ones.

The map of the environment L can be partitioned into partial maps L_i. The partitioning function p can be geometric (for example all features within a specific area, temporal (for example all features extracted in a certain time interval) or a combination of both. Such a set of landmarks will be referred to as a partial map.

$$p : L \rightarrow \{L_i, i \in I\} \tag{1}$$

Since our robot uses a map of geometric landmarks, the recognition problem is defined as matching two different partial maps originating from different points in time.

A partial map match $\mu \subset L_a \times L_b$ is a relation which is one-to-one over a suitable subset $D_a \times D_b, D_a \subset L_a, D_b \subset L_b$. On the set M of partial map matches, the match disparity $| \cdot |$ is given by ν (to be defined later on). The optimal partial map match μ_0 is then given by

$$|\mu_0| = min(|\mu|, \mu \in M) \tag{2}$$

Since the robot can approach the same region in different ways, landmarks which are present in one partial map are not necessarily part of another partial map of the same environment. Furthermore, the parameters of the same landmark can differ considerably from one partial map to another. This means that when comparing two partial maps factors such as map incompleteness as well as map rotation and translation have to be specifically dealt with.

The straight forward approach of selecting and matching two subsets of landmarks is as follows:

- select n landmarks from the N_a landmarks of L_a
- select n landmarks from the N_b landmarks of L_b
- for each one-to-one assignment of landmarks compute the disparity of the match.

The computational complexity of this approach is characterised by the $\binom{N_a}{n}$ and $\binom{N_b}{n}$ possibilities for subset selection and the up to $n!$ possible assignments of these subsets. The advantage of this approach is that it encaptures the local structure of the environment and therefore yields more robust results in the face of large uncertainties in the parameters of the single landmarks.

At the other end of the spectrum, a comparison of single landmarks yields a complexity of only $N_a \times N_b$. However, single landmarks do not have enough structure to be comparable. The large uncertainty on the landmark parameters makes the comparison even more difficult.

Therefore it seems advantageous to combine both approaches to obtain a matching algorithm which on the one hand uses the local structure of the environment, and on the other hand approaches the computational complexity of single landmark comparison.

4.1   Landmark Signature

Our approach to using local structure, while at the same time limiting the complexity of the landmark comparison, is based on the concept of landmark signatures. The signature Σ_i of landmark l_i is a set of features which makes a single landmark more distinguishable within a set of landmarks. The signature can contain information about the landmark itself, such as colour, texture or other non-geometric properties Γ. On the other hand, information about the environment of the landmark, such as relative angles and distances w.r.t. other non-parallel landmarks Ω, distances to other parallel landmarks Π and other geometric features can be stored in the signature.

$$\Sigma_i = \{\Omega_i, \Pi_i \, \Gamma_i, \ldots\} \tag{3}$$

Since the loop closing problem deals with translational and rotational errors in parts of the map, care should be taken that the geometric properties of the signature are invariant w.r.t. these errors. Otherwise, the comparison of landmarks corresponding to the same environmental feature may fail. Examples of such geometric invariant constructs are

- distance between "parallel" landmarks $\pi \in \Pi$ (see figure 3)
- distance between and relative angles w.r.t. pairs of "orthogonal" landmarks $\omega \in \Omega$ (see figure 4)

By "parallel" we mean landmarks that have almost constant distance along their extension; in our case a threshold of 3° deviation from exactly parallel in orientation works well. By "orthogonal" we mean landmarks that allow for a robust calculation of the intersection point and intersection angle; in our case a threshold of 60° deviation from exactly orthogonal orientation works well.

The above constructs are not only invariant w.r.t. the rotation and translation errors of the map, but also w.r.t. the individual landmark lengths. Depending on the path traversed during map acquisition, the type of range sensor used and the landmark extraction process, the landmark lengths are subject to a large uncertainty. Therefore landmark lengths cannot be reliably used when comparing signatures.

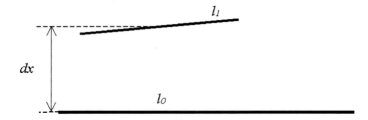

Fig. 3. A *paral* π. The distance between two parallel landmarks is invariant w.r.t. rotation and translation and the landmark lengths. For nearly parallel lines the dependency on landmark lengths is negligible.

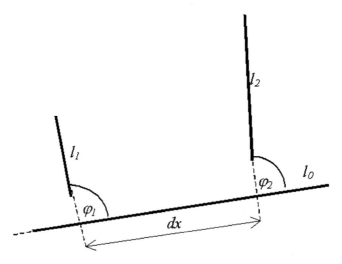

Fig. 4. An *orthogpair* ω. The distance between two orthogonal landmarks, and the relative angles between the base landmark and the orthogonal landmarks is invariant w.r.t. rotation and translation and the landmark lengths.

4.2   Comparison of Signed Landmarks

Let L_a and L_b be two different partial maps and let $\{\Sigma\}_a$ and $\{\Sigma\}_b$ be the corresponding sets of signatures. The disparity ν assigns a real number to each pair of signatures:

$$\nu : (\{\Sigma\}_a \times \{\Sigma\}_b) \to R \tag{4}$$

It is based on the disparities of the individual signature elements, given by

$$\nu_\Omega : (\{\Omega\}_a \times \{\Omega\}_b) \to R \tag{5}$$

$$\nu_\Pi : (\{\Pi\}_a \times \{\Pi\}_b) \to R \tag{6}$$

$$\nu_\Gamma : (\{\Gamma\}_a \times \{\Gamma\}_b) \to R \tag{7}$$

The value of the disparity is calculated from the signature elements. For example, let $\pi_m \in \Pi_i \in \Sigma_i$ be one entry in the parals of landmark $l_i \in L_a$ and π_n be one

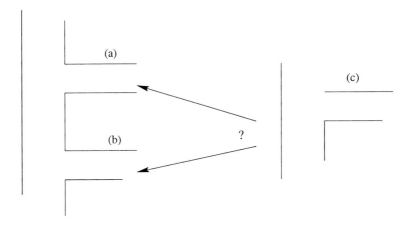

Fig. 5. The assignment dilemma: Is (a) = (c) or (b) = (c)?

entry in the parals of landmark $l_j \in L_b$, the disparity between π_m and π_n can be determined by

$$\nu_\pi(\pi_n, \pi_m) = |dx_n - dx_m| \tag{8}$$

The disparity for the orthog pair ω can be determined accordingly.

Since the individual signature elements can only be compared amongst equal types, the signature disparity between two landmarks is given by

$$\nu(\Sigma_i, \Sigma_j) = \sum_{k \in \{\Omega, \Pi, \Gamma\}} \alpha_k \nu_k(k_i, k_j) \tag{9}$$

where $\alpha_k \in [0, 1]$.

The disparity measure $\nu_k, k \in \{\Omega, \Pi, \Gamma\}$ is based on the best global assignment of the individual elements of type k. This assignment is usually not unique. As an example, consider the case when there are several adjacent corridors of the same widths in the environment (see figure 5). There is no one-to-one mapping between all orthog pairs, so that some possible assignments have to be left out. The problem of finding the best global assignment can be solved by dynamic programming [4]. To apply this technique, the sets of features to be assigned have to be linearly ordered. A linear order on the signature elements can be defined in a canonical way; for example, the parals can be ordered by the size of d_x.

Dynamic Programming. For the classic dynamic programming algorithm [1], a tableau is built with the ordered signature elements $\sigma_m \in \Sigma_{k,i}$ spanning one axis and $\sigma_n \in \Sigma_{k,j}$ spanning the other axis. The dimensions of the tableau are $M = |\Sigma_{k,i}|$ and $N = |\Sigma_{k,j}|$.

The calculation of the best global assignment of these signature elements is divided into a forward step and a backtracking step. During the forward step,

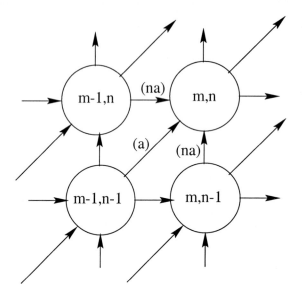

Fig. 6. The cost of assigning two signature elements.

the cost of assigning or skipping individual signature elements is computed. At node (m, n), the decision is made whether the element σ_m of signature Σ_i should be assigned to the element σ_n of signature Σ_j (see figure 6). The cost at node (m, n) is given by

$$c(m, n) = min(c(m - 1, n - 1) + c^a, c(m - 1, n) + c^{na}, c(m, n - 1) + c^{na}) \quad (10)$$

where c^{na} is the cost for not assigning the signature elements and c^a is the cost for assigning the elements. Therefore the cost at any node contains the cost for all the assignments made up to that time, plus the cost of assigning the individual signature elements represented by the node. During the forward step, the branch to each node that actually did yield the minimum cost is stored. This information is later needed for backtracking.

In the backtracking step, the path in the node tableau with the minimal cost is searched for. This is done by starting at the top right node (M, N) and then selecting the node that has lead to the minimum cost at this node. This is repeated until the path arrives at node $(0, 0)$. The contribution of signature elements of class k to the match disparity between the two landmarks l_i and l_j is the cost $v_k^{i,j} = c(M, N)$. The overall match disparity is then computed according to equation (9).

Because the orthog pairs and the parals are derived from landmarks (see figures 3 and 4), the above calculation not only yields the disparity $v^{i,j}$ between individual landmarks l_i and l_j, but also generates a partial map match $\mu^{i,j}$. In this partial map match, in addition to the landmarks l_i and l_j also those landmarks are mapped onto each other that correspond to the signature elements that are part of the best global assignments made in the dynamic programming stage.

Neglecting geometric aspects of the map building process, the best match μ_0 is then given by

$$|\mu_0| = min(\nu^{i,j}|l_i \in L_a, l_j \in L_b) \tag{11}$$

However, each map match $\mu^{i,j}$ yields a geometrical transformation $g^{i,j}$ with which the landmarks of the one set are transformed into the corresponding landmarks of the other set. Some geometrical transformations are more plausible than others (a 10m translation is more unlikely after a path length of 30m, than a 50cm translation). Therefore, based on a model of the geometrical map building error U, the best match is selected whose geometrical transformation $g^{i,j}$ is small based on the Mahalanobis Distance $d^{i,j}$ of the match.

Taking the geometric aspects into account, the optimal match μ_0 between two partial maps is given by

$$|\mu_0| = min(\nu^{i,j}|l_i \in L_a, l_j \in L_b, d^{i,j} = \frac{(g^{i,j})^2}{U} < \delta) \tag{12}$$

If a match μ_0 exists, the robot has recognised a previously mapped area. The geometrical transformation $g^{i,j}$ gives an indication of the map building error that has accumulated between leaving the area and re-entering it again.

The partial map match states that at least two apparently different landmarks actually originate from the same real world object with a high plausability. As a rule, a partial map match maps not only one pair of landmarks onto each other, but several:

$$L'_a \subset L_a \equiv L'_b \subset L_b \; and \; |L'_a| = |L'_b| > 1 \tag{13}$$

This equivalence is the starting point for the correction of the map.

5   Map Consistency

When a partial map match has been detected (with $g > 0$), it means that although the map is locally geometrically correct, its global geometric properties are incorrect. When correcting the global properties, care should be taken that the local properties are maintained as much as possible at the same time. For this purpose the concept of a flexible truss is introduced[1].

A truss is defined as a set of nodes Q, where the nodes are connected by a set of links Λ. A node is given by

$$q = (x, y, \beta) \tag{14}$$

where (x, y) represent the position and β the orientation of the node in a global coordinate system. A link between two nodes q_i and q_j is given by

$$\lambda(i, j) = (d_{ij}, \alpha_{ij}, \alpha_{ji}) \tag{15}$$

---

[1] Earlier attempts at modeling the map by means of a set of relative coordinate transformations similar to Lu and Milios turned out to be numerically unstable.

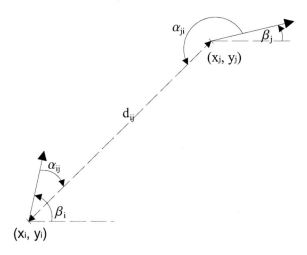

Fig. 7. A link between two nodes in the truss.

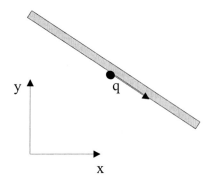

Fig. 8. A landmark represented by a node

where d_{ij} is the distance between the two nodes, and α_{ij}, α_{ji} are the angles between the node orientations and the link respectively (see also Figure 7).

A landmark can be represented by a truss node, consisting of the middle point of the line segment and the orientation of the segment (see also Figure 8). The nodes q_i and q_j of two landmarks which form part of a partial map match should therefore be replaced by a common node q'_{ij} (see also Figure 9). All links connecting to nodes q_i and q_j are therefore deformed by the partial map match.

The energy of the truss is given by the sum of the energy contained in all the links of the truss. This is given by

$$e = \sum_{(i,j)\in N} \left(\frac{(d_{ij} - \hat{d}_{ij})^2}{w_{ij}^d} + \frac{\Delta\alpha_{ij}^2}{w_{ij}^\alpha} + \frac{\Delta\alpha_{ji}^2}{w_{ji}^\alpha} \right) \tag{16}$$

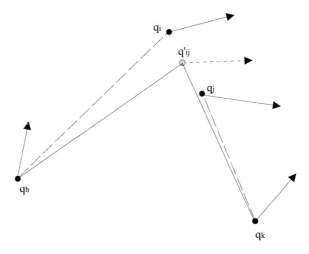

Fig. 9. The link deformation as a result of a partial map match

where

$$\Delta\alpha_{ij} = \arctan\left(\frac{-\Delta x \, sin\gamma_i + \Delta y \, cos\gamma_i}{\Delta x \, cos\gamma_i + \Delta y \, sin\gamma_i}\right) \qquad (17)$$

$$\Delta\alpha_{ji} = \arctan\left(\frac{\Delta x \, sin\gamma_j - \Delta y \, cos\gamma_j}{-\Delta x \, cos\gamma_j - \Delta y \, sin\gamma_j}\right) \qquad (18)$$

$$\Delta x = x_j - x_i \qquad (19)$$

$$\Delta y = y_j - y_i \qquad (20)$$

$$\gamma_i = \beta_i + \alpha_{ij} \qquad (21)$$

$$\gamma_j = \beta_j + \alpha_{ji} \qquad (22)$$

and $\hat{d}_{ij}$ is the original link length, and $w_{ij}^d, w_{ij}^\alpha, w_{ji}^\alpha$ are the respective weights.

Before the partial map match, the truss contains no energy, since all links are in their original state. After the partial map match, all links connecting to the transformed nodes introduce energy into the truss. By minimising the total energy on the truss, the mapping error is propagated throughout the truss, trying to keep the local geometric map information as correct as possible. When minimising the total energy, it is important that equation (16) is differentiable and that the derivative can be calculated explicitly, so that efficient optimisation algorithms can be applied.

Optimisation. As pointed out by Lu and Milios, numerical optimisation can be used to minimise the energy in a set of links and nodes. It turns out that the energy equation (16) has many local minima, which makes it difficult to obtain the optimal solution, i.e. the best possible map. Global optimisation could circumvent the numerous local minima. However since it would literally take

Fig. 10. Cleaning robot.

hours or days to optimise a map (typically > 2000 variables), this option is not
feasible. Therefore, we have to be careful about the starting point for the optimisation. This starting point is found by distributing the mapping error along
the truss. To each link in the truss the same part of the overall translation and
rotation error is assigned. It turns out that with such a heuristic, local optimisation leads to good results. For the local optimisation, different algorithms
were tested (steepest descent, quasi-Newton and conjugate gradient). In the end
all algorithms yielded almost identical results. However the conjugate gradient
method [11] turned out to be the fastest.

6   Experimental Results

The algorithms were tested extensively in simulation and on cleaning robot prototypes, and have been proven in the field on about twenty Hefter cleaning robots
(see Figure 10), equipped with the the SIemens NAvigation System (SINAS).
SINAS features an industrial PC with a Pentium 233 MHz processor, 32 MB of
DRAM and 20 MB of solid state hard disk. It uses a SICK laser range scanner
and a Siemens sonar system. Furthermore it is equipped with an KVH fibre optic
gyroscope and encoders on its wheels.

Fig. 11. Partial maps to match: new (grey) and old (black) part of the map.

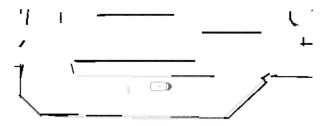

Fig. 12. Partial maps after match and first correction.

The result of this experience is that the robot can very reliably build maps of supermarkets and corridor environments of more than 2000 m^2 size and localise itself in these maps to about $5 - 10$ cm accuracy. The matching algorithm typically takes 100 ms and the subsequent optimisation step takes about 15 s, depending on the complexity of the environment.

The robot builds the map autonomously. Starting at the point denoted by ∘ in Figure 1, it is driven along the plotted route to the point denoted by ▷. When the robot gets back to the area of the starting point, it encounters the situation seen in the bottom left corner of Figure 1; Figure 11 shows both partial maps enlarged.

It should be noted that the partial maps may differ considerably in size. Furthermore not all features need to have counterparts in the other partial map. This clearly illustrates that our partial map matching algorithm is quite robust w.r.t. to missing landmarks and varying landmark lengths. It should be kept in mind that the algorithm is not designed to make all possible landmark assignments.

At this point, a partial map match is made and some of the landmarks are identified. Based on these identified landmarks a transformation is found that ideally puts these landmarks on top of each other. The result of this local partial map match and map modification is shown in Figure 12.

The consistency of the map can now locally be restored, but the geometrical properties of the rest of the map have to be maintained as much as possible. In order to do this, these properties are captured in the truss shown to the left in Figure 13.

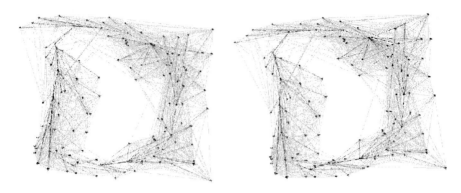

Fig. 13. Truss containing local geometric information before (left) and after (right) optimisation.

After optimisation, the gap in the truss visible in the lower left corner is closed and the other nodes in the truss are moved accordingly (see the right truss Figure 13. Note that the absolute positions and orientations of the nodes are not part of the target function: the map floats freely.

These steps of partial map match and subsequent global optimisation to restore consistency while at the same time maintaining local geometric correctness are repeated as the robot is driven through the entire supermarket. In Figure 14, the resulting map of the supermarket is shown.

Some remarks should also be made on the limits of this map building procedure:

The map building algorithm is designed for on-line operation, with limited need for memory. Therefore, line primitives are used which makes the algorithm suited for environments that exhibit enough walls. This holds true for many indoor environments like office buildings, hospitals and supermarkets.

Both in the landmark extraction stage which is based on Kalman filtering and in the partial map match stage, mismatches will irrecoverably make the system fail. However, due to good odometry and pose information from the gyroscope, these cases almost never occur in practice.

The robot does not match all landmarks that are in the environment. Since the laser range scanner we use only has a 180° field of view, the path along which the robot is driven determines which landmarks are mapped. Since it is very difficult to reproduce an exact path in such a complicated environment, two maps taken from different runs differ in terms of their content in some areas.

From the experience with many robots in continuous productive use we can say that the robustness and performance of the method meet the requirements. Very rarely a second teach run is needed. Since ground truth about areas of hundreds and thousands of square meters is not available, we can only talk about local geometric errors. Due to the nature of the optimization problem exhibiting many local minima, these occasionally occur, but as a rule are on the order of a couple of centimeters. This is largely sufficient in practical applications.

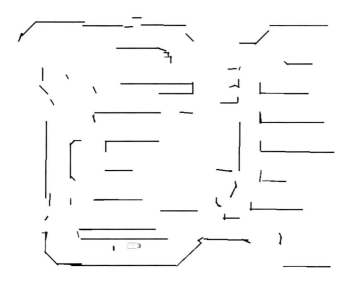

Fig. 14. The final map of the supermarket.

7   Conclusions

In this paper an approach to the construction of large consistent geometric landmark maps was presented. This approach consists of three main steps:

- extract landmarks
- recognise previously visited areas
- restitute consistency

For the recognition of previously visited areas, an algorithm, based on partial map matching, has been presented which also yields a one to one mapping
between a subset of landmarks. The partial map match is robust w.r.t. missing
landmarks, rotational and translational errors of the map building process and
the considerable uncertainty in individual landmark lengths.

For the restitution of consistency, an algorithm has been presented based
on modeling the map as a flexible truss and minimizing the energy contained
therein by means of numerical optimization.

Cycling through these three steps, the robot is able to map large environments *on-line*, which represents a significant advance in the field of map learning.

References

1. Bellman R., *Dynamic Programming*, Princeton University Press, Oxford University
   Press, London, 1957.
2. Borenstein. J, Everett H.R., Feng L., *Navigating Mobile Robots, Systems and Techniques*, AK Peters, Wellesley Massachusetts, 1996.

3. Chatila R., Laumond J.-P., Position referencing and consistent world modeling for mobile robots *ICRA Proceedings*, 1985.
4. Crowley J.L., Mely S., Kurek M., Mobile Robot Perception Using Vertical Line Stereo, Kanade, T. et al (ed), *Intelligent Autonomous Systems 2*, 1989, 597–607.
5. Elfes A., Robot Navigation: Integrating Perception, Environmental Constraints and Task Execution Within a Probabilistic Framework, L. Dorst et al (ed), *Reasoning with Uncertainty in Robotics* Lecture Notes in Artificial Intelligence No. 1093, Springer-Verlag, Berlin Heidelberg New York, 1995, 93–130.
6. Gutmann J.-S., Konolige K., Incremental Mapping of Large Cyclic Environments, *International Symposium on Computational Intelligence in Robotics and Automation* (CIRA'99), Monterey, 1999, 318–325.
7. Iyengar S.S., Elfes A. (ed): *Automous Mobile Robots, Perception, Mapping, and Navigation*, IEEE Computer Society Press, Los Alamitos California, 1991.
8. Karch O., Noltemeier H., Robot Localisation - Theory and Practice, *IROS Proceedings*, 1997, 850–856.
9. Leonard J.J., Durrant-Whyte H.F., Cox I.J., Dynamic Map building for an autonomous mobile robot, *Internal Journal of Robotics Research* **11(4)**, 1992, 89–96.
10. Lu F., Milios E., Globally consistent range scan alignment for environment mapping, *Autonomous Robots*, Vol. 4, 1997, 333–349.
11. Press W.H., Flannery B.P., Teukolsky S.A., Vetterling W.T., *Numerical Recipes in C, The Art of Scientific Computing*, Cambridge University Press, Cambridge, New York, Melbourne, 1988, 317–324.
12. Rencken W.D., Concurrent localisation and map building for mobile robots using ultrasonic sensors, *IROS Proceedings*, 1993, 2129–2197.
13. Rencken W.D., Feiten W., Zöllner R., Relocalisation by Partial Map Matching, *Modelling and Planning for Sensor Based Intelligent Robot Systems*, Lecture Notes in Artificial Intelligence Vol. 1724, Springer-Verlag, Berlin, Germany, 1999, 21–35.
14. Rencken W.D., Feiten W., Soika M., Building Large Consistent Maps, *KI-Zeitung* to appear
15. Thrun S., Burgard W., Fox D., A Probabilistic Approach for Concurrent Map Aquisition and Localization for Mobile Robots, *Technical Report, School of Computer Science, Carnegie Mellon University*, CMU–CS–97–183.
16. Thrun S., Burgard W., Fox D., A Real-time Algorithm for Mobile Robot Mapping with Applications to Multi-Robot and 3D Mapping. *ICRA Proceedings*, 2000, 321–328.
17. Weiss, G., von Puttkamer, E., A Map based on Laserscans without Geometric Interpretation, U. Rembold et al (ed), *Intelligent Autonomous Systems*, IOS Press, 1995, 403–407.

Tactile Man-Robot Interaction for an Industrial Service Robot

Steen Kristensen[1], Mathias Neumann[2], Sven Horstmann[1],
Frieder Lohnert[1], and Andreas Stopp[1]

[1] DaimlerChrysler Research and Technology, Cognition and Robotics Group
Alt-Moabit 96A, D-10559 Berlin, Germany
{steen.kristensen,sven.horstmann,frieder.lohnert,
andreas.stopp}@daimlerchrysler.com
[2] Technische Universität Berlin
Institut für Technische Informatik
Franklinstraße 28/29
D-10587 Berlin, Germany
mn@cs.tu-berlin.de

Abstract. In this paper, research towards a cooperative robotic assistant for manufacturing environments is presented. The aim of this research is to develop a robotic assistant which can easily be instructed how to either perform tasks autonomously or in cooperation with humans. Focus of this paper is on the tactile interaction between man and robot which is used for teaching as well as for direct cooperation in tasks jointly performed by worker and robot.

1   Introduction

In this paper we describe past and ongoing research efforts at DaimlerChrysler Research and Technology's Cognition and Robotics Group where over the last years work has been conducted on human-friendly robots for space, office, and factory automation.

A major goal of this work has been (and still is) to develop robots that can assist, co-exist with, and be taught by humans. Therefore, apart from developing the "standard" mobile robot capabilities such as landmark recognition, path planning, obstacle avoidance etc. our research effort has been aimed at the development of learning capabilities that will allow the user to quickly and intuitively teach the robot new environments, new objects, new skills, and new tasks. We believe this is the only viable way of creating robotic assistants that can be flexible enough to function robustly in the very diverse habitats of humans and thus be accepted as truly helpful devices. In this paper we present some of the results of this work.

Current research is aimed towards improving the man-machine interaction by adding more communication and cognition capabilities. This has the purpose of further simplifying the teaching of the robot but also to make it more "cooperative" by having it interpret human commands and behaviour in the

G.D. Hager et al. (Eds.): Sensor Based Intelligent Robots, LNCS 2238, pp. 177–194, 2002.

given context, allowing it to make better decisions about when, how, and where to assist the human co-worker(s). An important criterion is, however, that the robot can also perform tasks autonomously once instructed/taught by a human worker. Additionally it should be able to learn incrementally, i.e. to improve its performance during task execution by "passively" receiving or actively requesting information (the latter could for example be in the case where the robot detects ambiguities which it cannot resolve autonomously).

A typical scenario for a new robotic assistant in an industrial setting would be:

- The robot is led through the factory halls and is shown important places (stores, work stations, work cells etc.),
- the robot is shown relevant objects, e.g. tools, workpieces, and containers,
- the robot is shown how to dock by work cells, containers etc. in order to perform the relevant manipulation tasks,
- the robot is taught how to grasp various objects and how (and possibly in what sequence) to place them in corresponding containers or work cells,
- in case of a cooperation task, the robot is shown when and how to assist the human worker.

For service robots in manufacturing it is important that the man-robot communication channels are robust and fault-tolerant rather than high bandwidth (in the information theoretic sense). The need for high robustness is motivated by the fact that mistakes/misinterpretations on the side of the robot can have severe consequences economically as well as for the safety of the human workers. On the other hand the environments are typically more ordered and the range of tasks more restricted than in the case for domestic service robots. This means that robustness can partly be achieved through using a set of low bandwidth channels. Tactile sensing using a force/torque sensor (FTS) is an example of such a channel with which a limited amount of information can robustly be conveyed to the robot.

In Section 2 some more background in terms of related work and project specific constraints are outlined. In Sections 3–6 various methods for interaction and cooperation based on force/torque sensing are presented. Finally in Section 7, the results are discussed and summarised.

2   Background

In previous work [1], we have described how we can interactively teach new environment models in a quick and robust manner. These models are represented as topological graphs [2, 3] well-known from various mobile robot applications [4–7]. The topological graphs we use are extended with metric and symbolic information about objects such as work stations and places such as rooms or stores. An example of a model is shown in Figure 1.

In general, the world model can be said to contain elements which are relatively static over longer time spans. This has the obvious advantage that the

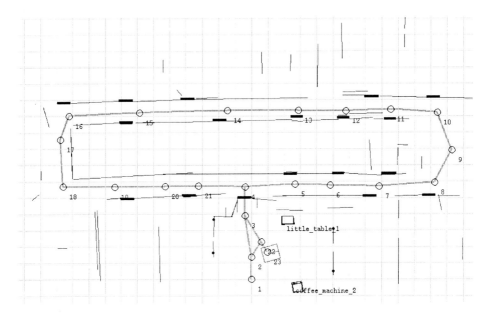

Fig. 1. World model taught by leading the robot through a previously unvisited environment. Grid size is 1 meter. The thick lines and the circles they are connecting form the basic topological graph. The thin lines are the walls, the black bars are the doors. The squares represent special objects, such as work stations, which are modelled in 3D.

model only needs to be occasionally updated. Furthermore, the symbolic nature of the model means that it forms a good basis for planning of high level missions like "fetch object A in store room 12". However, this kind of model does not form a sufficient basis for planning and executing detailed manipulation and cooperation tasks. For this purpose we have developed a number of sensor-based skills which allow the user to interact with the robot in order to control its posture, to hand over objects, and to interactively teach the robot how to grasp and manipulate objects.

The robot, Clever, which we use for the development of various interaction schemes is shown in Figure 2. Clever can be seen as a technology demonstrator which serves as the basis for later "real" industrial service robots which will be able to carry and manipulate heavier loads.

On Clever, a force/torque sensor is mounted in the wrist of the manipulator, i.e. between the arm and the gripper. This has the advantage that the same sensor can be used for sensor based cooperation, manipulation and for compliant motion, where the user "takes the robot by the hand" which we have found to be an intuitive and straightforward way for humans to interact with the robot.

The drawback of using only one sensor is that the force/torque measurement system cannot distinguish the user induced forces and torques from those resulting from the payload of the manipulator. In service robots for domestic use where the payloads are typically light (mass < 1kg) the user may well be able

Fig. 2. The robot "Clever" used in the experiments. The robot consists of a differential drive platform equipped with a 6-DOF Amtec manipulator and a Sick lidar plus a set of colour CCD cameras mounted on a pan/tilt unit. The arm, shown in more detail to the right, is equipped with a force/torque sensor between the wrist and the gripper, which is a standard parallel jaw gripper with force sensors in the "fingers".

to compensate for the load of the workpiece when guiding the robot using compliant motion. For an industrial service robot, potentially carrying significantly heavier loads, this is not an option, though. Therefore it is necessary for the robot to be able to distinguish between user and payload induced forces and torques. How we do this is outlined in Section 3.

The mobile manipulator as a whole has nine degrees of freedom. Therefore the problem of redundancy has to be solved, i.e. a method is needed to determine the posture of the mobile manipulator. Often mobile manipulators are modelled as one kinematic chain, but in our case the nonholonomic mobile platform used has a slower response to movement commands than the manipulator[1]. Thus, in Section 4 an approach to force guidance is presented which takes the different characters of both subsystems into account.

The special problem of force guidance is that the direction of the movement is controlled by the operator without the robot knowing what the actual purpose of the movement is. This means that the trajectories cannot be adjusted to the task. Therefore the control algorithm as described in Section 4 tends to result in the platform following the tool in a quite sub-optimal manner. Situa-

---

[1] For a foreseeable future we expect this to be characteristic of all industrial service robots although we hope that some day more agile industrial platforms may be at our disposal.

tions where the platform gets in the way of itself are possible. In particular, the nonholonomic platform makes it difficult to change the position to one parallel to the forward direction of the robot, causing for example manoeuvring to be quite time-consuming.

In order to solve this problem in a pragmatic way, the raw force/torque signal from the FTS is surveyed. The operator should be enabled to independently control both tool location and posture of the mobile manipulator. An approach to implement a system for this purpose is presented in Section 4.5.

3   Calibration of the Force/Torque Measurement System

As argued above, it is necessary to be able to (re-)calibrate the force/torque measurement system on-line to account for any payload held in the gripper when moving the manipulator using compliant motion.

If the robot carries a (rigid) workpiece the payload induced force/torques are nearly constant[2] for a fixed orientation of the tool centre point (TCP). However, when the orientation of the TCP is changed as a result of a user guiding the arm, the forces and torques caused by the gripper and the payload change. To compensate for those effects it is necessary to know the mass and the vector to the centre of gravity (relative to the FTS) of the combined gripper/payload object. Having such a model, it is possible to calculate the forces and torques induced by the gripper/payload for any given TCP orientation and thus to determine the forces and torques applied externally to the manipulator.

To calculate the mass and the vector to the centre of gravity of the object we sample the forces and torques measured by the sensor and then rotate the object a few degrees around the axis of the torques. A second sample compared with the first gives the needed information. Because the object only needs to be moved a little, and thus only a limited amount of freespace must be available, it is normally not a problem to do the calibration "on-site" where the objects was grasped.

This approach has been shown to work well for slow movements (where centrifugal forces can be ignored) like force guidance controlled by humans.

4   Coordination of the Manipulator and the Mobile Platform

Clever's manipulator can change its position freely as desired within its work space limits. Due to this and to the slower and kinematically more constrained performance of the platform it appears reasonable that the force/torque guided motion is performed mainly relative to the platform by moving the TCP [8].

---

[2] If the platform moves with accelerations which are small compared to gravity.

4.1 Force/Torque Guided Motion of the Manipulator

Usual force guidance is implemented in the way that the velocity of the tool is a linear function of forces $\boldsymbol{F}(t) = (F_x(t), F_y(t), F_z(t))^T$ and torques $\boldsymbol{M}(t) = (M_x(t), M_y(t), M_z(t))^T$ [9].

However, the manipulator arm used cannot be controlled by velocities. Instead the goal configuration and a time interval that results from the control rate f_c are used to control the arm in position mode. Furthermore the noisy sensor signal requires filtering to ensure that only reasonable large signals are taken into account.

The following non-linear function is proposed:

$$^P\boldsymbol{r}(t + \frac{1}{f_c}) = {}^P\boldsymbol{r}(t) + \frac{^Pv_{max}}{f_c} \begin{pmatrix} \text{sign}(F_x(t)) \; e^{(1 - \frac{F_{max}}{|F_x(t)|})} \\ \text{sign}(F_y(t)) \; e^{(1 - \frac{F_{max}}{|F_y(t)|})} \\ \text{sign}(F_z(t)) \; e^{(1 - \frac{F_{max}}{|F_z(t)|})} \end{pmatrix} \tag{1}$$

The desired position $^P\boldsymbol{r}$ of the tool at time $t + \frac{1}{f_c}$ is reached from the current position at time t in compliance with $0 < |F_{x,y,z}(t)| \le F_{max}$. $^Pv_{max}$ specifies the maximal permitted translation velocity of the tool with respect to the platform frame.

The desired orientation $^P\boldsymbol{A}_R$ at time $t + \frac{1}{f_c}$ can be described by:

$$^P\boldsymbol{A}_R(t + \frac{1}{f_c}) = \boldsymbol{R}_{XYZ}(w_x, w_y, w_z) \; {}^P\boldsymbol{A}_R(t) \tag{2}$$

where $\boldsymbol{R}_{XYZ}$ is a matrix representing the rotations of the TCP around the axes of the platform frame. The three rotation angles are defined as:

$$w_{x,y,z} = \frac{^P\omega_{max}}{f_c} \; \text{sign}(M_{x,y,z}(t)) \; e^{(1 - \frac{M_{max}}{|M_{x,y,z}(t)|})} \tag{3}$$

$0 < |M_{x,y,z}(t)| \le M_{max}$ has to hold and $^P\omega_{max}$ specifies the maximal permitted rotation velocity of the tool with respect to the platform frame.

4.2 Compensation of Platform Movements

The platform as the second subsystem is able to move simultaneously and independently. But the tool should only be moved with respect to the fixed world frame according to the forces applied to the FTS. Therefore the manipulator has to compensate the movements of the platform.

Let $^W\boldsymbol{p}(t) = ({}^Wp_x(t), {}^Wp_y(t), {}^Wp_\phi(t))^T$ denote the position (a reference point in the centre, Figure 3) and the orientation of the platform with respect to the world frame. $^W\boldsymbol{p}(t)$ is known from the platform's odometry sensing.

The transformation

$$^Wp_x(t - \frac{1}{f_c}) + {}^Pr_x(t) \cos({}^Wp_\phi(t - \frac{1}{f_c})) - {}^Pr_y(t) \sin({}^Wp_\phi(t - \frac{1}{f_c})) = \\ {}^Wp_x(t) + ({}^Pr_x(t) + {}^P\Delta r_x) \cos({}^Wp_\phi(t)) - ({}^Pr_y(t) + {}^P\Delta r_y) \sin({}^Wp_\phi(t)) \tag{4}$$

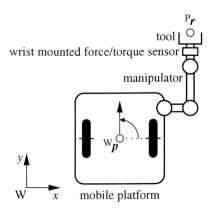

Fig. 3. The mobile manipulator and its reference points. $^W p$ is the reference point of the platform between the wheels (black areas) and $^P r$ is the reference point of the arm at the end of the tool (TCP). The orientation is given by the angle between the x-axis of the fixed world frame and the driving direction (arrow).

$$
\begin{aligned}
&^W p_y(t - \tfrac{1}{f_c}) + {}^P r_x(t) \sin({}^W p_\phi(t - \tfrac{1}{f_c})) + {}^P r_y(t) \cos({}^W p_\phi(t - \tfrac{1}{f_c})) = \\
&^W p_y(t) + ({}^P r_x(t) + {}^P \Delta r_x) \sin({}^W p_\phi(t)) + ({}^P r_y(t) + {}^P \Delta r_y) \cos({}^W p_\phi(t))
\end{aligned}
\tag{5}
$$

of the current x-y-coordinates of the tool with respect to the platform into the world frame on the basis of the current and the last configuration of the platform results in the offsets

$$
\begin{aligned}
^P \Delta r_x = &-{}^P r_x(t) \\
&+ {}^P r_x(t) \cos({}^W p_\phi(t) - {}^W p_\phi(t - \tfrac{1}{f_c})) \\
&+ {}^P r_y(t) \sin({}^W p_\phi(t) - {}^W p_\phi(t - \tfrac{1}{f_c})) \\
&+ (-{}^W p_x(t) + {}^W p_x(t - \tfrac{1}{f_c})) \cos({}^W p_\phi(t)) \\
&+ (-{}^W p_y(t) + {}^W p_y(t - \tfrac{1}{f_c})) \sin({}^W p_\phi(t))
\end{aligned}
\tag{6}
$$

$$
\begin{aligned}
^P \Delta r_y = &-{}^P r_y(t) \\
&- {}^P r_x(t) \sin({}^W p_\phi(t) - {}^W p_\phi(t - \tfrac{1}{f_c})) \\
&+ {}^P r_y(t) \cos({}^W p_\phi(t) - {}^W p_\phi(t - \tfrac{1}{f_c})) \\
&+ ({}^W p_x(t) - {}^W p_x(t - \tfrac{1}{f_c})) \sin({}^W p_\phi(t)) \\
&+ (-{}^W p_y(t) + {}^W p_y(t - \tfrac{1}{f_c})) \cos({}^W p_\phi(t))
\end{aligned}
\tag{7}
$$

for moving in the x-y-plane and

$$\Delta w_z = ^W p_\phi(t) - ^W p_\phi(t - \frac{1}{f_c})$$ (8)

for rotating around the z-axis of the world. These offsets have to be added in equations 1 and 2.

4.3   Preferred Configurations of the Manipulator

Based on the computed target position $^P r(t + \frac{1}{f_c})$ and the associated configuration it is possible to determine a preferred configuration close to $^P r(t + \frac{1}{f_c})$ which provides more manipulability [10]. To reach this configuration the mobility of the platform should be used without moving the tool with respect to the world frame.

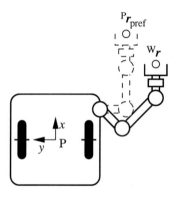

Fig. 4. The preferred point $^P r_{pref}$ is fixed relatively to the platform. The platform has to move to match $^P r_{pref}$ with $^W r$.

The mobile platform is able to move only in the x-y-plane of the world which is parallel to the x-y-plane of the platform frame. Therefore a new position of the tool

$$\begin{pmatrix} ^P r_{pref,x}(t + \frac{1}{f_c}) \\ ^P r_{pref,y}(t + \frac{1}{f_c}) \end{pmatrix}$$ (9)

can be found with respect to the platform which yields a more favourable configuration of the manipulator (Figure 4). The z-coordinate and the orientation in the world frame must be constant.

4.4   Closed Loop Platform Controller

The only task of the platform is to move in the next interval so that the arm will be able to take on its preferred configuration [11].

The nonholonomic mobile platform has the following decoupling matrix $\boldsymbol{\Phi}$ which describes the correlation between the velocities of the preferred point in work space and in configuration space.

$$\boldsymbol{\Phi}(^{\mathrm{W}}\boldsymbol{p}(t)) = \begin{pmatrix} \Phi_{11}(^{\mathrm{W}}\boldsymbol{p}(t)) & \Phi_{12}(^{\mathrm{W}}\boldsymbol{p}(t)) \\ \Phi_{21}(^{\mathrm{W}}\boldsymbol{p}(t)) & \Phi_{22}(^{\mathrm{W}}\boldsymbol{p}(t)) \end{pmatrix} \tag{10}$$

$$\Phi_{11}(^{\mathrm{W}}\boldsymbol{p}(t)) = \cos(^{\mathrm{W}}p_\phi(t))$$

$$\Phi_{12}(^{\mathrm{W}}\boldsymbol{p}(t)) = -^{\mathrm{P}}r_{\mathrm{pref},x}(t + \frac{1}{f_c})\sin(^{\mathrm{W}}p_\phi(t)) - ^{\mathrm{P}}r_{\mathrm{pref},y}(t + \frac{1}{f_c})\cos(^{\mathrm{W}}p_\phi(t))$$

$$\Phi_{21}(^{\mathrm{W}}\boldsymbol{p}(t)) = \sin(^{\mathrm{W}}p_\phi(t)) \tag{11}$$

$$\Phi_{22}(^{\mathrm{W}}\boldsymbol{p}(t)) = ^{\mathrm{P}}r_{\mathrm{pref},x}(t + \frac{1}{f_c})\cos(^{\mathrm{W}}p_\phi(t)) - ^{\mathrm{P}}r_{\mathrm{pref},y}(t + \frac{1}{f_c})\sin(^{\mathrm{W}}p_\phi(t))$$

With the transformation

$$^{\mathrm{W}}r_{\mathrm{pref},x}(t + \frac{1}{f_c}) = {}^{\mathrm{W}}p_x(t)$$
$$+ {}^{\mathrm{P}}r_{\mathrm{pref},x}(t + \frac{1}{f_c})\cos(^{\mathrm{W}}p_\phi(t))$$
$$- {}^{\mathrm{P}}r_{\mathrm{pref},y}(t + \frac{1}{f_c})\sin(^{\mathrm{W}}p_\phi(t))$$

$$^{\mathrm{W}}r_{\mathrm{pref},y}(t + \frac{1}{f_c}) = {}^{\mathrm{W}}p_y(t) \tag{12}$$
$$+ {}^{\mathrm{P}}r_{\mathrm{pref},x}(t + \frac{1}{f_c})\sin(^{\mathrm{W}}p_\phi(t))$$
$$+ {}^{\mathrm{P}}r_{\mathrm{pref},y}(t + \frac{1}{f_c})\cos(^{\mathrm{W}}p_\phi(t))$$

and the gain g_{P} the control vector

$$\begin{pmatrix} v_{\mathrm{lin}} \\ v_{\mathrm{rot}} \end{pmatrix} = \boldsymbol{\Phi}^{-1}(^{\mathrm{W}}\boldsymbol{p}(t))\boldsymbol{v} \tag{13}$$

with

$$\boldsymbol{v} = \begin{pmatrix} ^{\mathrm{W}}\dot{r}_x + g_{\mathrm{P}}\left(^{\mathrm{W}}r_x(t + \frac{1}{f_c}) - ^{\mathrm{W}}r_{\mathrm{pref},x}(t + \frac{1}{f_c}) \right) \\ ^{\mathrm{W}}\dot{r}_y + g_{\mathrm{P}}\left(^{\mathrm{W}}r_y(t + \frac{1}{f_c}) - ^{\mathrm{W}}r_{\mathrm{pref},y}(t + \frac{1}{f_c}) \right) \end{pmatrix} \tag{14}$$

can be computed considering the desired velocity of the tool.

In the next step the arm has to compensate this movement as mentioned above.

4.5   Controlling the Posture According to the User's Intention

When the preferred point is fixed with respect to the platform the redundancy resulting from the 9 degrees of freedom is eliminated. The drawback of this

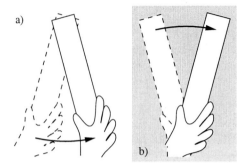

Fig. 5. Two possible strategies for moving large objects with one hand on a frictional surface.

simple scheme is of course that the (rather arbitrary) position of the preferred point may not be optimal for the given tasks. In other words, the solution of the redundancy problem depends on the task which is known to the operator. Therefore the user's intention has to be conveyed to the robot.

The following situation might occur: The tool is at the desired location in the world. However, the platform has the wrong orientation (not to the purpose of the user's intention). It should move sideways to turn the tool.

Corresponding forces and torques can be applied by the user but then the machine displaces the tool because of the force guidance algorithm as mentioned above. The user's intention is not regarded.

Inspired by the way humans are moving large objects on a surface with friction the following approach has been developed.

Consider the bar in Figure 5. In the case of turning it around the grasp point (Figures 5b and 6b) the ratio of the force and the torque is easily to be distinguished from that of the rotation about the opposite point (Figures 5a and 6a).

If the bar is representing the mobile manipulator, then the movement, as shown at Figure 5a, is equivalent to the known mode of force guidance. In case of Figure 5b a new mode is defined leading to sideways movements.

It is quite obvious from Figure 6 that only F_y and M_z are important for moving sideways. The ratio of both signals determines the desired mode of motion.

Figure 7 shows the regions which can be characterised as modes of moving. The gray shaded regions correspond to the cases 6b and 6d and will lead to changes of the posture. Selecting the ranges it has to be ensured that displacement along the y-axis (dashed ellipse) and orientation change of the tool around the z-axis (scored ellipse) remain possible. The modes shown in Figure 6a and 6c have to be allowed too. These regions of pairs of (F_y, M_z) are called force/torque symbols of modes of movement.

The expression

$$\kappa = \frac{-\arctan(\varrho(F_y(t)M_z(t) - \varepsilon))}{\pi} + \frac{1}{2} \tag{15}$$

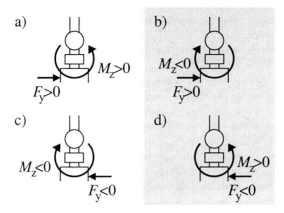

Fig. 6. Possibilities to apply significant values of force and torque to the tool.

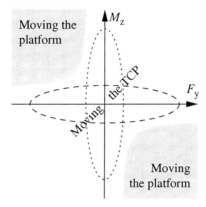

Fig. 7. Regions of force-moments pairs (F_y, M_z). Each area represents a simple mode of moving.

determines the degree of membership of a measured pair of (F_y, M_z) to the gray shaded regions of Figure 7. ϱ specifies the smoothness and ε specifies the threshold of the regions. The values for ϱ and ε have been determined empirically.

4.6   Moving the Virtual Preferred Point

As described above, forces and torques induced by the operator may be interpreted as an intention to move sideways (gray shaded regions in Figure 7). But performing sideways movements implies complicated manoeuvres because of the platform's nonholonomic constraints. Therefore, the sideways force component was instead chosen for overlaying the preferred point (9) with a supplementary offset $^{P}\Delta r_{\mathrm{pref},y}$. This offset is computed iteratively by

$$^{P}\Delta r_{\mathrm{pref},y}(t + \frac{1}{f_c}) = {^{P}}\Delta r_{\mathrm{pref},y}(t) + \kappa\, h\, \mathrm{sign}(F_y(t))\, e^{(1 - \frac{F_{\max}}{|F_y(t)|})} \qquad (16)$$

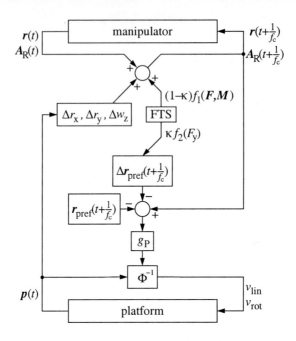

Fig. 8. Simplified block diagram of the controller. Both (platform and arm) controllers and the relations between them are shown.

That does in itself not change the position of the platform but the changed relative orientation of the arm enables a different manoeuvring behaviour.

The relation of force, moment and velocity of the preferred point displacement is quite complex and can thus be difficult to handle by the operator. Therefore the factor h has to be carefully determined which we have done experimentally.

The offset $^{P}\Delta r_{pref,y}$ has to be integrated in the platform controller as described in Section 4.4. The force guided movement of the tool (Section 4.1) has to be scaled by $1 - \kappa$ respectively.

Of course this method makes it impossible for the system to change the orientation and position of the tool independently and freely. But mostly such movements can be emulated in two steps, e.g. by first rotating the tool to the desired orientation and then moving it to the target position.

4.7   Experiments

A good example to show the advantages of the proposed method is to try to achieve a sideways displacement by moving on a zigzag course. This manoeuvre is expected to be used frequently when humans cooperate with robotic assistants.

For the experiments, the system shown in Figure 2 was used.

A simplified control architecture, as shown in Figure 8, was used. Due to the communication structure and driver software the control rate is limited to 4Hz.

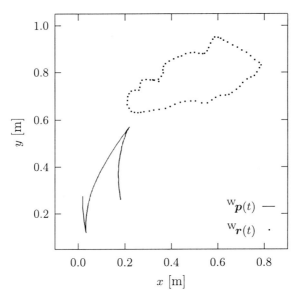

Fig. 9. Positions of the platform ($^{\mathrm{W}}p(t)$) and the tool ($^{\mathrm{W}}r(t)$) when moving without changing the preferred point (classical force guided motion).

To demonstrate the functionality, the mobility of the arm was limited to that of the platform (x-y-plane). $^{\mathrm{W}}\dot{r}$ was set to 0 and g_{P} was set to 1 because we wanted the platform to move slowly in the experiments, ρ was set to 1 and ε to $-50\mathrm{N}^2\mathrm{m}$ (determined experimentally).

Figure 9 and Figure 10 show roughly the same movement of the platform. The starting points are equivalent to the situation shown in Figure 3. In the second case the operator had the possibility to displace the preferred point. It can be seen that in order to achieve a greater sideways displacement of the platform, a smaller movement of the tool is required. Figure 11 shows the user taking Clever for a walk.

5   Teaching of Arm Movements

For teaching the arm movements necessary to perform manipulation tasks we have pursued two directions; teaching via a graphical user interface and teaching by directly moving the arm around using compliant motion. Normally, the "crude" movements are taught using the GUI while finer movements are better taught by directly moving the arm to the desired position and orientation.

In order to be able to generalise grasping and manipulation movements to cases where the objects and/or positions to place them are different from when they were taught, the movements are recorded relative to the objects and not relative to the robot's frame of reference. This means that initially the object of interest has to be located with respect to the robot.

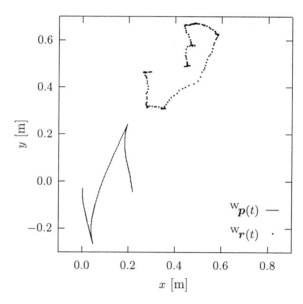

Fig. 10. Positions of the platform and the tool by manoeuvring the robot with the special meaning of the force/torque signal.

Fig. 11. Human guiding the robot using compliant motion.

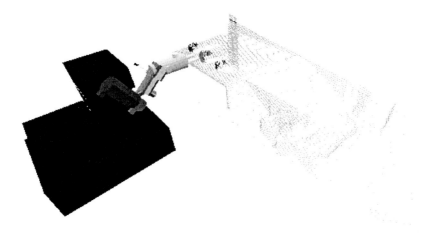

Fig. 12. Example of the localisation of an object on a table. The scene consisting of a table with three objects on it has been scanned by the robot's laser range finder resulting in the 3D point cloud in the right half of the image. The ICP algorithm has converged to the object in the middle (the lighter model points cover the darker scene points). On the basis of this localisation it is calculated how to grasp the object which is shown in the display as feedback to the user.

For the recognition and localisation of objects not easily describable with geometric primitives we use the *iterative closest point* (ICP) algorithm [12]. This has the advantage that object models can be generated simply by scanning the object with a laser range finder and cutting out the relevant part of the resulting 3D point cloud. The algorithm iteratively seeks to match model points to scene points (using a nearest neighbour metric) and—based on this match—to register the model and the scene objects. When the iteration has converged we compare the mean square error between scene and model points, and if this is below a given threshold, we assume the scene object to be of the same type as the model object and that these furthermore coincide. Due to its simplicity, the algorithm works rather quickly, but due to the fact that it is basically a gradient descent, it has problems with local minima. However, for scenes where foreground and background can easily be segmented, it has been shown to work well. In Figure 12 an example of the localisation of a workpiece on a table is shown.

Knowing the pose of the object relative to the robot, it is straightforward to transform and store the arm movements performed by the user in a teaching "run" in object centred coordinates. Thus, when the robot operates autonomously, it can transform the taught trajectories according to the given current pose of the object and so perform the manipulation at this new, relative position. We have demonstrated the validity of this method with various tasks such as picking up workpieces and placing them into workstations.

We are currently investigating how to generalise taught grasping trajectories to different classes of objects. Also we are developing methods for interactively extending and improving already existing trajectories.

6   Handing over Objects

Finally, we would like to present an example of a simple skill showing how tactile sensing combined with acoustic feedback (i.e. two low bandwidth channels) can be combined to achieve a very useful and robust functionality for man-robot interaction.

The most elementary and simple cooperation between robot and man is the robot delivering a workpiece to a worker. This means that the manipulator presents the load so that it can be easily taken by a human if the gripper of the manipulator is releasing it. For domestic applications with light objects, an object can simply be released when the user signals that he/she has taken hold of it (e.g. using speech input or by shaking the object a little). The problem in an industrial context, where objects as heavy as 10 kg can be handled (by humans), is that the human cannot sense the weight of the object until it is released by the robot. If the object is heavier than expected, this may lead to the human dropping it, possibly causing injury.

To ensure that such accidents are avoided, we have implemented a skill that makes the robot release the object only when the user already supports its full weight. Knowing the weight of the gripper and the object from calibration this is easy to measure using the force/torque sensor. As feedback to the user, signalling if he/she has to lift more or less, a tone is produced whose pitch increases to a certain level at which two short tone pulses a generated, indicating that the user is now supporting the full weight of the object. The robot then waits a short period, and if the weight is still supported by the user, the gripper is opened. Figure 13 shows Clever handing over a workpiece to the user.

Albeit extremely simple (or maybe exactly because of that), this skill has proven to be intuitively easy to understand and thus very useful in our cooperation experiments.

7   Discussion and Conclusion

In this paper, we have described research done for a service robot for manufacturing at DaimlerChrysler Research and Technology.

In particular, we presented methods developed for tactile interaction for cooperation and for teaching the robot manipulation skills and trajectories using a single wrist mounted force/torque sensor.

First we outlined how we calibrate the payload of the manipulator on-line which enables the user to teach and cooperate using compliant motion even when the robot is carrying a heavy object in its gripper.

Second we presented results from current research on extending "classical" methods for coordinated arm/platform movements with methods for the estimation of the user's intentions. Knowing these intentions allows the robot to better determine how to move (i.e. how to solve the redundancy) as a result of the forces and torques induced by the user, thus making the cooperation faster, less straining, and more intuitive for the user. First results from manoeuvring

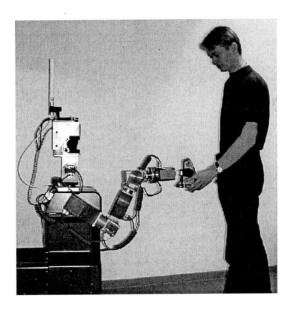

Fig. 13. Clever handing over an object.

experiments were presented. However, more work is required on how to define preferred configurations and how to select one if several exist. Furthermore we will investigate to what extent force/torque "symbols" can be used to initiate manoeuvres, possibly in combination with other communication channels such as speech and vision.

Then we outlined how we teach various manipulation trajectories in a sensory context, allowing for generalisation to cases where the poses of objects as well as the target positions where to place them vary.

Finally, as an example of a simple but very efficient and robust cooperation skill, we presented a method using acoustic feedback to ensure that the worker is not surprised and thus possibly injured by the weight of a workpiece when this is handed over by the robot.

We would like to emphasise the fact that the presented methods are developed in the context of a larger system which is used as a research platform for human-robot interaction in the manufacturing process. This provides us with a unique opportunity to test and evaluate these techniques together with other learning and interaction methods (based on speech, vision, and laser sensors). The lesson we have learned so far is that a combination of simple, robust and partly redundant interaction mechanisms leads to a safer and faster to learn interaction than do more sophisticated methods based on a single sensor. Therefore we will continue to explore such systems since safety and ease of use (which also has a positive effect on safety) are of paramount importance in manufacturing where on the other hand the cost of adding another sensor is not as critical as for robots aimed at the mass consumer market.

Acknowledgements

This research was partly sponsored by the German Ministry for Education and Research under the projects NEUROS, Neural Skills for Intelligent Robot Systems, and MORPHA, Intelligent Anthropomorphic Assistance Systems.

References

1. Kristensen, S., Hansen, V., Horstmann, S., Klandt, J., Kondak, K., Lohnert, F., Stopp, A.: Interactive Learning of World Model Information for a Service Robot. In: Sensor Based Intelligent Robots. Lecture Notes in Artificial Intelligence (1724), Springer (1999) 49–67
2. Kuipers, B.J., Buyn, Y-T.: A Robust, Qualitative Method for Robot Spatial Learning. In: Proceedings of the Seventh National Conference on Artificial Intelligence. AAAI Press, Menlo Park, Calif. (1988) 774–779
3. Kuipers, B.J., Buyn, Y-T.: A Robot Exploration and Mapping Strategy Based on a Semantic Hierarchy of Spatial Representations. Robotics and Autonomous Systems, Vol. 8 (1991), 47–63
4. Gutmann, J–S., Nebel, B.: Navigation mobiler Roboter mit Laserscans. In: Autonome Mobile Systeme. Springer (1997), in German.
5. Koenig, S., Simmons, R.G.: Xavier: A Robot Navigation Architecture Based on Partially Observable Markov Decision Process Models. In: Kortenkamp, D., Bonasso, R.P., Murphy, R.: Artificial Intelligence and Mobile Robots. AAAI Press/The MIT Press (1998) 91–122
6. Thrun, S., Bücken, A., Burgard, W., Fox, D., Fröhlinghaus, T., Hennig, D., Hofmann, T., Krell, M., Schmidt, T.: Map Learning and High–Speed Navigation in RHINO. In: Kortenkamp, D., Bonasso, R.P., Murphy, R.: Artificial Intelligence and Mobile Robots. AAAI Press/The MIT Press (1998) 21–52
7. Jensfelt, P., Kristensen, S.: Active Global Localisation for a Mobile Robot Using Multiple Hypothesis Tracking. In: Workshop on Reasoning with Uncertainty in Robot Navigation (Workshop ROB-3 at the International Joint Conference on Artificial Intelligence), Stockholm, Sweden (1999) 13–22
8. Chung, J.H., Velinsky, S.A., Hess, R.A.: Interaction Control of a Redundant Mobile Manipulator. The International Journal of Robotics Research, Vol. 17(12) (1998) 1302–1309
9. Kazanzides, P., Zuhars, J., Mittelstadt, B., Taylor, R.H.: Force Sensing and Control for a Surgical Robot. In: Proceedings of the 1992 IEEE International Conference on Robotics and Automation (1992) 612–617
10. Yoshikawa, T.: Foundations of Robotics: Analysis and Control. The MIT Press, Cambridge, Massachusetts (1990)
11. Yamamoto, Y.: Control and Coordination of Locomotion and Manipulation of a Wheeled Mobile Manipulator. Ph.D.-Thesis, Department of Computer and Information Science School of Engineering and Applied Science, University of Pennsylvania (1994)
12. Besl, P.J., McKay, N.D.: A Method for Registration of 3-D Shapes. IEEE Transactions on Pattern Analysis and Machine Intelligence, Vol. 14(2) (1992) 239–256

Multiple-Robot Motion Planning
=
Parallel Processing + Geometry

Susan Hert[1] and Brad Richards[2]

[1] Max-Planck-Institut für Informatik, Saarbrücken, Germany
`hert@mpi-sb.mpg.de`
[2] Vassar College, Poughkeepsie, NY, USA
`richards@cs.vassar.edu`

Abstract. We present two problems in multiple-robot motion planning
that can be quite naturally solved using techniques from the parallel processing community to dictate how the robots interact with each other and
techniques from computational geometry to apply these techniques in the
geometric environment in which the robots operate. The first problem
we consider is a load-balancing problem in which a pool of work must
be divided among a set of processors in order to minimize the amount
of time required to complete all the work. We describe a simple polygon partitioning algorithm that allows techniques from parallel processor
scheduling to be applied in the multiple-robot setting in order to achieve
a good balance of the work. The second problem is that of collision
avoidance, where one must avoid that two (or more) processors occupy
the same resource at the same time. For this problem, we describe a
procedure for robot interaction that is derived from procedures used in
shared-memory computers along with a geometric data structure that
can efficiently determine when there are potential robot collisions.

1 Introduction

There is a wealth of research devoted to the problems that arise in environments
where sets of tasks are to be processed in parallel. In most cases, the overarching
goal is to accomplish the tasks most efficiently while avoiding or minimizing interference among the processors. The problems appear in many different settings,
involving different types and numbers of processors and different definitions of
interference, but, regardless of the setting, many of the basic problems are the
same. Given the similarities of the problems and the goals, it is is reasonable
to expect that techniques for addressing the problems in one setting may be
applicable in other settings as well. This is true whether the processors are stationary computer processors or robots moving about in a workspace. However,
since robots, unlike computer processors, interact with each other and the environment in ways that are determined by their geometry and the geometry of
the environment, applying the techniques of parallel processing is possible only
after the corresponding geometric problems that arise have been addressed. In

G.D. Hager et al. (Eds.): Sensor Based Intelligent Robots, LNCS 2238, pp. 195–215, 2002.
© Springer-Verlag Berlin Heidelberg 2002

this paper, we provide efficient solutions to two basic problems in multiple-robot motion planning using techniques from parallel processing to determine how the robots interact with each other and techniques from computational geometry to transfer these techniques into the geometric robotics environment.

In Section 2 we consider the problem of terrain covering using a team of robots. We show how this problem can be viewed as a variant of a load balancing problem in parallel processing and thus can be addressed using similar methods. Put simply, load balancing is the problem of dividing a given amount of work among a set of processors such that each processor does roughly the same amount of work and thus the total time required to complete all the work is minimized. This is a fundamental problem in parallel processing that has been well studied in various forms. In this setting, the work generally takes the form of a program that is to be collectively executed by the processors. We consider the case in which dividing the work among the processors corresponds to assigning certain portions of the data to each processor. A natural counterpart of this in the robotics setting arises when a team of robots is assigned a terrain-covering task. That is, the robots must collectively execute an algorithm that assures that each part of an environment is visited by at least one robot. In this setting, the amount of work to be done roughly corresponds to the area of the region to be covered and thus to assign each robot a certain amount of work one must be able to divide the environment (which we model as a polygon) into pieces of given areas.

Though algorithms exist for partitioning a given polygon into pieces of given areas [3, 11] these algorithms are appropriate only when all the work is to be divided among the robots before any processing is done. This corresponds to applying the so-called *static scheduling* load balancing techniques from parallel processing. These techniques have the disadvantage that they must divide the work based on only an estimate of the amount of time required to finish the work and cannot adjust the loads of the processors based on the actual time required. This can lead to a large imbalance in the work load and thus a far-from-optimal completion time. This disadvantage is not present when applying purely dynamic scheduling techniques where work is assigned to idle processors in small pieces. However, these techniques suffer from the amount of overhead required in the on-line assignment of the work. Thus hybrid techniques have been developed [19] in which some significant portion of the work is assigned to each processor before computation begins and then, as processors finish their statically assigned work, other portions of the work are assigned dynamically. To apply such a technique in the robotics setting, a variation of the partitioning problems addressed in [3] and [11] must be solved. We present this problem in Section 2.2 and provide a simple algorithm to solve it. Notice that the technique is built on the assumption that not everything about the environment is known and thus is well suited for the case when the polygon being divided is only an approximation of the robots' actual environment.

The second problem we consider is one of the most fundamental ones in multiple-robot motion planning: produce a collision-free coordinated motion plan for a set of robots working in a common environment. We illustrate that this can be addressed using one of the most basic mechanisms in parallel processing for avoiding "collisions" between parallel processors. To clarify what is meant by "colliding processors", consider the analogy that each robot corresponds to a processor and each path a robot follows corresponds to a part of a program that is being executed by the corresponding processor. Thus, it naturally follows that the parts of the workspace occupied by a robot as it moves along its path correspond to the parts of a parallel computer's memory that are accessed when a processor executes its piece of a program. To make the analogy most clear, we consider that it is only the pieces of memory that a processor writes that correspond to the locations a robot occupies[1]. Given this, the definition of colliding processors is immediate: two processors attempting to write to the same piece of memory at the same time collide with each other.

A basic technique for avoiding such processor collisions is to use a monitor to protect regions of shared memory where collisions may occur. A processor wishing to modify data within the region must first acquire access to it through the monitor. Once modifications have been made, the region is released and other processors may access the region. One can easily see that this corresponds to the basic technique for avoiding collisions in sensor-based robot motion planning: when one robot senses another robot at a certain location, it does not try to move there. In Section 3, we extend this notion of a monitor beyond the purely sensor-based setting and illustrate that by maintaining a data structure in which the current locations as well as the next planned path segments of the robots are stored, collision-free motion plans can be produced that have potentially shorter waiting times for the robots than those produced by purely off-line planners and can lead to better performance than purely sensor-based planners.

2   Terrain Covering with a Team

2.1   The Problem and Related Work

A terrain-covering algorithm is an algorithm that produces a path for a robot that causes it to visit each point in its environment at least once. Such algorithms could be used by lawn-mowing or vacuum-cleaning robots. Though it has been shown that finding a minimal such path is NP-hard [2], a number of polynomial-time algorithms have been developed that produce reasonably efficient paths for a single robot [2, 12, 24, 28].

When the lawn to be mowed or the floor to be vacuumed is quite large, one naturally thinks of using more than one robot to accomplish the mowing or vacuuming task. Though algorithms have been developed for the related, but ultimately quite different, problem of terrain or model acquisition by a team of robots (e.g., [21, 25]), we are unaware of any algorithms that directly address the terrain-covering problem for more than one robot. The reason for this may

---

[1] Reading a piece of memory can be seen to correspond to the robot sensing another part of its environment, but this is not relevant for the problem we currently consider.

be simply that the solution to this problem looks quite obvious: partition the environment into non-overlapping pieces and assign each piece to one robot that executes a single-robot terrain-covering algorithm. The partition should ideally be done such that each robot does the same amount of work, or, if the robots are of different abilities, such that all robots finish their work at the same time. Using terminology from the field of parallel computation, we wish to produce an optimal *load balance* for the robots.

This goal, though easy to state, is far from easy to attain. There are a number of reasons for this, not the least of which is that it is unclear how large the non-overlapping pieces should be in order for all robots to be busy for the same length of time. The amount of work to be done by a robot in a certain region depends not only on the shape and size of the region but also on the algorithm being executed by the robot. Thus, a precise quantification of the amount of work a region requires is difficult to obtain in general. However, since the amount of work to be done when covering a certain region is roughly proportional to the area of the region, area can be used as a reasonable measure when producing the partition. Thus, a reasonably balanced partitioning of the work among the terrain-covering robots can be provided by solving the nontrivial geometric problem of partitioning a polygon into pieces of given areas.

The so-called area partitioning problem has been considered in a number of other works. Two algorithms have been presented that produce partitions into connected pieces of given, possibly unequal, sizes [3, 11] under different constraints. Though most other works that have considered the problem of partitioning polygons based on areas consider specifically the case of producing a cut that bisects the area of the polygon [4, 26] (or simultaneously bisects the area of two disjoint polygons [7, 29]), the methods generalize for producing partitions of other proportions as well. However, the methods that are formulated for convex polygons [5, 7, 10, 29] do not generalize for nonconvex polygons and the methods developed to handle nonconvex polygons or polygons with holes [4, 26] do not guarantee that the cuts will result in connected polygons. This is obviously an important consideration for the terrain-covering application. Covering many disconnected pieces of an environment generally requires more work than covering a single connected piece since the robot must travel among the disconnected pieces.

In Section 2.2, we argue that, as in the field of parallel processing, a good balance of work among the team of terrain-covering robots can be produced by initially leaving some of the environment unassigned to any robot and then parceling this remaining area out as the robots finish their initial portions of the work. To accomplish this, a new variant of the area partitioning problem must be solved. We present this problem in Section 2.3 and show how the algorithms presented in [3] and [11] can be combined to solve the problem.

2.2   Load Balancing

In the field of parallel processing, the problem of load balancing, where one wishes to divide work among a set of p processors to achieve a minimal completion time for all the work, has been studied in many different forms. Methods for addressing these problems vary from completely static methods, in which the work is divided into p large chunks and assigned to the processors before any computation begins [19] to fully dynamic methods in which the work is assigned to processors in smaller chunks as computation proceeds and the processors finish their assigned tasks (e.g., [14, 23]). The major advantage of the former method, the amount of overhead incurred in assigning work to processors, is the largest disadvantage of the latter method. Similarly, while the latter method can achieve a better balance of work (and thus a lower total time for the work) since it can adapt the balance based on the actual time required for particular tasks, the former method suffers from having to produce the division of work based on only estimates, which may be quite inaccurate, of the time required. Thus hybrid methods have been developed [19] in which some (possibly large) portion of the work is assigned to each processor before computation begins and the remaining work is assigned in smaller chunks as necessary to keep the processors busy. These methods attempt to achieve the best of both worlds – low overhead with a good balance of work and have been observed to perform quite well in comparison to other methods [19].

It is reasonable to believe that similarly good results will result from applying these hybrid load balancing techniques to the problem of terrain covering with a team of robots. We therefore need to be able to divide the work of the robots (i.e., their polygonal environment) into pieces that will allow these techniques to be applied. Though the algorithms presented in [3] and [11] can be used to partition polygons into pieces of arbitrary, given areas, and can thus be used to achieve a static scheduling of the terrain-covering work, these algorithms are not ideal for use with the hybrid scheduling methods. To be most effectively applied, these methods require that the portion of the work left initially unassigned to any robot be easily accessible to all robots. In other words, this unassigned part of the environment should be connected to all the assigned portions. In this way, whichever robot finishes its assigned work first can be assigned a new portion of the environment to cover as easily as any other. This leads us to a new variant of the area partitioning problem which we call *connected area partitioning*.

2.3   Connected Area Partitioning

Given a polygon $\mathcal{P}$ and $p+1$ values $a_1, \ldots, a_{p+1}$ such that $\sum_{i=1}^{p+1} a_i = Area(\mathcal{P})$, we wish to produce a partition of $\mathcal{P}$ into $p+1$ pieces $P_1, \ldots, P_{p+1}$ such that $Area(P_i) = a_i$ for all i, and pieces $P_1, \ldots, P_p$ are each connected to P_{p+1}. We solve this problem using variants of the techniques presented in [3] and [11].

Both algorithms begin with a polygon that has been partitioned into convex pieces (e.g., [9, 13, 16, 30]) and produce the desired area partition by dividing the convex pieces in an appropriate manner to satisfy the constraints of the problem. The major advantage of beginning with the convex decomposition is that the partitioning algorithm can then capitalize on this work and produce a natural bias toward partition polygons that, while not necessarily convex, maintain to some extent the good shape characteristics of these convex pieces. We assume a similar preprocessing step here. That is, polygon $\mathcal{P}$ is assumed to have been partitioned into q convex pieces $\mathcal{CP}_1, \ldots, \mathcal{CP}_q$ without introducing any extra vertices [9, 13, 16].

We also assume here, as in [3], that the polygon $\mathcal{P}$ is simply connected. Though this is often not the case in robot environments, the algorithm we describe can easily be applied to nonsimply connected polygons by artificially connecting each hole to the outer boundary of the polygon and thus converting such polygons into simply connected polygons.

As in the previous work, we build a connectivity graph corresponding to the convex decomposition of $\mathcal{P}$. Because we assume the convex decomposition of $\mathcal{P}$ produced no new vertices and $\mathcal{P}$ is simply connected, the connectivity graph is a tree T. With each node N_j in the tree, we associate the value $Area(N_j)$, which is the sum of the areas of the convex pieces corresponding to the nodes in the subtree rooted at N_j. The value $Area(N_j)$ is also referred to as the *size* of the node (Figure 1). We assume the nodes of T and the convex pieces $\mathcal{CP}_1, \ldots, \mathcal{CP}_q$ are numbered in the same way. For brevity, we sometimes refer to the nodes and pieces interchangeably. For example, we may speak of the vertices of a node N in the tree, by which we mean the vertices of the convex piece that corresponds to N.

Because we wish to have the unassigned portion of $\mathcal{P}$ easily accessible to every robot, we attempt to ensure that the polygon P_{p+1} will be in the "middle" of the polygon $\mathcal{P}$ by choosing as the root of T the largest node of maximum degree (Figure 1). This node is denoted N_r and the corresponding convex piece is $\mathcal{CP}_r$. We then construct the partition polygons $P_1, \ldots, P_p$ beginning with the children of N_r, attempting to work from the leaves of T inward. In what follows, let $c_1, \ldots, c_k$ denote the children of N_r in clockwise order.

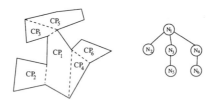

Fig. 1. A convex decomposition and the corresponding connectivity tree.

To create a balanced area partition, we first choose a set of partition polygons to be created from a subset of the child subtrees. We then carve out a polygon of exactly the size required for this set of partition polygons. Finally, we divide this polygon into the required number of pieces, taking care that each of the resulting polygons is connected to the unassigned area. For the first partition, we employ a modified version of the sweep-line technique developed for the general area partitioning algorithm presented in [3] that carves out a polygon with a given area beginning with a certain node in the tree of convex pieces. For the second partition, in which we must assure that each polygon not only has the correct area but also that it contains some portion of a particular edge on its boundary, we use anchor points, as in the algorithm presented in [11], to dictate where the sweep-line begins.

Ideally, we would like to construct the partition polygons $P_1, \ldots, P_p$ from disjoint sets of root subtrees. This would result in the robots being well distributed around the environment and therefore less likely to interfere with each other when moving to the next portions of the environment assigned to them. Of course, this is not always possible. Furthermore, the following theorem shows that in order to find a best distribution of the area requirements among the child subtrees would require an exhaustive search of the possibilities unless P = NP:

Theorem 1. *Determining if there is a partition of the set $\{a_1, \ldots, a_p\}$ into k' subsets $s_1, \ldots, s_{k'}$, such that the sum of the values in each subset s_j is no more than the size of c_j is NP-complete.*

Proof. Reduction from BIN-PACKING.   □

We therefore rely on heuristics to decide which subtrees to use for creating the various partition polygons. We first order the area requirements $a_1, \ldots, a_p$ and children $c_1, \ldots, c_k$ in descending order. Then, we check for the following special case: $k \geq p$ and for each $i = 1, \ldots, p$ $Area(c_i) > a_i$. In this case, the problem is simplified to p instances of the area partitioning problem [3]. For each instance, we wish to partition a given polygon (the one corresponding to child subtree c_i) into two pieces of given areas (one of area a_i and the other of area $Area(c_i) - a_i$). We solve each of these problems independently and have then produced the desired connected area partition.

In the more general case, the children of N_r are either not sufficiently large or not sufficiently numerous to allow the partition polygons to be produced independently of each other using disjoint subtrees of T. It is also not possible in general to decide first which polygons are to be constructed from which subtrees and then to do the partitioning work. This is possible only if the subtrees to be used are all bigger than the partition polygons to be created from them.

For this more general case, we use the children of N_r in clockwise order beginning with the largest, and assign partition polygons to the subtrees in order from largest to smallest. That is, we begin with child c_1 and then find the smallest value i_2 such that $\alpha = \sum_{i=1}^{i_2} a_i \geq Area(c_1)$. Partition polygons $P_1, \ldots, P_{i_2}$ will be created using child c_1 and perhaps some of its siblings. To determine which siblings are necessary, we use the sweep-line technique summarized in Figure 2 and illustrated in Figure 3 to create a *super-polygon SP* with area α. That is, we sweep a segment L around $\mathcal{CP}_r$ until enough area has been accumulated for super-polygon SP. If, during the sweep, an edge that connects $\mathcal{CP}_r$ to another convex piece, say the one corresponding to child c_2, is encountered, a comparison of the current accumulated area, α and $Area(c_2)$ is done to determine if all, some, or none of the sibling subtree is required to complete SP. The sweep then proceeds accordingly, recursively dividing the sibling subtree if necessary.

Input: N_j node in which to sweep; p anchor point of sweep-line $L = (L_s, L_e)$;
                    a area required; P current polygon
Output: polygon P with area a
Complexity: $O(n)$
Procedure $DivideConvexPiece$

$L_s = L_e = p$
if N_j is a leaf and $a < Area(N_j)$ then
   Rotate L clockwise around CP_j until area to the right of L equals a.
else
  while $(a > 0)$ do
      Sweep L CW in N_j until area to right of L is equal to a
        or vertex shared with a neighboring convex piece is found
      e_{cw} = next CW edge in N_j
      N_{cw} = convex piece sharing edge e_{cw}
      $a = a-$ area to the right of L in N_j
      if $(a = 0)$ return
      if area of next CW triangle in N_j to left of L is $> a$
        pt = exact area interpolation point on e_{cw}
        add triangle (L_s, L_e, pt) to P
        return
      else if e_{cw} is not a partition edge and $Area(N_{cw}) < a$
        assign nodes in subtree rooted at N_{cw} to P
        $a = a - Area(N_{cw})$
      else
        $midPt$ = midpoint of edge e_{cw}
        $a' = a'-$ area of triangle $(L_s, L_e, midPt)$
        $DivideConvexPiece(N_{cw}, a, P)$
        return
      endif
  enddo
endif

Fig. 2. Summary of the sweep-line procedure used to produce a polygon P of a given area. This is a modified version of the procedure used in [11].

When SP has been constructed, we remove SP from $\mathcal{P}$, remove any nodes that were included in SP from T and adjust the sizes of any partially used nodes (including the root node). The process is then repeated starting with area a_{i_2+1} and the next subtree of N_r that was not completely used by SP. This continues until all partition polygons have been assigned to some super-polygon constructed from a child of the root node (*i.e.*, $i_2 + 1 > p$) or there are no more children left.

If there are no more children left, then the area for the remaining polygons is carved out of the remaining portion of CP_r. Care must be taken here because the edges of the unassigned portion of CP_r are the edges that are to be used to assure that the partition polygons remain connected to polygon P_{p+1}. Thus, if an edge created during the construction of a previous super-polygon is swept

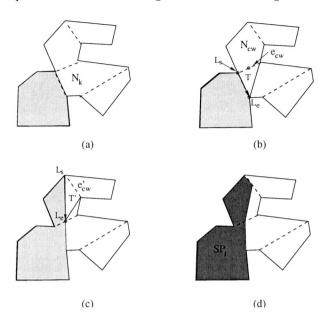

Fig. 3. Illustration of the sweep-line procedure described in Figure 2. (a) The shaded piece is assigned as the basis for superpolygon SP_1 (b) Sweep in N_k stops with sweep line at $L_s L_e$. The area triangle T is too small to complete polygon SP_1 so recur with $N_{cw} = N_k$ (c) Sweep until next neighbor is reached. The area of triangle T' is larger than the remaining area required by SP_1. Find the point on e'_{cw} that results in a triangle with correct area. (d) The complete superpolygon SP_1.

over during the construction of the last super-polygon, then the super-polygon to which this edge belongs is simply merged with the last super-polygon and thus this final super-polygon will be split into more pieces in the end.

Note that, by construction, the following lemma holds.

Lemma 1. *The unassigned polygon P_{p+1} is simply connected and shares an edge of non-zero length with each super-polygon.*

Once the super-polygons have been created, we must divide each into its set of partition polygons. We assure that each partition polygon is connected to the remaining unassigned portion of the original polygon by placing *anchor points* on the boundaries between its super-polygons and polygon P_{p+1}. These anchor points are used to anchor the ends of the sweep lines used to create the final partition polygons. Thus, for a super-polygon SP that is to be split into p' partition polygons, $p' - 1$ anchor points are spaced evenly along the boundary between SP and P_{p+1}. Once the anchor points have been placed, the procedure outlined in Figure 2 is used again, once for each anchor point and thus the connected area partition is produced.

A summary of the algorithm for producing a connected area partition is given in Figure 4 and the steps are illustrated in Figure 5. Panel in Figure 5(a) shows

Input: $CP_1, \ldots, CP_q$, a convex partition of $\mathcal{P}$; area requirements $a_1, \ldots, a_{p+1}$, with $a_1, \ldots, a_p$ in descending order.

Output: set of polygons $P_1, \ldots, P_{p+1}$ that form a partition of $\mathcal{P}$ with $Area(P_i) = a_i$ for all i and such that for $1 < i < p+1$, P_i is connected to P_{p+1}.

Complexity: $O(n + q \log q + pq)$

Procedure:

1. Build connectivity tree T of convex partition with the root being the largest node with maximum degree
2. if $k \geq p$ and for each $i = 1, \ldots, p$ there exists a unique c_i such that $Area(c_i) \geq a_i$

   for $i = 1, \ldots, p$ do
   produce P_i from c_i using area partitioning algorithm.

   else
   2.1 $c = c_1$; $i_1 = 1$;
   2.2 While $i_1 \leq p$
       2.2.1 Find smallest $i_2 \geq i_1$ such that $\alpha = \sum_{i=i_1}^{i_2} a_i > Area(c)$;
       2.2.2 Beginning with node c, construct a super-polygon SP with area α using the sweep-line procedure outlined in Figure 2. Let j' be the index of the next subtree not completely used in creating SP.
       2.2.3 Remove SP from $\mathcal{P}$, and adjust the size of $c_{j'}$
       2.2.4 $i_1 = i_2 + 1$; $c = c_{j'}$
   2.3 Distribute anchor points around the boundary of the unassigned polygon P_{p+1}
   2.4 Use sweep-line procedure of Figure 2 again starting sweep at the anchor points to create the polygons with areas a_i

Fig. 4. Summary of the connected area partitioning algorithm.

the convex decomposition and its connectivity tree (with the root shown as an open box). In panel (b), the first super-polygon, SP_1, for partition polygons P_1 and P_2 has been constructed. The second super-polygon, SP_2, for P_3 and P_4 is shown in panel (c) and anchor points (open circles) have been placed on the boundaries shared by polygon P_5 and SP_1 and SP_2. In panel (d), the final partition has been produced by splitting the super-polygons into polygons of the desired sizes.

Run-Time Analysis.   In this section, we prove the following theorem

Theorem 2. *Given a polygon with n vertices, together with a decomposition of the polygon into q convex pieces and given $p + 1$ areas, the connected area partitioning algorithm outlined in Figure 4 requires $O(n + q \log q + pq)$ time in the worst case.*

Proof. In step 1, we build the connectivity tree T corresponding to the convex decomposition. This can be done in $O(n)$ time via a simple depth-first or breadth-first traversal of the pieces, assuming, as is reasonable, that the output of the convex decomposition pre-processing step includes information about

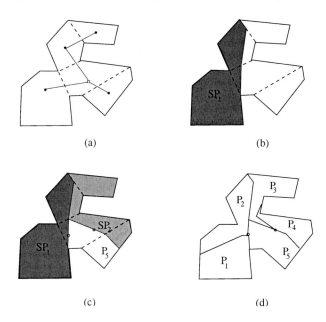

(a) (b)

(c) (d)

Fig. 5. Example of a polygon beign partitioned into five pieces of given areas.

which pieces neighbor which others. If the initial tree built in this manner has a root node other than the largest node of highest degree, this node can be identified and the tree re-rooted in $O(q)$ time.

In step 2 of the algorithm, we first test in $O(q\log q + p)$ time whether each polygon $P_1,\ldots,P_p$ can be built from a separate child subtree of the root. If this is possible, then each partition polygon can be constructed in time linear in the number of convex pieces in its subtree and the number of vertices. Thus, since the subtrees are disjoint, all the partition polygons can be computed in $O(n+q)$ in this case.

When the special case does not hold, we must first construct the super-polygons and from these the partition polygons. Step 2.2.1, in which the sizes of the super-polygons are determined, requires $O(p)$ time in total. Each super-polygon is constructed in step 2.2.2 using the sweep-line algorithm outlined in Figure 2, which in general requires time linear in the number of vertices of the super-polygon [3]. A closer look reveals that the time required is linear in the number of complete subtrees used for the polygon (which can be added to the polygon in constant time) plus the number of vertices of any partially used subtree (since the vertices of the subtree must be visited in order to cordon off the correct portion of the subtree for the polygon). Because any partially used subtree for one super-polygon will be completely used by the next super-polygon, the total number of vertices visited to create all the super-polygons is $O(n)$ and thus the time required to compute all super-polygons is $O(n)$. Removal of a super-polygon from the original polygon $\mathcal{P}$ (step 2.2.3) can be done in time linear in the number of convex pieces divided to create this polygon. Since no

convex piece, other than possibly $\mathcal{CP}_r$, is divided into more than two pieces, and
the root node is divided into no more pieces than there are super-polygons, this
means the time required to remove all the super-polygons is $O(q+p)$. Step 2.2.4
obviously requires constant time.

To distribute the anchor points along the boundary of P_{p+1} requires time
linear in the number of anchor points, which is $O(p)$, and the size of the boundary
of P_{p+1}, which is $O(n)$. Finally, computing the partition polygons once the anchor
points have been placed requires at most $O(n+pq)$ time in total in the worst
case [3]. □

Discussion. The algorithm we have presented for solving this new version of the
area partitioning problem, while simple and efficient, does have its drawbacks.
The partition polygons produced may be shaped in ways that hamper the robots'
efficiency (*e.g.*, with sharp corners or narrow regions in which it is difficult to
maneuver). Thus two interesting questions for further investigation are whether
other constraints can be imposed on the polygons to be produced and whether
local modifications to the polygons can be performed as a post-processing step
to achieve shorter overall completion time.

Another issue to address arises when one considers what happens after the
initial partitioning is done. After each robot has finished its initial portion of
the work, it will be assigned a new section of the environment, which is part
of the initially unassigned polygon P_{p+1}. Though our algorithm guarantees that
each robot has easy access to this polygon, it is still possible that a robot will
have to pass through a portion of P_{p+1} assigned to another robot to get to its
newly assigned region of P_{p+1}. One would generally like to avoid this sort of
interference and thus the question arises whether it is possible to produce a
partition that not only allows portions of the region to be assigned dynamically
but also guarantees that no robot will trespass on another's region.

Many of these questions are best addressed via experimentation. Work is
currently underway on an implementation of this algorithm in a simulated environment.

3   Multiple-Robot Collision Avoidance

3.1   The Problem and Related Work

Collision avoidance is, of course, one of the most fundamental problems that
must be addressed when working in a multiple-robot environment. While sensor-
based robots can avoid collisions simply by heeding what their sensors tell them,
purely reactive motion is not sufficient to produce a set of paths in a constrained
dynamic environment that will lead each robot to a designated target position.
This has led to the study of motion planners for multiple robots, which attempt
to build plans for a set of robots that both avoid collisions and cause the robots
to reach their target positions. Complete planners such as the one of Parsons
and Canny [22] are theoretically interesting because they guarantee a solution

will be found if one exists. However, these planners generally compute plans in the combined configuration space of the set of robots and thus are inherently computationally expensive. More efficient incomplete planners [1, 6, 8, 31] have been developed that use a decoupled approach, first introduced by Kant and Zucker [15]. In these planners, paths of the robots are computed independently of each other and then velocity profiles along these paths are computed that assure the robots do not collide.

The so-called path coordination problem concerns itself only with the second part of the decoupled planning approach. That is, it is assumed that the paths of the robots have been computed and thus the problem is to produce a coordinated motion plan along these paths. In the work of O'Donnell and Lozano-Pérez [20], techniques from concurrent database processing were adapted to produce a method for computing deadlock-free and collision-free coordinated motion plans for two robots. There, the notion of a coordination diagram was introduced. A coordinated motion plan for the two robots corresponds to a path through this two-dimensional space in which obstacles correspond to mutual positions where the robots collide. This method has been adapted to handle the case of more than two robots by exploiting the fact that obstacles in a coordination diagram for more than two robots are cylindrical in shape [17, 18].

Here, we also consider the path coordination problem. The specific setting we consider is the following: Each robot R_i has a starting point S_i, a target point T_i, and a method for getting from one to the other independently of other robots in the environment. We wish to produce a coordinated motion plan for the robots. That is, we want to produce a plan that assures the robots will not collide with each other as they move along their computed paths. It is assumed here that each robot can perfectly localize itself and communicate its location and intended future motion to the other robots.

In Section 3.3, we propose a general framework in which to formulate the path coordination problem. As in the work of O'Donnell and Lozano-Pérez, we borrow ideas from parallel computing to address this problem. In particular, the framework is built around the idea of using *monitors* (Section 3.2) to avoid that two processors have access to the same resource (in our case, space) at the same time. Our approach differs significantly from coordination diagram-based approaches, which take into account the entire path of each robot when producing the motion plan. We seek to localize collision detection to smaller path segments and thus potentially reduce the computation time required as well as the time required to execute the motion plan. In Section 3.3, we illustrate that the framework we propose is general enough to encompass a wide range of planners, including purely reactive motion planners and off-line, decoupled planners. We then present a new on-line coordinated motion planner built out of this framework.

3.2   Locks and Monitors

Locks are one of the simplest ways for parallel programs to ensure exclusive access to regions of shared memory. Before modifying shared data, a processor first acquires a lock that protects that region. A processor that wishes to acquire a lock on a region must wait until the previous lock holder releases its lock before it can proceed (*i.e.*, it *blocks* at the lock acquisition attempt). The parallel system itself does not associate a lock with any particular region of memory; it is left up to the parallel application to enforce some discipline that associates locks with collections of shared data.

Locks can be difficult to use correctly since programmers must manually associate them with appropriate regions of memory, and must remember both to lock before modifying and to release the lock when modifications are complete. The exact semantics of the lock also vary from system to system, in particular with respect to the order in which competing processors are granted access.

Monitors have evolved from locks to address these shortcomings and are directly supported by a number of modern programming languages (most notably, Java). In contrast to a lock, a monitor encapsulates both the data to which exclusive access must be guaranteed and the routines that attempt to modify the shared data. Thus, the data to be protected is explicitly associated with a particular monitor. Calls to monitor routines block until previous invocations have completed (or until a certain variable in the monitor reaches a certain state), so there is no need to acquire or release locks and therefore no errors can arise due to their misapplication. Finally, monitors efficiently support waiting processors, keeping waiting processors in a queue structure by default. Waiting processors are awakened (notified) when the processor that currently has access to the monitor's data completes.

3.3   Space Monitoring

To apply the monitoring mechanism used in parallel computing to our robotics problem we require a method of keeping track of which parts of the shared resource are currently in use, a means of deciding whether resource requests are in conflict or not and, if so, how the conflict should be resolved. The resource to be managed in the robotics setting is space, and resource requests take the form of path segments the robots wish to follow. We therefore require geometric data structures and decision procedures to apply this technique. Before going into these geometric details, we first describe the general procedure used by the robots in the presence of these details to get from their starting points to their target points.

Motion Planning with Monitors.   Each robot R_i follows a path in the workspace from S_i to T_i. Before the robot can move along a segment of its path it must acquire access to this path segment, which guarantees that the robot will not collide with any other robots along this segment. This is done by using a monitor to protect accesses to each segment. A given robot uses an access routine within the monitor to claim exclusive access to the path segment, and will relinquish the segment when it moves on to the next segment. When access to a particular segment is obtained, the robot begins to move along this segment and releases its hold on its previous path segment. All robots that may have been waiting to move along a path segment that intersected this previous segment are notified that the segment is free. Then the robot issues a pre-request to the central planner for its next path segment. This means that the planner can begin the potentially costly geometric computations required to determine if there are potential collisions along this segment before the robot reaches it. In this way, some or all of the latency that may be caused by this computation can be hidden[2].

Notice that we do not designate here how the path segments are determined or how or when access to particular segments is granted since the general technique does not depend on any particular path structure or access method. For example, it can easily be seen that this framework incorporates a simple reactive, sensor-based motion planner in the following way: each path segment is a location of a robot; using the access routine of the monitor corresponds to a robot moving to a particular location; and notifying robots of a freed segment simply means that the robots can detect with their sensors that the robot has moved out of its previous location. One can also see that by breaking each path into segments around its intersection points with other paths and granting access to the segments based on a pre-computed priority ordering of the robots we arrive at a decoupled planning method similar to those suggested by Kant and Zucker [15] and Buckley [6].

1. `next_seg` = first path segment for R_i
2. while R_i is not at its target
   2.1 `next_seg.acquire()` /* blocking request */
   2.2 `curr_seg.release()`
   2.3 `curr_seg = next_seg`
   2.4 `next_seg` = path segment after end of `curr_seg`
   2.5 send pre-request for `next_seg` /* nonblocking request */
   2.6 move to the end of `curr_seg`

Fig. 6. Procedure each robot R_i uses to get from S_i to T_i

This simple procedure used by each robot is summarized in Figure 6. In Steps 2.1, 2.2, and 2.5, the robot interacts with the central controller. The procedures for monitoring a segment and notifying about the release of a path segment are trivial procedures summarized in Figure 7. The pre-request procedure is a little more involved because it is with this procedure that the bulk of the work is done including determining where potential collisions are and deciding how to avoid them. Care must be taken here to assure that deadlocks are also avoided. After describing the data structure we employ, we discuss the pre-request procedure in general and discuss a particular instance for use in on-line path coordination.

---

[2] This technique of pre-fetching to hide latency is a common one used not only in parallel programming but also in networking and hardware design.

Data Structures. The only thing we need to keep track of during this process is the set of path segments that are either currently active (robots are currently occupying part of them) or pending (they are the result of pre-requests). For each segment we need to know: whether it is a current or pending segment; a description of the area swept out by the robot as it moves along this path segment (the *trace* of the robot on this segment); and the bounding box of this trace. For each segment there is also an associated monitor like the one shown in Figure 7. Through its condition variable, `blocked`, the monitor keeps track of which segments are waiting for a particular segment to be freed. Typically this is handled through a queue, but this is not necessary and other structures for the waiting robots may be employed as appropriate.

The segments are stored in two orthogonal sorted lists. These lists are sorted based on the coordinates of their bounding boxes, with one list sorted according to the boxes' x-coordinate intervals and the other according to the y-coordinate intervals. The intervals are sorted lexicographically. Such a structure of orthogonal lists allows for $O(\log n_s)$ query, insertion, and deletion time, where $n_s = O(n)$ is the number of segments.

Decisions that Avoid Collisions. When a robot sends a pre-request for its next path segment, it supplies to the controller a simple description of the path segment it wishes to follow. The controller uses this, together with information about the size and shape of the robot's body, to compute the trace Tr of the robot along the path segment. This can be done, for example, using the method described in [27]. To determine if there are potential collisions, the controller then checks if this trace intersects the traces of any other segments currently in its data structure. This is done by first computing the bounding box of this trace, and then using the coordinates of the bounding box to search through the two orthogonal lists of segments to find the segments whose bounding boxes intersect this one. These are the only segments whose traces could intersect Tr. Each such trace is checked for an intersection with Tr in turn. If there is an intersection, then it must be determined which robot will have priority and be able to move through the area first. This robot is then given priority over the lower priority robot for the two intersecting path segments. The procedure is summarized in Figure 8.

As pointed out above, the decision about which robot has higher priority can be done in any number of ways. The easiest way, however, is to adhere to a complete priority ordering of the robots as is done, for example, in [6, 8]. If this ordering is strictly followed, the robots can reach their goals without getting deadlocked. However, when the robots' paths are each broken into more than one path segment and thus decisions are made on a local basis, one must do more than simply adhere to the priority ordering in each local decision to avoid deadlock. For example, the situation depicted in Figure 9 could quite easily occur. Each robot has arrived at its current location without interfering with any other robots and each waits for another to release its current segment before it can proceed. When situations such as this are detected (which is easily done by maintaining

```
Monitor {

      ID_list acquire_order;      // Controller sets this up
      cond blocked;               // Sleeping robots
      bool in_use = FALSE;        // Someone using segment?

   // segment is acquired if not already in use and my_ID is at the
   // head of the list; otherwise the robot sleeps until it reaches
   // the head of the list
    procedure acquire( my_ID )
    {
        if ( in_use ||
              acquire_order.is_empty() ||
              my_ID != acquire_order.head() )
          wait(blocked);
        in_use = TRUE;
    }

    procedure release()
    {
        acquire_order.delete_head();  // get out of the way
        in_use = FALSE;
        signal(blocked);              // awaken front robot
    }

    procedure extend_list( ID )
    {
        acquire_order.insert(ID);     // inserts the ID in proper order
        signal(blocked);              // awaken front robot that may be
                                      // waiting on an empty list for
                                      // processing to finish
    }
}
```

Fig. 7. A monitor for a segment contains two data items and three procedures.

1. Compute trace Tr_i of robot R_i along path segment s_i.
2. Compute the bounding box of trace Tr_i.
3. Find the set of segments $\mathcal{S}$, for which the trace bounding boxes intersect the bounding box of Tr_i
4. For each segment $s_j \in \mathcal{S}$, if Tr_i intersects the trace Tr_j of s_j then
   4.1 Let $low \in \{i, j\}$ be the index of the lower priority robot and $high$ the index of the higher priority robot.
   4.2 Use the **extend_list** method for the monitors of s_{low} and s_{high} to give R_{high} priority over R_{low} for both segments.

Fig. 8. Procedure used to process a pre-request for a path segment.

a graph of dependencies among the path segments), some re-planning of at least one robot's path is required. For example, the robot with highest priority could be instructed to reroute its next path segment in the presence of new obstacles, which are the other robots. Alternatively, one of the lower priority robots could be instructed to backtrack along its current path segment to a point that gives free access to the highest priority robot.

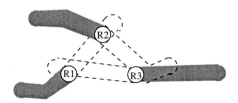

Fig. 9. Robots R_1, R_2, and R_3 are deadlocked. Their current path segments are shown shaded in grey; their next segments are shown outlined with dashed lines. None can move since the next path segment of each is occupied by another robot.

Both these methods avoid deadlock as well as collisions. The former method, while simpler to state, has the disadvantage of requiring re-planning of some path segments, but it is more generally applicable. The latter approach is appropriate only when the path segments are sufficiently long to allow a robot to remain within its current path segment (where it will not collide with another robot) while moving out of the way of another robot. This approach also requires that it be possible to modify a segment while it is being monitored, which would require some modifications and additions to the procedures outlined in Figure 7.

Discussion. We have mentioned that the general framework we propose here incorporates the purely on-line reactive method of collision avoidance. In fact, the on-line method of path coordination we proposed in the previous section can be seen as an extension of the reactive method. It therefore shares some of the advantages and disadvantages of this method. In particular, the method can adapt to the reality of the current situation and therefore potentially produce more efficient motion plans than off-line path coordination methods since the length of time robots will wait for each other is based on more accurate information than for the off-line methods. But, as with purely reactive planning, potential deadlock situations can be reached. Because the method is not purely reactive, the deadlocks can be avoided at the possible cost of some re-planning. By using pre-requests, the cost of this re-planning can be minimized since much of the computation can go on as the robots move along other path segments. Our method can also be seen to be superior to purely reactive methods since decisions can be made on a larger, possibly varying, scale and thus can incorporate more information without any visible decline in performance.

Extensions of the framework that are worth considering include having distributed control in which robots communicate only with those that are in their vicinity to negotiate ownership of certain parts of the space and also making the monitor a bit more flexible so it can accommodate segments that are modified by the controller. This latter extension could make for a more dynamic and efficient path coordination mechanism.

4   Conclusion

We have illustrated with two basic problems from multiple-robot motion planning that the area of parallel processing provides a set of well-tested tools with which solutions to multiple-robot motion planning problems can be built out of geometric building blocks. The first problem we considered was a problem of dividing work among a set of robots working in a common environment. There we applied ideas used to solve load-balancing problems in parallel processing to motivate a new polygon partitioning algorithm that allows similar ideas to be implemented in the robotics setting. The second problem we considered was the path coordination problem. Here we applied a basic technique used in parallel systems to avoid processors having conflicting access to resources simultaneously to arrive at a general framework into which several existing collision-avoidance systems fit. We also proposed a new on-line method for path coordination that fits into this framework. The similarities in the types of problems considered in parallel computing and multiple-robot motion planning are striking, thus further investigation into the parallels between these two fields is warranted to see if there are other places where one field might benefit from results already attained in the other.

References

1. R. Alani, F. Robert, F. Ingrand, and S. Suzuki. Multi-robot cooperation through incremental planmerging. In *Proceedings 1995 IEEE International Conference on Robotics and Automation*, pages 2573–2578, 1995.
2. E. M. Arkin, S. P. Fekete, and J. S. B. Mitchell. Approximation algorithms for lawn mowing and milling. Technical report, Mathematisches Institut, Universität zu Köln, 1997.
3. H. Bast and S. Hert. The area partitioning problem. In *Proceedings of the 12th Annual Canadian Conference on Computational Geometry*, pages 163–172, 2000.
4. K. F. Böhringer, B. R. Donald, and D. Halperin. On the area bisectors of a polygon. *Discrete Computational Geometery*, 22:269–285, 1999.
5. P. Bose, J. Czyzowicz, E. Kranakis, D. Krizanc, and D. Lessard. Near-optimal partitioning of rectangles and prisms. In *Proceedings 11th Canadian Conference on Computational Geometry*, pages 162–165, 1999.
6. S. Buckley. Fast motion planning for multiple moving robots. In *Proceedings 1989 IEEE International Conference on Robotics and Automation*, pages 322–326, May 1989.
7. M. Díaz and J. O'Rourke. Ham-sandwich sectioning of polygons. In *Proceedings 2nd Canadian Conference on Computational Geometry*, pages 282–286, 1991.

8. M. Erdmann and T. Lozano-Pérez. On multiple moving objects. *Algorithmica*, 2:477–521, 1987.
9. D. H. Greene. The decomposition of polygons into convex parts. In F. P. Preparata, editor, *Computational Geometry*, volume 1 of *Advances in Computing Research*, pages 235–259. JAI Press, London, England, 1983.
10. R. Guàrdia and F. Hurtado. On the equipartitions of convex bodies and convex polygons. In *Proceedings 16th European Workshop on Computational Geometry*, pages 47–50, 2000.
11. S. Hert and V. Lumelsky. Polygon area decomposition for multiple-robot workspace division. *International Journal of Computational Geometry and Applications*, 8(4):437–466, 1998.
12. S. Hert, S. Tiwari, and V. Lumelsky. A terrain-covering algorithm for an AUV. *Journal of Autonomous Robots*, 3:91 – 119, 1996.
13. S. Hertel and K. Mehlhorn. Fast triangulation of simple polygons. In *Proceedings of Conference on Foundations of Computation Theory*, pages 207–218, New York, 1983. Springer-Verlag.
14. S. F. Hummel, E. Schonberg, and E. L. Flynn. Factoring: A method for scheduling parallel loops. *Communications of the ACM*, 35(8):90–101, 1992.
15. K. Kant and S. W. Zucker. Toward efficient trajectory planning: the path-velocity decomposition. *International Journal of Robotics Research*, 5(3):72–89, 1986.
16. J. M. Keil. Decomposing a polygon into simpler components. *SIAM Journal on Computing*, 14:799–817, 1985.
17. S. LaValle and S. Hutchinson. Optimal motion planning for multiple robots having independen t goals. In *Proceedings IEEE International Conference on Robotics and Automation*, pages 2847–2852, April 1996.
18. S. Leroy, J.-P. Laumond, and T. Siméon. Multiple path coordination for mobile robots: a geometric algor ithm. In Dean Thomas, editor, *Proceedings of the 16th International Joint Conference on Artif icial Intelligence (IJCAI-99-Vol2)*, pages 1118–1123. Morgan Kaufmann Publishers, 1999.
19. J. Liu, V. A. Saletore, and T. G. Lewis. Safe self-scheduling: A parallel loop schedule scheme for shared-memory multiprocessors. *International Journal of Parallel Programming*, 22:589–616, 1994.
20. P. A. O'Donnell and T. Lozano-Pérez. Deadlock-free and collision-free coordination of two robot manipulators. In *Proceedings 1989 IEEE International Conference on Robotics and Automation*, pages 484–489, May 1989.
21. B. J. Oommen, S. S. Iyengar, N. S. V. Rao, and R.L Kashyap. Robot navigation in unknown terrains using learned visibility graphs. Part I: The disjoint convex obstacle case. *IEEE Journal of Robotics and Automation*, RA-3(6):672 – 681, 1987.
22. D. Parsons and J. Canny. A motion planner for multiple mobile robots. In *Proceedings 1990 IEEE International Conference on Robotics and Automation*, pages 8–13, May 1990.
23. C. Polychronopoulous and D. J. Kuck. Guided self-scheduling: A practical scheduling scheme for parallel supercomputers. *IEEE Transactions on Computers*, 36(12):1425–1439, 1987.
24. N. S. V. Rao, S. S. Iyengar, B. J. Oommen, and R. L. Kashyap. On terrain model acquisition by a point robot amidst polyhedral obs tacles. *International Journal of Robotics and Automation*, 4(4):450–455, 1988.
25. N. S. V. Rao, V. Protopopescu, and N Manickam. Cooperative terrain model acquisition by a team of two or three point-robots. In *Proceedings 1996 International Conference on Robotics and Automation*, volume 2, pages 1427–1433, 1996.

26. T. C. Shermer. A linear time algorithm for bisecting a polygon. *Information Processing Letters*, 41:135–140, 1992.

27. T. Siméon, S. Leroy, and J. P. Laumond. A collision checker for car-like robots coordination. In *Proceedings of the IEEE International Conference on Robotics and Automation (ICRA-98)*, pages 46–51, May 16–20 1998.

28. J. M. Smith, C. Y. Choo, and N. M. Nasrabadi. Terrain acquisition algorithm for an autonomous mobile robot with finite-range sensors. In *Proceedings of SPIE Applications of Artificial Intelligence IX*, volume 1468, pages 493–501, Orlando, FL, 1991.

29. I. Stojmenović. Bisections and ham-sandwich cuts of convex polygons and poly-hedra. *Information Processing Letters*, 38:15–21, 1991.

30. S. B. Tor and A. E. Middleditch. Convex decomposition of simple polygons. *ACM Transactions on Graphics*, 3:244–265, 1984.

31. C. W. Warren. Multiple robot path coordination using artificial potential fields. In *Proceedings 1990 International Conference on Robotics and Automation*, pages 500–505, May 1990.

Modelling, Control and Perception for an Autonomous Robotic Airship

Alberto Elfes[1,2], Samuel S. Bueno[1], Josué J.G. Ramos[1], Ely C. de Paiva[1],
Marcel Bergerman[1], José R.H. Carvalho[1], Silvio M. Maeta[1],
Luiz G.B. Mirisola[1], Bruno G. Faria[1], and José R. Azinheira[3]

[1] Information Technology Institute, Campinas, SP 13081, Brazil
`{ssbueno,josue,ely,mbergerman,regis,maeta,mirisola,faria}@iti.br`
[2] Research Institute for Applied Knowledge Processing – FAW, 89073 Ulm, Germany
`elfes@faw.uni-ulm.de`
[3] Instituto Superior Técnico, IDMEC/IST, 1049-001 Lisbon, Portugal
`jraz@dem.ist.utl.pt`

Abstract. Robotic unmanned aerial vehicles have an enormous potential as observation and data-gathering platforms for a wide variety of applications. These applications include environmental and biodiversity research and monitoring, urban planning and traffic control, inspection of man-made structures, mineral and archaeological prospecting, surveillance and law enforcement, communications, and many others. Robotic airships, in particular, are of great interest as observation platforms, due to their potential for extended mission times, low platform vibration characteristics, and hovering capability. In this paper we provide an overview of Project AURORA (Autonomous Unmanned Remote Monitoring Robotic Airship), a research effort that focusses on the development of the technologies required for substantially autonomous robotic airships. We discuss airship modelling and control, autonomous navigation, and sensor-based flight control. We also present the hardware and software architectures developed for the airship. Additionally, we discuss our current research in airborne perception and monitoring, including mission-specific target acquisition, discrimination and identification tasks. The paper also presents experimental results from our work.

1 Introduction

Unmanned aerial vehicles (UAVs) have a very wide spectrum of potential applications. In addition to their use as military intelligence gathering and surveillance platforms, UAVs have enormous potential in civilian and scientific applications. Civilian applications include traffic monitoring and urban planning, inspection of large-scale man-made structures (such as power transmission lines, pipelines, roads and dams), agricultural and livestock studies, crop yield prediction, land use surveys, planning of harvesting, logging and fishing operations, law enforcement, humanitarian demining efforts, disaster relief support, and telecommunications relay, among many others. Scientific applications cover areas such as mineral and archaeological site prospecting, satellite mimicry for ground

G.D. Hager et al. (Eds.): Sensor Based Intelligent Robots, LNCS 2238, pp. 216–244, 2002.

truth/remote sensor calibration, and environmental, biodiversity, and climate research and monitoring studies. These in turn include sensing and monitoring of forests, national parks and endangered ecological sites, limnological studies, and characterization of biodiversity spectra, as well as monitoring of air composition and pollution above cities and industrial sites, identification of sources of atmospheric and limnological pollutants, and monitoring of long-term climate change indicators.

Today, some of these applications are pursued based on data and imagery obtained from remote sensing satellites, balloon-borne sensors, or manned aircraft-based and field surveys. Each one of these sources of information has significant drawbacks, however. Balloons are not maneuverable and consequently the area to be surveyed cannot be precisely controlled. Satellite imagery available for civilian applications is limited in terms of the spatial (pixel) resolution and the spectral bands available, as well as in terms of the geographical and temporal swaths provided by the satellite (Figs. 1, 2). Manned aerophotogrammetric or aerial inspection surveys, while giving the mission planner control over many of the variables mentioned above, are very costly in terms of aircraft deployment, crew time, maintenance time, etc., and their extensive use is therefore beyond the financial scope of many governments and international agencies.

We see robotic unmanned aerial vehicles — UAVs with significant levels of autonomy, able to plan and execute extended missions with a significant degree of autonomy — as the answer to many of the scientific and civilian data acquisition needs outlined above. We believe that the development of unmanned, substantially autonomous robotic aerial vehicles will ultimately allow the airborne acquisition of information in such a manner as to give the user the ability to choose the spatial and time resolution of the data to be acquired, to define the appropriate geographical coverage, and to select the sensor systems of relevance for a specific data-gathering mission, while doing so at more readily affordable costs. This will lead directly to an expansion of scientific and civilian uses of aerial data and to significant social and economic benefits deriving from this expansion.

The advances that occurred over the last decade in the areas of sensors, sensor interpretation, and control and navigation systems have supported the increasing use of unmanned semi-autonomous ground and underwater robotic vehicles in a variety of applications. However, comparatively less progress has been made in the development and deployment of substantially autonomous robotic aerial vehicles. As mentioned earlier, UAVs play an important role in military reconnaissance and surveillance missions [26,5]; additionally, agencies such as NASA are developing airborne systems as platforms for environmental and climate research [27]. Many of these systems are flown using a combination of remote control during critical phases of the mission and onboard navigation systems for flight path execution. As far as is known to the authors, relatively little work has been done towards the development of substantially autonomous onboard analysis and control capabilities, such dynamic replanning of flight patterns in response to the sensor imagery being obtained.

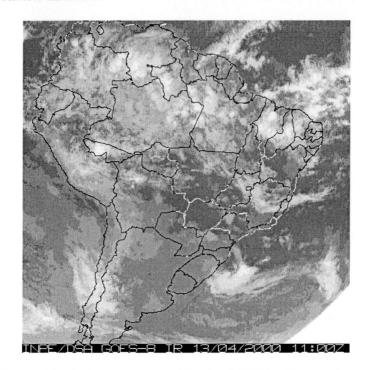

Fig. 1. Image of South America captured by the GOES Satellite, made available by the National Space Research Institute (INPE), Brazil. To provide an idea of the scale of the data gathering problem for environmental research and monitoring applications, it suffices to mention that the country of Brazil covers an area of 8.5 million km^2, while the Amazon basin, which lies mostly in Brazil but extends beyond its borders, has a size of approximately 6.5 million km^2.

In the research discussed in this paper, our main goal is the development of the underlying technologies for substantially autonomous airborne vehicle operation. This includes the ability to perform mission, navigation, and sensor deployment planning and execution, failure diagnosis and recovery, and adaptive replanning of mission tasks based on real-time evaluation of sensor information and constraints on the airborne system and its surroundings. At the same time, our work also involves development of prototype UAVs for testing the technologies being developed. Our current driving applications involve environmental, biodiversity, and climate research and monitoring [17], for which we have chosen Lighter-Than-Air (LTA) vehicles as the airborne vehicle technology of choice.

In this paper we provide a status report on Project AURORA (Autonomous Unmanned Remote Monitoring Robotic Airship). We discuss airship modelling and control, autonomous navigation, and sensor-based flight control. We also present the hardware and software architectures developed for the airship. Additionally, we discuss our current research in airborne perception and monitoring, including mission-specific target acquisition, discrimination and identification tasks. The paper includes experimental results from our work.

Fig. 2. Forest clearing through uncontrolled fires. Identification and control of illegal forest burning activities is largely beyond the resolution of currently available civilian satellite imagery.

2   Airships as Observation Platforms

Most of the environmental monitoring studies mentioned in Section 1 have profiles that require low speed, low-altitude airborne data gathering platforms. These vehicles should also be able to hover above an area and have extended airborne capabilities for long duration studies; generate low noise and turbulence, so as not to disturb the environment that is being measured and monitored; generate very low vibration, reducing sensor noise and hardware malfunction; be able to take off and land vertically, so that maintenance and refueling can be done without the need for runways, and also so that remote, difficult to access regions with limited logistics support can be monitored; be highly maneuverable; have a large payload to weight ratio; and have a low operational cost.

Of the four basic possible aerial vehicles types (airplanes, helicopters, airships, and balloons) the last ones are not considered here because they are not maneuverable. Table 1 compares the three remaining vehicles types with respect to the requirements set above, where high compliance with each requirement is indicated by three marks ($+++$), and low or no compliance by one mark ($+$).

As can be inferred from Table 1, airships are, on most accounts, better suited to environmental monitoring tasks than airplanes or helicopters. Fundamentally, this is due to two reasons. Firstly, airships derive the largest part of their lift from aerostatic, rather than aerodynamic, forces. Therefore, an airship is not required to spend significant amounts of energy to float in the air, but only to move between locations or to counteract the drift caused by wind. Consequently, airships need smaller engines than airplanes and helicopters for propulsion, which in turn produce less noise, vibration, and turbulence, and consume less fuel. Secondly, airships have a dynamic behaviour that is intrinsically of higher stability than

Table 1. Comparison of aerial vehicle technologies for airborne sensing and monitoring applications.

Requirement	Airplane	Helicopter	Airship
Low speed, low altitude flight	+	+++	+++
Hovering capability	+	+++	+++
Long endurance	++	+	+++
Vertical take-off/landing	+	+++	+++
Good maneuverability	++	+++	+
Payload to weight ratio	++	+	+++
Safe operation	++	+	+++
Low noise and turbulence	+	+	+++
Low vibration	++	+	+++
Low operational cost	++	+	+++
Simplicity of operation	++	+	+++

other airborne vehicles, making them ideally suited as low-vibration observation platforms.

It should be noted that, among UAV aircraft used today, by far the most commonly employed are reduced-scale fixed-wing vehicles (airplanes), followed by rotary-wing (helicopter) aircraft. Airships are only recently becoming a focus of interest in the UAV world [5], although their advantages are recognized in other areas [25, 28].

3   Project AURORA

Project AURORA (Autonomous Unmanned Remote Monitoring Robotic Airship) focusses on the development of the technologies required for substantially autonomous airborne vehicle operation. As part of this effort, we also build prototype UAVs, both for testing the technologies being developed and for exploring the use of UAVs in novel applications. As mentioned earlier, our current driving applications are largely related to environmental, biodiversity, and climate research and monitoring. The work presented here builds on previous research done by the authors on remotely piloted helicopters [32] and on autonomous underwater vehicles [9].

The project, conditioned on funding, aims at developing a sequence of prototypes capable of successively higher mission times and ranges, with increasing levels of autonomy, evolving from mainly teleoperated to substantially autonomous systems. The main tasks involved in autonomous flight are summarized for the various flight phases in Fig. 3. Details on various parts of the project can be found in [16, 7, 30, 2, 15].

3.1   The AURORA I Vehicle

For the current phase of the project, a first prototype, AURORA I, has been built and is used to test the autonomy technologies being developed (Fig. 4).

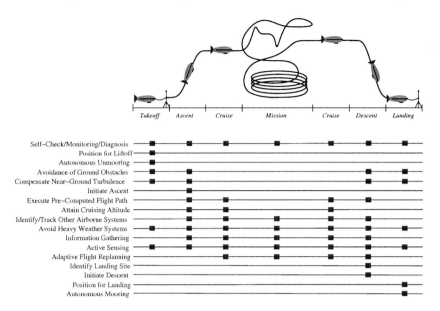

Fig. 3. Flight phases for a substantially autonomous UAV. The table shows the major tasks to be addressed at each flight phase.

The major physical subsystems of AURORA I include: the airship; the onboard control and navigation subsystems, including the vehicle sensors, hardware, and software; the communications subsystem; the mission sensors; and a base station (Fig. 5). By vehicle sensors we understand those atmospheric, inertial, positioning, and imaging sensors required by the vehicle to accomplish its autonomous navigation tasks. Mission sensors are those selected for specific aerial data-gathering needs, and are not discussed in detail here. The other subsystems listed are described in the sequence.

3.2 The Airship

AURORA I is conceived as a proof-of-concept system, to be used in low-speed, low-altitude applications. The LTA platform is the AS800 by Airspeed Airships [36, 1]. The vehicle is a non-rigid airship (blimp). It has a length of 9 m, a diameter of 2.25 m, and a volume of 24 m^3 (Fig. 4). It is equipped with two vectorable engines on the sides of the sensor and communications pod, and has four control surfaces at the stern, arranged in an "$\times$" configuration. The payload capacity of the airship is 10 kg at sea level, and its maximum speed is 50 km/h. Onboard sensors include DGPS, INS and relative wind speed systems for flight control, and video cameras for navigation and monitoring.

The onboard control and navigation subsystems are responsible for sensor data acquisition and actuator control, based on flight profiles uploaded from the ground. The airship hardware consists of an onboard computer, microprocessors,

Fig. 4. The AURORA I Robotic Airship. The airship, shown moored on the left and in flight on the right, is 9 m long, has a diameter of 2.25 m, and a volume of 24 m^3.

internal sensors, and actuators. The software consists of a 3-layer architecture. The communications subsystem is composed of radio links which transmit data and commands between the airship and the base station. This system includes also video links for the transmission of imagery captured by cameras mounted onboard. The mobile base station is composed of a processing and communications infrastructure, a portable mooring mast, and a ground vehicle for equipment transportation.

While in the sequence we mention specific commercial hardware products being used, we caution the reader that we frequently reevaluate these systems and replace them by others with better performance as these become commercially available.

3.3   Hardware

The internal sensor suite used for flight path execution purposes includes both inertial navigation sensors (compass, accelerometers, inclinometers, and gyroscopes) and a GPS (global positioning system) receiver. Cameras mounted on the airship's gondola have the purpose of providing aerial images to the operator on the ground, and serving for visual control and navigation based on geographical features of the terrain.

Fig. 5 presents the hardware architecture that integrates the airship's onboard equipment. The onboard subsystems include a CPU, sensors, actuators, and a communications subsystem. Most of the components are mounted inside the airship's gondola (Fig. 6). Control and navigation sensors (including compass, inertial measurement unit, airspeed sensor, barometric altimeter, GPS receiver and engine speed sensor), as well as vehicle state and diagnostic sensors

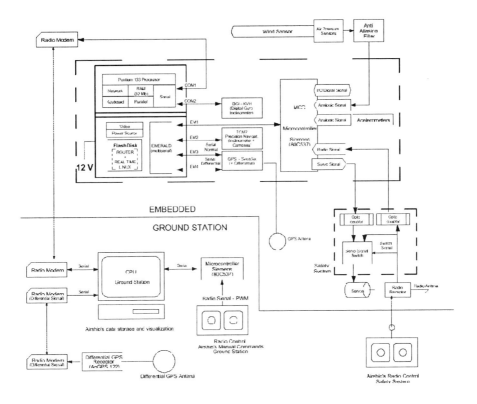

Fig. 5. Hardware architecture of the AURORA system.

Fig. 6. Onboard CPU and sensors.

(including control surface and vectorization position sensors, engine temperature, and fuel and battery levels) are connected to a PC 104 computer via serial ports or a CAN bus. All actuators (including engines, thruster vectorization and control surfaces) are connected to the PC 104 via a microcontroller.

The CPU is a PC104, a small size, low energy consumption computer widely utilized in embedded and industrial applications. In our case it includes: a MOPS lcd5 board with a Pentium 133 MHz processor, 32 MB RAM, two serial ports, one parallel interface, a 10base-T Ethernet network interface, and keyboard and IDE adapters; an Emerald MM multiserial board with four additional serial ports (yielding a total of six serial ports); a V104 power supply fed by a 12 V battery; and a VGA PCM-3510 video adapter to connect the CPU to a monitor for system development and testing.

3.4 Sensors

The followings sensor subsystems are currently used on the airship:

- *GPS with Differential Correction:* we utilize a Trimble Navigation SveeSix GPS receiver composed of a PC104-compatible board with digital output and an antenna mounted on the upper side of the airship. Another GPS receiver, located at the ground station, sends correction data to the onboard GPS.
- *Inertial Measurement Unit:* we use a Crossbow DMU-AHRS Dynamic Measurement Unit, which is composed of three magnetometers, three accelerometers and three gyroscopes, and provides roll, pitch, and yaw (heading) rates and attitude, thereby serving as an inclinometer and a compass as well.
- *Wind sensor:* a wind sensor built for Project AURORA by the IDMEC/IST [3] measures the relative airship air speed in all three axes as well as the barometric altitude.

The current control and navigation system uses the data from the above sensors directly. A data fusion module based on an Extended Kalman Filter is presently under development, and will be used to provide more accurate estimates of airship state variables.

Except for the onboard GPS receiver, which connects directly to the PC104 bus, all other sensors are connected to the CPU via serial ports and a SAB 80C517 8-bit CMOS microcontroller. The microcontroller also receives PWM signals from a radio control unit and generates PWM signals to the actuators. It communicates with the CPU via a RS232 interface.

3.5 Ground Station

The ground station is composed of a microcomputer, a differential GPS receiver, and a microcontroller board connected to a remote control unit (RCU) (Fig. 7).

The ground station computer is a Pentium II 300 MHz system with multiple serial ports to which the radio modem interface to the onboard system, the differential GPS, and the RCU are connected. The differential GPS system, a Trimble AgGPS 122, is positioned close to the ground control station and with an unobstructed view of the sky, and sends correction data to the onboard GPS receiver via the communication system.

Fig. 7. Ground station with PC, differential GPS and teleoperation unit.

A teleoperation control unit at the ground station allows us to manually control the airship while testing the entire hardware and software components of our system. A custom designed microcontroller board converts the PWM pulses of the RCU into digital data and sends them to the ground computer. From there, the data is sent to the onboard system via the communication architecture described in the sequel.

For safety purposes we developed a backup command system which allows the ground operator to take over control of the airship in case of a software or hardware failure. To this end we utilize one of the channels of the RCU. The backup safety system electrically decouples the onboard CPU from the RCU through optocouplers, which prevents electric disturbances at the CPU from spreading to the radio control circuit. Furthermore, the CPU/sensors subsystem and the RCU/actuators subsystem are connected to different power sources, which allows us to control the airship even in the event of electrical failure.

3.6 Telemetry System

Communication between the ground station and the airship occurs over two radio links. The first one operates in analog mode to transmit video imagery from the airship to the ground station. The second one operates in digital mode to transmit sensor telemetry and command data between the ground station and the onboard processor.

The video system is composed of a color CCD camera, a transmission antenna mounted onboard, and a Pulnix flat plate reception antenna mounted on the ground station.

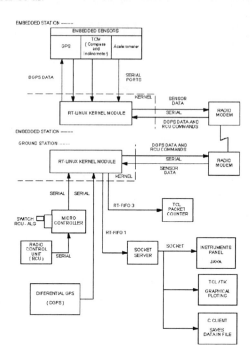

Fig. 8. Low-level software architecture utilized in AURORA.

The data transmission system is composed of a pair of spread spectrum radio modems, with a direct line-of-sight data transmission range of 30 km. An error detection scheme utilizing CRC and packet retransmission mechanisms insures data integrity.

3.7   Software Architecture

The software architecture consists of a 3-layer structure, combined with a high-level data flow programming method and system development environment.

As the underlying operating system for the AURORA project we have chosen to use RT-Linux. It is known to be reliable and robust, can be used under real-time requirements, and requires relatively little memory and disk space. As airships have relatively large time constants, this allows us to run the individual control and navigation modules as processes under RT-Linux (Fig. 8).

Processes running on the onboard system read and send sensor data to the ground station and execute the autonomous flight control strategies, sending commands to the actuators. These tasks are executed at a 100ms rate. This sampling interval was selected based on simulated flights and corroborated as sufficient in actual experimental flights.

The ground station sends commands and mission paths to the onboard control system. It also receives sensor data from the airship, displaying them in real time during simulated and actual flights. The ground system records all data re-

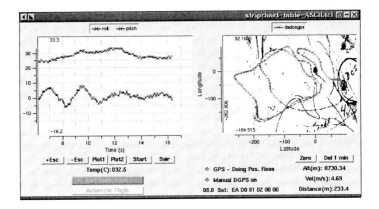

Fig. 9. Real-time flight data visualization. Hand sketches can be overlaid as well.

ceived from the airship for post-flight analysis and visualization. For pilot training and airship manual control we utilize the remote control unit, connected to one of the serial ports of the ground station through a microcontroller.

The complexity of the system being developed requires a deliberative-reactive intermediate-level process control and communications architecture, where different subsystems can run independently and as separate threads, while able to exchange information and activate or inhibit each other. To facilitate the management of the different processes in a distributed environment, our architecture is being built on top of the Task Control Architecture (TCA) [35], a set of primitives that helps structure process scheduling and interprocess communication.

As the higher-level, overarching control structure we are considering a layered, multi-rate approach similar to the ATLAS architecture [9, 13], which allows nested control cycles to be run at progressively lower speeds, but at increasing levels of competency.

3.8 Human-Machine Interface

The human-machine interface (HMI) that runs on the ground station provides the communication and visualization mechanism between the operator and the navigation system onboard the airship. In AURORA I this interface is used by the operator to specify a flight plan, composed of sets of locations, altitudes, and flight primitives (takeoff, cruise, hover, land). During flight, onboard sensor data and imagery received from the airship are displayed in real time to the operator.

Telemetry data visualization, particularly of GPS and inertial sensor data, both for simulated and actual flights, is done using a Tcl/Tk script (Fig. 9). All data received by the ground station is recorded for post-flight analysis, and the HMI is able to perform flight playback for mission evaluation purposes.

As part of the HMI we have developed a physical model-based virtual reality airship simulator[31]. The simulator is based on a very accurate dynamic model of the airship, outlined in Section 4.1, and incorporates real-world topograph-

Fig. 10. Airship simulation and visualization during a flight.

ical information of selected regions. The simulator is used to validate control strategies and navigation methods, for pilot training, and for mission planning and pre-evaluation. Since visualization of the airship flight and of the sensor data occurs within a model of a standard airplane-type cockpit (Fig. 10), the operator can easily follow the progress of the current mission, making decisions such as uploading new mission plans to the airship or aborting the mission in the case of unforeseen events. Additionally, since the simulator is built using VRML/Java, it is possible to "fly" the airship and perform experiments with the airship control system over the Internet, for all phases of flight, including take-off and landing.

In future work we plan to enhance the HMI, interfacing it to a geographical information system (GIS). This will allow the operator to define mission profiles directly on a geographical chart. The navigation system will then plan the airship's flight path and profile.

4   Airship Control

4.1   Dynamic Modeling and Control System

As the basis for the development of the control and navigation strategies, we have developed a 6-DOF physical model of the airship that includes the nonlinear flight dynamics of the system. We briefly review the model here, while a more detailed presentation is found in [21].

The dynamic model assumes that motion is referenced to a system of orthogonal body axes fixed in the airship, with the origin at the Center of Volume (CV), assumed to coincide with the gross Center of Buoyancy (CB) (Fig. 11).

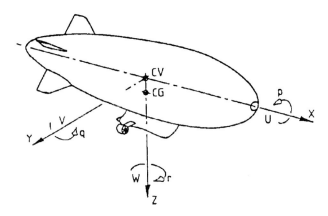

Fig. 11. Local reference frame for the airship.

The orientation of this body-fixed frame (X, Y, Z) with respect to an Earth-fixed frame (X_E, Y_E, Z_E) is obtained through the Euler angles (Φ, Θ, Ψ). The airship linear and angular velocities are given by (U, V, W) and (P, Q, R), respectively. Angular velocities (P, Q, R) are also referred to as the roll, pitch and yaw rates.

While developing an accurate mathematical model of airship flight dynamics, the following aspects were taken into account:

- The airship displaces a large volume of air and consequently its virtual mass and inertia properties are substantial when compared with those associated with the vehicle itself. The model incorporates the mass and inertia of the vehicle, the mass and inertia of the buoyancy air (the air displaced by the total volume of the airship), and the virtual mass and inertia of the air surrounding the airship that is displaced by the relative motion of the airship in the atmosphere.
- The airship mass may change in flight due to ballonet deflation or inflation. In our model we have assumed that the airship mass varies slowly, and the associated time derivatives are zero.
- To accommodate the constant change of the position of the Center of Gravity (CG), the airship motion has to be referenced to a system of orthogonal axes fixed relative to the vehicle with the origin at the Center of Volume (CV) (Fig. 11).
- The airship is assumed to be a rigid body, so that aeroelastic effects are ignored.
- The airframe is assumed to be symmetric about its vertical (XZ) plane, so that both the CV and the CG lie in the symmetry plane.

Taking into account the considerations given above, we have developed a dynamic model that is expressed as:

$$M\frac{dx_A}{dt} = F_d(x_A) + F_a(x_A) + P + G \tag{1}$$

where M is the 6×6 mass matrix and includes both the actual inertia of the airship as well as the virtual inertia elements associated with the dynamics of buoyant vehicles; $x_A = [U, V, W, P, Q, R]$ is the vector of airship state variables; F_d is the 6×1 dynamics vector containing the Coriolis and centrifugal terms; F_a is the 6×1 vector of aerodynamic forces and moments; P is the 6×1 vector of propulsion forces and moments, and G is the 6×1 gravity vector, which is a function of the difference between the weight and buoyancy forces.

The aerodynamic model we developed is based on the seminal work presented in [20], and takes advantage of information from a wind tunnel database built to model the Westinghouse YEZ-2A airship [20]. The adaptation was possible due to the same length/diameter ratio (4:1) of both airships. The aerodynamic coefficients available in the database are a function of the aerodynamic incidence angles (angle of attack and sideslip angle varying in the range of $[-25°, +25°]$), and of the three deflections of the tail surfaces (elevator, rudder and aileron, also varying in the range of $[-25°, +25°]$).

Using this 6-DOF non-linear model, a SIMULINK-based control system development environment was built to allow the design and validation of flight control and trajectory following strategies [29].

As control actuators, the AS800 has deflection surfaces and thrusters. The "×"-arranged deflection surfaces generate the equivalent rudder and elevator commands of the classical "+" tail, with allowable deflections in the range of $\pm 25°$. The engines can be vectorized from $0°$ to $+90°$ up. The rudder and elevator are responsible for directional control (left and right, up and down) and their effect is due to aerodynamic forces produced at medium to high speeds. The thrusters are responsible for generating the necessary forces for controlled airship motion. Their vectorization is used for vertical load compensation and for control of vertical displacements at low speeds. It is important to note that for large airships operating under the critical airspeed of 5 m/s, the control surfaces are considered to be ineffective [21, 20]. For the AS800, control is still possible at low airspeeds due to the flow generated by the main thrusters, which increases the efficiency of the deflection surfaces. The airship also has a stern thruster used mainly during hover.

Based on the model outlined above, we have analyzed the airship's motion and found three interesting control challenges: non-minimum phase behavior and oscillatory modes at low speeds, time-varying behavior due to altitude variations and fuel burning, and variable efficiency of the actuators depending on airship speed. All of these issues were taken into account in the design of the control system.

The airship control system is designed as a 3-layer control system [29]. At the bottom level, the actuators described above provide the means for maneuvering the airship along its course. At the intermediate level, two main control algorithms with different structures are available to command the actuators based on the mission profile. These two control algorithms implement longitudinal and lateral control. The longitudinal control algorithm, currently under development, is based on a feedforward/feedback structure, and controls the propulsion, vectoring and elevator deflection for take-off, cruise, hover, and landing maneuvering operations. The lateral control algorithm, discussed in more detail below, controls rudder deflection and the tail thruster for turning maneuvers. Finally, the top level of the control architecture is implemented as a supervisory layer that is responsible for failure detection and mission replanning in the presence of unexpected events.

4.2   Path Tracking

An important airborne vehicle autonomy problem is following a pre-computed flight path, defined by a set of points given by their coordinates (latitude and longitude), with given speed and altitude profiles. In this section we introduce the trajectory following problem and outline the approaches investigated in AURORA, as well as the trajectory error metrics.

We posit trajectory following as an optimal control problem, where we compute a command input that minimizes the path tracking error for a given flight path. Allowable flight paths are defined as sequences of straight line segments joining waypoints. The heading change at each waypoint (between consecutive segments) may vary in the $\pm 180°$ range, and the distance between the actual airship position and the precomputed trajectory path is to be continuously minimized. The longitudinal motion is maintained at a constant altitude and airspeed, and will be considered decoupled from the lateral motion; this is a common assumption in aerial vehicle control [22].

We start by presenting the airship lateral dynamics model. The dynamics of the airship in the horizontal plane is given by the fourth order linear state space system:

$$\dot{x} = Ax + Bu \tag{2}$$

where the state x includes the sideslip angle β, yaw rate R, roll rate P and yaw angle Ψ. The control input u is the rudder deflection ζ.

The path tracking error metric is defined in terms of the distance error δ to the desired path, the angular error ϵ, and the ground speed V (Fig 12). Assuming the airship ground speed to be a constant and the angular error ϵ to be small, the following linearized path tracking model results:

$$\begin{cases} \delta = V \sin \epsilon = V_0 \, \epsilon \\ \epsilon = R \end{cases} \tag{3}$$

where V_0 is the reference ground speed considered for design purposes.

In order to accommodate both the distance and angular errors in a single equation, a look-ahead error δ_a may be estimated some time ahead of the actual position:

$$\delta_a \approx \delta + V_0 \, \Delta t \, \epsilon \tag{4}$$

This strategy has already been successfully used for the guidance of both unmanned aircraft [24] and ground mobile robots [6].

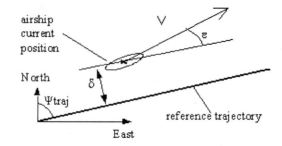

Fig. 12. Definition of Path Tracking Error.

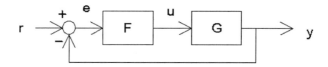

Fig. 13. H_∞ control block diagram.

It is important to note that the ground speed V will be kept constant only in the absence of wind (as the longitudinal controller maintains a constant airspeed). Therefore, the tracking controllers should present robustness properties in order to assure good tracking under wind incidence. This is the case for the control strategies presented below, whose robustness was verified through simulated examples and actual flight experiments.

H_∞ Control Approach

The first approach we developed for the nonlinear control problem presented above uses the linearized path tracking error dynamics (Eq. 3) around a trim condition for a constant fixed airspeed. A robust H_∞ design is used next to assure that a previously defined operation envelope is covered, taking into account the simplified model and unmodeled uncertainties.

The control problem may then be expressed as the search for a regulator F of the output y, where the error e is expressed in terms of the look-ahead track error δ_a (Fig. 13). The system G to be controlled is obtained as the series of the lateral airship model (Eq. 2) and the linearized path tracking model (Eq. 3). The lateral dynamic model of the airship used in the design is an approximation of Eq. 2, where only the second order horizontal motion (with states β and R) is considered. Here the control signal is fed directly to the rudder without the need for any lower level controller.

For the correct shaping of the controller, a mixed sensitivity H_∞ technique is used [8], allowing us to specify the characteristics of the closed loop in the frequency domain. Three weighting functions (W_1, W_2, W_3) are used, respectively, for the sensitivity function S_1 (performance), for the actuation sensitivity function S_2 and for the complementary sensitivity function S_3 (stability robustness)

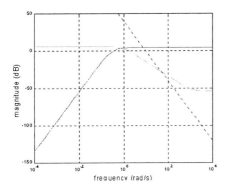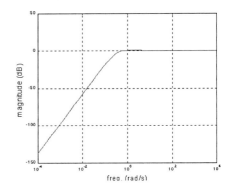

Fig. 14. The diagram on the left shows the inverse of the weighting functions for performance sensitivity W_1^{-1} (solid line), actuation sensitivity W_2^{-1} (dashed line), and complementary sensitivity W_3^{-1} (dotted line). The diagram on the right shows the closed-loop performance sensitivity of the H_∞ control system.

(Fig. 14). The look-ahead distance used for the controller design is chosen with a reference speed of 10 m/s and time of 2.5 s, so that $e = \delta_a \approx \delta + 25\epsilon$.

The sensitivity weighting function W_1 is chosen to permit good tracking at low frequencies. The complementary sensitivity weighting function W_3 is used for high frequency noise rejection. Finally, the actuation sensitivity weighting W_2 reduces the use of actuators for higher frequencies. The closed-loop behavior of the linearized system can be expressed in terms of the performance sensitivity function (Fig. 14), which shows excellent agreement with the frequency shaping: good command following at low frequencies (low error), and disturbance rejection at higher frequencies.

The controller is designed in continuous time, reduced to a fifth order controller and finally put in discrete form with a 10 Hz sampling rate. A detailed discussion of this approach is found in [29, 30, 34, 4].

PI Control Approach

The second approach investigated is based on a classical PI approach, with a heading control inner loop and a path-tracking outer loop (Fig 15).

The heading controller is based on a proportional-derivative (PD) controller. The path-tracking controller is a proportional-integral (PI) controller whose output, added to the trajectory heading angle Ψ_{traj}, yields the reference signal Ψ_{ref} for the heading controller. The PI controller input is the look-ahead path tracking error δ_a, given in equation Eq. 4. The PI controller uses the tracking error δ_a to correct the reference signal for the heading controller, with the necessary correction forcing the tracking error to decrease. It is interesting to note from this control structure that, in the absence of wind, the reference heading angle will be the same heading angle Ψ_{ref} of the trajectory. In the presence of wind, however, to compensate for the lateral forces, the airship heading angle will assume the value necessary to minimize the look-ahead tracking error δ_a, and the

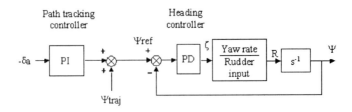

Fig. 15. PI control block diagram.

airship will consequently fly "sideways" under such conditions. Finally, we mention that the proportional and integral gains of the controller are obtained by trial and error, and that the PI controller uses an anti-wind up strategy to avoid saturation of the integral term.

The lateral dynamic model used is the fourth order transfer function from the yaw angle Ψ to the rudder deflection ζ, obtained from Eq. 2. This single controller presents good robustness properties for different dynamic models, covering a wide range of airspeeds and heavinesses (difference between airship weight and buoyancy forces). Again, details are found in [29, 30, 34, 4].

4.3 Case Study

To test the tracking controller developed, extensive simulations were made, covering airspeeds in the range from 5 to 15 m/s, with wind disturbances ranging from zero (no wind) to 70% of the vehicle airspeed. Both the PI and H_∞ controllers were used, in order to evaluate the path tracking methodology as well as the behavior of the two control schemes.

In all cases, the mission flight path chosen consists of 5 waypoints separated by approximately 200 m, with heading changes between 26° and 128°. These distances and angles pose difficult maneuverability conditions for the AS800 airship. The results of the case studies are only summarized below; a detailed discussion is found in [29, 30, 34, 4], while a sample tracking result is shown in Fig. 16.

For flights without wind, a trimmed airspeed of 10 m/s was used, which corresponds to the reference condition used in the trajectory error generation. The IAE criterion for the distance error δ as well as the variance of the control signal were computed for the PI and H_∞ approaches, and no substantial difference between their performances was noticed. Significant deviations occured only at very sharp angles, which make the maneuvers more difficult (Fig. 16).

For flights with wind, a trimmed airspeed of 5 m/s was used, together with a wind interference of constant speed at 3 m/s, blowing from the east. The flight path chosen for the simulation tests allows us to verify the robustness of the proposed methods for different flight situations: initially, the airship has to fly with the wind at 90°, then against the wind (which means that the ground speed is smaller), and then with a tail wind (which means that the ground speed is higher).

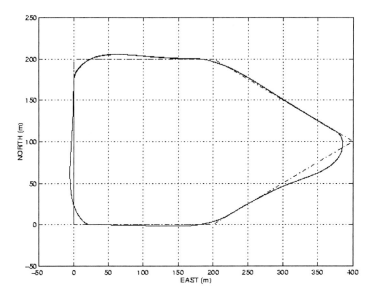

Fig. 16. Flight path reference and real trajectory with H_∞ controller (no wind).

The PI and H_∞ approaches follow the reference flight path well, with little influence from the presence of wind. In most situations, the H_∞ controller is more ambitious and makes the control signal saturate for longer periods of time. At higher airspeeds, this may cause the control actuator signal to oscillate slightly, corresponding to a higher controller gain. Despite this, the variances of the control signals are very close for the PI and H_∞ approaches. However, the H_∞ controller shows a remarkable advantage over the PI controller when the airship is subjected to step disturbances [4].

Overall, it can be said that the trajectory error minimization strategy leads to PI and H_∞ approaches exhibiting similar path tracking behavior and robustness characteristics, while the H_∞ controller presents superior performance under wind disturbances.

5   Autonomous Flight

Initial experimental validation of the modelling and control work presented above was done by testing the PI guidance control method, which was implemented on the onboard computer using the C language. The PI approach was chosen for the first test flight experiments due to the simplicity of the control structure and its implementation. The H_∞ approach has been implemented and will be undergoing testing shortly. Airship position and heading were obtained from DGPS and compass data.

Our first successful autonomous flight occurred on March 4th, 2000. The AURORA airship was flown over the CIACOM military field outside of the city

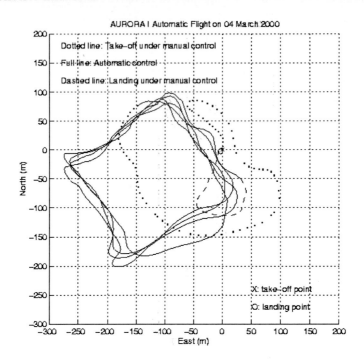

Fig. 17. Autonomous flight of the AURORA I airship, following a predefined mission trajectory.

of Campinas, Brazil (Figs. 17 and 18). In this flight, take-off and landing of the airship were done manually. The mission path, flown autonomously by the airship, was defined as a square with sides of 150 m length. Wind speed during the experiment stayed in the range of 0 to 10 km/h, blowing approximately from the northeast. Airship path following was controlled automatically by the onboard system, while altitude was controlled manually by the ground pilot [34, 33]. The look-ahead distance used for the controller design was chosen for a reference speed of 10 m/s and a look-ahead time of 2.5 s, leading to:

$$e = \delta_a = \delta + 25 \; \epsilon \tag{5}$$

The results of this experimental flight are presented in Figs. 17 and 18. In Fig. 17, the dotted line represents the airship motion under manual control from take-off until hand-over to autonomous control. The continuous line represents the airship motion under PI trajectory tracking control. Finally, the dashed line shows the motion of the airship after hand-back to manual control for the final landing approach. The plot clearly shows the adherence of the airship trajectory to the mission path, as well as overshoots due to the disturbing winds when the airship turns from southwest to northwest.

Fig. 18 presents one of the loops performed by the airship around the square. The dots represent the airship position and the lines represent its heading. As

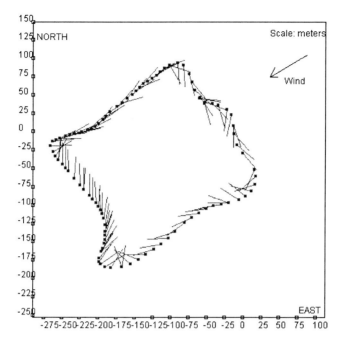

Fig. 18. AURORA I position and heading along a loop.

remarked before, the control method composed by the tracking and heading controllers automatically adjusts the airship heading to compensate for wind disturbances; for example, in the lower left part of the square loop, the airship navigates "sideways", while in the upper left it navigates mostly facing towards the trajectory.

We conclude this section by mentioning that we are currently testing the new version of our automatic control system for both airship altitude and attitude.

6   Perception

Extending the capabilities of a robotic aircraft so that it is able to dynamically respond to its state estimate and to the information obtained from its internal and mission sensors, thereby closing the high-level control cycle and allowing the robot to adjust its mission and flight plans accordingly, is one of the key challenges towards broader UAV deployment.

Sensor-based adaptive navigation of a robotic aircraft requires several perceptual competencies. Our work in perception-based navigation and control for the AURORA airship is still in an initial phase. It is currently focussed on two sets of issues: visual-based servoing for autonomous take-off, hovering, tracking, and landing purposes; and autonomous target recognition and tracking mechanisms for finding and identifying man-made structures (such as roads or pipelines),

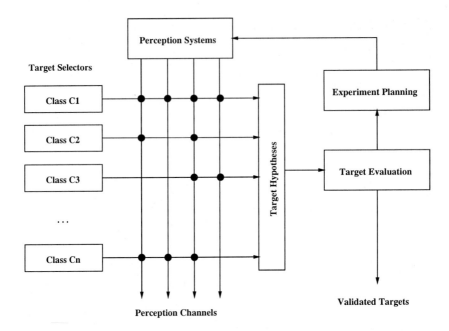

Fig. 19. System architecture for dynamic target identification: a target selection, validation and classification cycle is implemented.

geographical structures (such as rivers), air and water pollution sources, and biological targets of interest.

In this section, we report on preliminary work done towards the development of an approach to dynamic target recognition that is based on a cycle of hypothesis formulation, experiment planning for hypothesis validation, experiment execution, and hypothesis evaluation (Fig. 19). We also present some initial results obtained from adaptive aerial target identification and classification using visual imagery from airborne cameras [15, 14].

6.1 Adaptive Target Identification

As a representational framework, we encode sensor observations using stochastic visual lattice models [23] that draw on our previous work on the use of Markov Random Field (MRF) models [37] in robot perception and control [11, 10, 12].

For target identification and classification we use a classical hypothesis testing approach [19]. Consider a set of sensor observations, X. We want to classify the observation vector X into one of c classes, designated $\omega_1, \ldots, \omega_c$. For the two-class case, given X and the two classes ω_1 and ω_2, the use of Bayes' formula allows us to define the *discriminant function* $h(X)$ as:

$$h(X) = -\ln p_1[X|\omega_1] + \ln p_2[X|\omega_2] \underset{\omega_2}{\overset{\omega_1}{\gtrless}} \ln \frac{p_1[\omega_1]}{p_2[\omega_2]} \qquad (6)$$

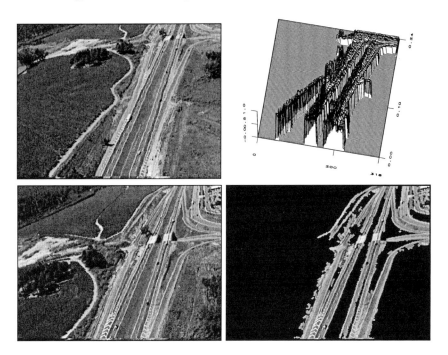

Fig. 20. Identification and tracking of a paved road using an airborne camera. The road classification probabilities for the upper left image are shown on the upper right, while for the lower left image the segmented image is shown on the lower right.

In the generalization to the c-class case, we assign the observation X to class k if the posterior distribution for k is the largest of all posterior distributions:

$$p_k[\omega_k|X] = \max_i p_i[\omega_i|X] \quad \Longrightarrow X \in \omega_k \qquad (7)$$

The Bayes classification error depends fundamentally on the conditional density functions $p_i[X|\omega_i]$. We affect the shape of these functions by explicitly controlling the position of the robot vehicle and its sensor parameters, thereby improving the classification error [12].

The architecture of the adaptive target recognition system we are developing is shown in Fig. 19. The target selectors, which determine what classes of targets are being sought, are switched on or off depending on the type of mission being executed. The selectors, in turn, are used to identify candidate target hypotheses for further evaluation. This may lead to an outright rejection or validation of the targets, or to controlled acquisition of additional imagery to increase the discriminatory capability of the system. Experimental results obtained from applying these concepts to aerial imagery are shown in section 6.3.

6.2 Optimal Design of Experiments

To control the acquisition of new data in an optimal way, we use an approach derived from the theory of optimal design of experiments [18] to discriminate

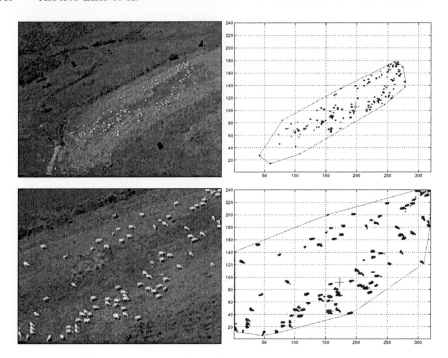

Fig. 21. Identification and tracking of Nelore cattle using an airborne camera. The aerial imagery is shown on the left, and the results of the target classification procedure on the right.

hypotheses based on the entropy measure. For c classes, we have a set of prior probabilities, $p_0[H_j], j = 1, \ldots, c$, that correspond to the hypotheses of the target belonging to the classes $\omega_1, \ldots, \omega_c$. Assuming that a new experiment $\mathcal{E}$ has been conducted in the form of a sensor observation, we obtain the posterior probabilities $p[H_j]$. We compute the information obtained from the observation using:

$$\Delta I(\mathcal{E}) = \eta(p[H]) - \eta(p_0[H]) = -\sum_{j=1}^{c} p_0[H_j] \ln p_0[H_j] + \sum_{j=1}^{c} p[H_j] \ln p[H_j] \quad (8)$$

For a finite-horizon problem and a finite set of sensing options (obtained from the tesselation of the representational space and a discretization of the sensor pose and parameter alternatives, see [12]), we can compute the expected value of ΔI with respect to the results of the observations. The sequence of observations (experiment) that maximizes the expected mean increment of information $E[\Delta I(\mathcal{E})]$ will be an optimal experiment.

6.3   Target Identification and Tracking Using Aerial Imagery

We now present some initial experimental results from the application of the concepts discussed above to autonomous identification, recognition and tracking of man-made and biological targets using aerial imagery. Fig. 20 shows results from paved road identification and tracking. Identification and segmentation of the roads in the images was done using probabilistic measures based on the spectral characteristics of the targets in the visible RGB bands. Atmospheric conditions and sensor limitations lead to a higher correct classification rate for road portions closer to the airborne camera, while some parts of the imagery that are further away from the airship are misclassified. As the airborne vehicle comes closer to the new target regions, however, the change in the distributions of the observations leads to a correct reclassification.

A second set of results shows a herd of Nelore cattle being identified from the air against pasture and forest backgrounds, and again being tracked over several image frames (Fig. 21). It is important to note that, as the airborne vehicle maneuvers to approach the herd, the correct classification rate rises to 100% in the last frame of the sequence. Identification and segmentation of the cattle were again done using its spectral characteristics in the visible RGB bands.

7   Conclusions

In this paper, we have argued that robotic unmanned aerial vehicles have an enormous potential as observation and data-gathering platforms for a wide variety of applications. We have presented an overview of Project AURORA, including the physically-based dynamic model developed for the system, the control approach used, and the hardware and software architectures. We also discussed our results in autonomous flight control, and preliminary work towards autonomous perception-based flight planning and execution. Our current work focusses on extending the models to a larger airship envelope, to extend the autonomous control algorithms, and to expand our perception-related research.

Acknowledgments

The work presented in this paper was partially supported by the Information Technology Institute (formerly called the Automation Institute) of the Brazilian Ministry of Science and Technology, by the National Council for Scientific and Technological Development (CNPq) under grant 68.0140/98-0, and by the Research Support Foundation of São Paulo (FAPESP) under grants 97/13384-7, 98/13563-1, 99/04631-6, 99/04645-7 and 00/01000-4. A. Elfes was supported during the year 2000 by a Mercator Professorship of the German Research Foundation (DFG), and by the Research Institute for Applied Knowledge Processing (FAW), Ulm, Germany. The views and conclusions contained in this document are those of the authors and should not be interpreted as representing the official policies, either expressed or implied, of the sponsoring organizations.

References

1. Airspeed Airships. *The AS800 Airship.* http://airship.demon.co.uk/airspeed.html, February 1998.
2. J. R. Azinheira, E. C. de Paiva, J. J. G. Ramos, and S. S. Bueno. Mission Path Following for an Autonomous Unmanned Airship. In *Proceedings of the 2000 IEEE International Conference on Robotics and Automation,* Detroit, MI, April 2000. IEEE.
3. J. R. Azinheira, H. V. Oliveira, and B. F. Rocha. Aerodynamic measurement system for the Project AURORA airship: Calibration Report and User Manual. Research report, Mechanical Engineering Institute, Instituto Superior Técnico, Lisbon, Portugal, 1999.
4. J. R. Azinheira, E. C. Paiva, J. J. G. Ramos, and S. S. Bueno. Mission Path Following for an Autonomous Unmanned Airship. In *Proceedings of the 2000 IEEE International Conference on Robotics and Automation,* San Francisco, USA, April 2000. IEEE.
5. J. H. Boschma. The development progress of the U.S. Army's SAA LITE Unmanned Robot Airship. In *Proceedings of the AIAA Lighter-Than-Air Systems Technology Conference.* AIAA, September 1993.
6. A. Botto, J. R. Azinheira, and J. S. Costa. A comparison between robust and predictive autonomous guidance controllers. In *Proceedings of the Eighteenth IASTED International Conference,* Innsbruck, Austria, February 1999. IASTED.
7. E. C. De Paiva, S. S. Bueno, S. B. V. Gomes, and M. Bergerman. A Control System Development Environment for the Aurora Semi-Autonomous Robotic Airship. In *Proceedings of the 1999 IEEE International Conference on Robotics and Automation,* Detroit, MI, May 1999. IEEE.
8. J. C. Doyle, B. Francis, and A. Tannenbaum. *Feedback Control Theory.* Macmillan Publishing Company, New York, 1992.
9. A. Elfes. *Autonomous Underwater Vehicle (AUV) Architecture.* IBM T. J. Watson Research Center Internal Report, 1990.
10. A. Elfes. Dynamic Control of Robot Perception Using Stochastic Spatial Models. In G. Schmidt, editor, *Information Processing in Mobile Robots,* Berlin, July 1991. Springer-Verlag.
11. A. Elfes. Multi-Source Spatial Fusion Using Bayesian Reasoning. In M. A. Abidi and R. C. Gonzalez, editors, *Data Fusion in Robotics and Machine Intelligence,* San Diego, CA, 1992. Academic Press.
12. A. Elfes. Robot Navigation: Integrating Perception, Environment Constraints and Task Execution Within a Probabilistic Framework. In L. Dorst, M. van Lambalgen, and F. Voorbraak, editors, *Reasoning With Uncertainty in Robotics,* volume 1093 of *Lecture Notes in Artificial Intelligence,* Berlin, Germany, 1996. Springer-Verlag.
13. A. Elfes. Incorporating Spatial Representations at Multiple Levels of Abstraction in a Replicated Multilayered Architecture for Robot Control. In R. C. Bolles, H. Bunke, and H. Noltemeier, editors, *Intelligent Robots: Sensing, Modelling, and Planning,* New York, 1997. World Scientific Publishers. Invited paper, 1996 International Dagstuhl Seminar on Intelligent Robots, Schloß Dagstuhl, Germany.
14. A. Elfes, M. Bergerman, and J. R. H. Carvalho. Dynamic Target Identification by an Aerial Robotic Vehicle. In G. Baratoff and H. Neumann, editors, *Dynamic Perception,* Ulm, Germany, September 2000. AKA, Berlin.

15. A. Elfes, M. Bergerman, and J. R. H. Carvalho. Towards Dynamic Target Identification Using Optimal Design of Experiments. In *Proceedings of the 2000 IEEE International Conference on Robotics and Automation*, San Francisco, CA, April 2000. IEEE.

16. A. Elfes, S. S. Bueno, M. Bergerman, and J. J. G. Ramos. A Semi-Autonomous Robotic Airship for Environmental Monitoring Missions. In *Proceedings of the 1998 IEEE International Conference on Robotics and Automation*, Leuven, Belgium, May 1998. IEEE.

17. A. Elfes, M. F. M. Campos, M. Bergerman, S. S. Bueno, and G. W. Podnar. A Robotic Unmanned Aerial Vehicle for Environmental Research and Monitoring. In *Proceedings of the First Scientific Conference on the Large Scale Biosphere-Atmosphere Experiment in Amazonia (LBA)*, Belém, Pará, Brazil, June 2000. LBA Central Office, CPTEC/INPE, Rod. Presidente Dutra, km 40, 12630-000 Cachoeira Paulista, SP, Brazil.

18. V. V. Fedorov. *Theory of Optimal Experiments*. Academic Press, New York, 1972.

19. K. Fukunaga. *Introduction to Statistical Pattern Recognition*. Academic Press, New York, 2nd edition edition, 1990.

20. S. B. V. Gomes. *An Investigation of the Flight Dynamics of Airships with Application to the YEZ-2A*. PhD thesis, College of Aeronautics, Cranfield University, 1990.

21. S. B. V. Gomes and J. J. G. Ramos. Airship Dynamic Modeling for Autonomous Operation. In *Proceedings of the 1998 IEEE International Conference on Robotics and Automation*, Leuven, Belgium, May 1998. IEEE.

22. B. G. Kaempf and K. H. Well. Attitude Control System for a Remotely-Controlled Airship. In *Proceedings of the 11th AIAA Lighter-Than-Air Systems Technology Conference*, Clearwater, USA, 1995. AIAA.

23. T. Kämpke and A. Elfes. Markov Sensing and Superresolution Images. In *Proceedings of the 10th INFORMS Applied Probability Conference (AP99)*, Ulm, Germany, July 1999.

24. P. Lourtie, J. R. Azinheira, J. P. Rente, and P. Felício. ARMOR Project — Autonomous Flight Capability. In *Proceedings of the AGARD FVP95 Specialist Meeting: Design and Operation of Unmanned Air Vehicles*, Turkey, October 1995. AGARD.

25. E. Mowforth. *An Introduction to the Airship*. The Airship Association Ltd., United Kingdom, 1991.

26. K. Munson. *Jane's Unmanned Aerial Vehicles and Targets*. Jane's Information Group Limited, Surrey, UK, 1996.

27. NASA. *Theseus Project*. http://www.hq.nasa.gov/office/mtpe/Theseus.html, October 1997.

28. O. J. Netherclift. *Airships Today and Tomorrow*. The Airship Association Ltd., United Kingdom, 1993.

29. E. C. Paiva, S. S. Bueno, S. B. V. Gomes, J. J. G. Ramos, and M. Bergerman. Control System Development Environment for AURORA's Semi-Autonomous Robotic Airship. In *Proceedings of the 1999 IEEE International Conference on Robotics and Automation*, Detroit, USA, May 1999. IEEE.

30. E. C. Paiva, A. Elfes, and M. Bergerman. Robust Control of an Unmanned Airship for Cooperative Robotic Applications. In *Proceedings of the 1999 International Workshop on Dynamic Problems in Mechanics and Mechatronics (EURODINAME '99)*, Schloß Reisensburg, Ulm, Germany, July 1999. Springer-Verlag.

31. J. J. G. Ramos, S. S. Maeta, M. Bergerman, S. S. Bueno, A. Bruciapaglia, and L. G. B. Mirisola. Development of a VRML/JAVA unmaned airship simulation environment. In *Proceedings of the 1999 International Conference on Intelligent Robots and Systems (IROS'99)*, Kyongju, South Korea, October 1999. IEEE/RSJ.

32. J. J. G. Ramos and O. Neves. Environment for unmanned helicopter control system development: application examples. In *Proceedings of the 4th IEEE Conference on Control Applications*, Albany, NY, September 1995. IEEE.

33. J. J. G. Ramos, E. C. Paiva, J. R. Azinheira, S. S. Bueno, S. M. Maeta, L. G. B. Mirisola, M. Bergerman, B. G. Faria, and A. Elfes. Flight Experiment with an Autonomous Unmanned Robotic Airship. In *Proceedings of the 2001 International Conference on Field and Service Robotics (FSR 2001)*, Helsinki, Finland, June 2001. IEEE.

34. J. J. G. Ramos, E. C. Paiva, S. M. Maeta, L. G. B. Mirisola, J. R. Azinheira, B. G. Faria, S. S. Bueno, M. Bergerman, C. S. Pereira, C. T. Fujiwara, J. P. Batistela, R. R. Frazzato, R. P. Peixoto, G. C. Martins, and A. Elfes. Project AURORA: A Status Report. In *Proceedings of the 3rd International Airship Convention and Exhibition (IACE 2000)*, Friedrichshafen, Germany, July 2000. The Airship Association, UK.

35. R. Simmons. Structured control for autonomous robots. *IEEE Transactions on Robotics and Automation*, 10(1), 1994.

36. N. Wells. Practical operation of remotely piloted airships. In *Proceedings of the 11th AIAA Lighter-than-Air Systems Technology Conference*, Clearwater Beach, FL, May 1995. AIAA.

37. G. Winkler. *Image Analysis, Random Fields and Dynamic Monte Carlo Methods*. Springer-Verlag, Berlin, 1995.

On the Competitive Complexity
of Navigation Tasks

Christian Icking[1], Thomas Kamphans[2], Rolf Klein[2], and Elmar Langetepe[2]

[1] FernUniversität Hagen, Praktische Informatik VI, D-58084 Hagen, Germany
[2] Universität Bonn, Institut für Informatik I, D-53117 Bonn, Germany

Abstract. A strategy S solving a navigation task T is called competitive with ratio r if the cost of solving any instance t of T does not exceed r times the cost of solving t optimally. The *competitive complexity* of task T is the smallest possible value r any strategy S can achieve. We discuss this notion, and survey some tasks whose competitive complexities are known. Then we report on new results and ongoing work on the competitive complexity of exploring an unknown cellular environment.

1 Introduction

Getting a robot to autonomously navigate an unknown environment is a challenge that comprises many different types of problems. From a theoretician's point of view, one could try to group them in the following way.

From the beginning, great effort has been made in coping with the *technical complexity* of designing, building, and running robots. This includes many difficult questions like how a system should be structured, how sensor readings can be evaluated, and how robots can be controlled.

While major progress was being made in this area, the *computational complexity* became important, too. A robot's actions should not be slowed down by internal computations necessary for recognizing objects and accordingly planning its motions. This naturally leads to the question of how data should be organized and processed, that is, to the design of suitable data structures and efficient algorithms.

Let us assume, for a moment, that all these problems were already solved to our satisfaction. Then we could put the robots to work, and the remaining challenge would be to do this in the most efficient way. Some navigation tasks would now be very simple. For example, if the robot were to move from a start position, s, to a target at a known position, t, in a known environment, it could quickly compute the optimum path from s to t [1], and safely follow it.

But if the location of t is not known? Then the robot has to search for the target in a systematic way. Only with luck will it discover t immediately; in general, when the robot eventually arrives at t, its path will be more expensive than the optimum path from s to t. Our interest is in keeping the extra cost

---

[1] See e. g. the survey article [37] for efficient algorithms for computing shortest paths in known two-dimensional environments.

G.D. Hager et al. (Eds.): Sensor Based Intelligent Robots, LNCS 2238, pp. 245–258, 2002.
© Springer-Verlag Berlin Heidelberg 2002

as small as possible, by designing a clever search strategy. The best result that can be achieved in principle is called the *competitive complexity* of the search problem at hand[2].

This notion is best illustrated by the following well-known example. Suppose a robot is facing a very long wall. There is a door leading to the other side, but the robot has no clue as to the location of the door (left or right, distance). The cost of moving is proportional to the length of the robot's path, i. e. there is no charge for turning. The robot will detect the door when it arrives there[3].

In order to find the door, the robot has to explore the left part and the right part of the wall alternatingly, each time increasing the exploration depths before returning to the start. If the depths are doubled each time, the total path until the door is found can be at most 9 times as long as the distance from the start point to the door; this is not hard to verify. Hence, this "doubling" strategy is competitive with ratio 9. It is quite surprising, and not easy to prove, that no (deterministic) strategy can achieve a ratio smaller than 9. Therefore, the wall search problem is of competitive complexity 9.

As opposed to the technical complexity, which lies entirely with the robot, and the computational complexity, which depends partially on the models used by the robot and partially on general properties of algorithms, the competitive complexity is a characteristic property of the navigation task alone.

At first glance the above example may appear a bit contrived. After all, a single, extremely long wall is an environment rarely encountered. But the above approach can be generalized to situations where an upper bound to the distance of the door is known in advance [19], or where m, rather than 2, alternatives must be searched [23], like corridors leading off a central room. This indicates that results first obtained for very basic situations may later be generalized to fit more realistic settings.

We think that knowing about the competitive complexities of basic navigation tasks is useful for another reason, too. Computer scientists learn that sorting n keys is of computational complexity $n \log n$. If they ever choose to implement a simple quadratic sorting algorithm they can rest assured that they are not too far from the optimum, for small values of n [4]. A similar observation may hold in robotics, i.e., we may be sure that a simple greedy strategy works not too bad in general.

Our final argument in favor of analysing the competitive complexity of navigation tasks is the following. Even when complete information about environment and task is available, computing the optimal solution may not be feasible. For example, computing the shortest inspection tour of a known environment

[2] The term "competitive analysis" was coined by Sleator and Tarjan [42] in the wider area of online algorithms, to express that a player is competing with an adversary who tries to make his job hard; see the survey volume [15].

[3] This search problem was first discovered in the sixties [5], generalized and analyzed by Gal [16], and "re-invented" by Baeza-Yates et al. [4]. Good improvements can be found in Schuierer [39].

[4] For small problem sizes the quadratic algorithm might even be better because of hidden constants.

with obstacles contains the traveling salesperson problem[5] and is, therefore, NP-hard. A good competitive exploration strategy could also be used as an approximation in this case.

How much do we know about the competitive complexities of basic navigation tasks? Apart from the wall problem mentioned above, there are only very few tasks whose complexities are precisely known. One of them is the *street problem* originally presented by Klein [26], whose complexity has finally been determined in Icking et al. [22] and independently by Semrau and Schuierer [40]. This problem is briefly reviewed in Sect. 2 of this paper.

In general it is quite difficult to precisely determine the competitive complexity of a task. Then we can at least try to put in an interval $[l, u]$, by proving that each possible strategy can be tricked into causing at least l times the minimum cost, and by providing some strategy that never causes more than u times the minimum cost, in solving any given instance of the task. A good survey of such results can be found in P. Berman's chapter in [15].

Even proving an upper bound for the cost incurred by a particular strategy can be a complicated matter. For example, we have shown in [20] that any unknown obstacle-free environment can be explored by a tour at most 26.5 times longer than the shortest possible inspection tour that could be computed offline, knowing the whole environment. Based on experiments we believe that our strategy performs much better, but it seems hard to prove a smaller upper bound.

In this paper we present new results and ongoing research on the competitive complexity of the exploration task for cellular environments with obstacles. The environment's free space consists of quadratic cells of equal size. At each time, the robot is located in some cell. Witin one step, it can realize which of the four adjacent cells are free, and move to one of them. On its way, it can maintain a partial map.

The robot's task is to start from a given cell on the boundary, visit each cell of the environment, and return to the start. The number of steps on this tour is compared with the length of the shortest possible offline tour that visits each cell at least once. However, computing this optimum tour is known to be NP-hard, see Itai et al. [25].

In Sect. 3.1 we show that this problem is of competitive complexity 2, by proving that any exploration strategy can be forced to make twice as many cell visits as necessary, even if returning to the start is *not* required.

However, the environments used in the proof are very special, in that they consist only of long and narrow corridors encircling a rectangular obstacle. The question arises if for environments that are fleshy, rather than skinny, the competitive complexity of the exploration problem is smaller. Clearly, when there are large areas of free cells the robot should be able to do better than visit each cell twice, on the average.

---

[5] We can simulate the cities to be visited by little spiral-shaped obstacles the robot must enter for inspection.

This question is investigated in Sect. 3.2. We present an exploration strategy URC[6] that is able to explore fleshy environments quite efficiently, and discuss its performance. Finally, in Sect. 4, we mention some open problems for further research.

2 An Optimal Online Strategy for Searching a Goal in a Street

2.1 History of the Street Problem

Suppose a point-shaped mobile robot equipped with a 360° vision system is placed inside a room whose walls are modeled by a simple polygon. Neither the floor-plan nor the position of the target point are known to the robot. As the robot moves around it can draw a partial map of those parts that have so far been visible. The robot is searching for a goal in the unknown environment and it will recognize the targetpoint on sight.

In arbitrary simple polygons no strategy can guarantee a search path a constant times as long as the shortest path from start to goal at the most. Therefore the following sub-class of polygons for which a constant performance ratio can be achieved was introduced by Klein [26, 27].

A polygon P with two distinguished vertices s and t is called a street if the two boundary chains leading from s to t are *mutually weakly* visible, i. e. if each point on one of the chains can see at least one point of the other; see Fig. 3 for an example. The first competitive strategy [26, 27] for searching a street had a competitive ratio not bigger than 5.72, i.e., the first upper bound for the ratio was given. Besides, it was shown that no strategy can achieve a competitive ratio of less than $\sqrt{2} \approx 1.41$, the ultimative lower bound to the ratio as turned out later.

Since then, street polygons have attracted considerable attention concerning structural properties, generalizations and applications in related fields, see Tseng et al. [43], Das et al. [9], Datta and Icking [12, 13], Datta et al. [11], López-Ortiz and Schuierer [32, 35], Ghosh and Saluja [17], Bröcker and Schuierer [7], Carlsson and Nilsson [8]. Bröcker and López-Ortiz [6] have shown that a constant performance ratio can be achieved for arbitrary start- and endpoints inside a street.

The main research has focussed on search strategies for improving the upper bound of 5.72 toward the $\sqrt{2}$ lower bound, see Icking [21] 4.44, Kleinberg [28] 2.61, López-Ortiz and Schuierer [31] 2.05, López-Ortiz and Schuierer [33] 1.73, Semrau [41] 1.57, Icking et al. [24] 1.51, Dasgupta et al. [10], Kranakis and Spatharis [29], López-Ortiz and Schuierer [34]. The gap between the upper and lower bound, also mentioned in Mitchell [37], was finally closed by Icking et al. [22] and independently by Semrau and Schuierer [40].

---

[6] A java implementation of this strategy is available at
  http://www.informatik.uni-bonn.de/I/GeomLab/Gridrobot .

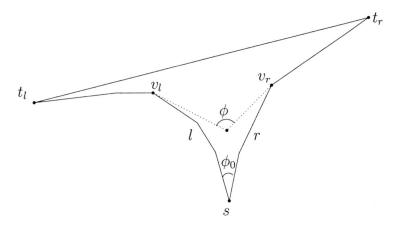

Fig. 1. A funnel.

2.2 Funnels

The competitive complexity of the street problem can be shown to be the same as for a subclass of streets called *funnels* [26]. A funnel consists of two convex chains starting from s at an opening angle of ϕ_0, that extend outwards; see Fig. 1. Inside the funnel the robot sees two foremost vertices, v_l and v_r, at an angle $\phi > \phi_0$. It knows that v_l belongs to the left, and v_r to the right boundary chain, but it does not know behind which of them the target is hidden. As the robot proceeds, v_l and v_r may change. Before the robot gets to the lid of the funnel it will be able to see the target.

Since the walking direction should be within the opening angle, ϕ, this angle is always *strictly increasing* but it never exceeds π. By this property, it is quite natural to use the opening angle ϕ for *parameterizing a strategy*.

For an arbitrary angle ϕ, let

$$K_\phi := \sqrt{1 + \sin \phi}.$$

2.3 An Optimal Strategy

In a funnel with opening angle π the goal is visible and there is a trivial strategy that achieves the optimal competitive factor $K_\pi = 1$. So we look backwards to decreasing angles.

Let us assume for the moment that the funnel is a triangle, and that we have a strategy with a competitive factor of K_{ϕ_2} for all triangular funnels of opening angle ϕ_2. How can we extend this to opening angles ϕ_1 with $\pi \geq \phi_2 > \phi_1 \geq \frac{\pi}{2}$?

Starting with an angle ϕ_1 at point p_1 we walk a certain path of length w until we reach an angle of ϕ_2 at point p_2 from where we can continue with the known strategy; see Fig. 2. We assume, for a moment, that the current left and right reflex vertices, v_l and v_r, see also Fig. 1, do not change.

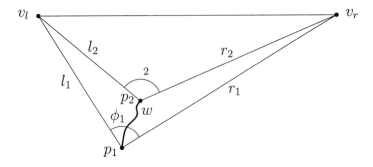

Fig. 2. Getting from angle ϕ_1 to ϕ_2.

Let l_1 and l_2 denote the distances from p_1 resp. p_2 to v_l at the left side and r_1 and r_2 the corresponding distances at the right. If $t = v_l$, the path length from p_1 to t is not greater than $w + K_{\phi_2} l_2$. If now $K_{\phi_1} l_1 \geq w + K_{\phi_2} l_2$ holds and the analogous inequality $K_{\phi_1} r_1 \geq w + K_{\phi_2} r_2$ for the right side, which can also be expressed as

$$w \leq \min(K_{\phi_1} l_1 - K_{\phi_2} l_2, K_{\phi_1} r_1 - K_{\phi_2} r_2)\,, \tag{1}$$

we have a competitive factor not bigger than K_{ϕ_1} for triangles with initial opening angle ϕ_1. Surprisingly, (1) is also *sufficient* for an overall competitive factor not bigger than K_{ϕ_1}, even if v_l or v_r change.

So the remaining question is, how should we satisfy (1)? Let us be bold and require

$$K_{\phi_1} l_1 - K_{\phi_2} l_2 = K_{\phi_1} r_1 - K_{\phi_2} r_2$$

or, equivalently

$$K_{\phi_2}(l_2 - r_2) = K_{\phi_1}(l_1 - r_1)\,. \tag{2}$$

This condition describes a nice curve that fulfils (1), see Icking et al. [22]. If the starting angle ϕ_0 is smaller than $\frac{\pi}{2}$ it suffices to walk along the angular bisector of ϕ this also fulfils (1) as shown in Langetepe [30].

To summarize, our strategy for searching a goal in an unknown street works as follows: for an example see Fig. 3.

Strategy WCA:
If the initial opening angle is less than $\frac{\pi}{2}$, walk along the current angular bisector of v_l and v_r until an opening angle of degree $\frac{\pi}{2}$ is reached. Depending on the actual parameters ϕ_0, l_0, and r_0, walk along the corresponding curve defined by condition (2) until one of v_l and v_r changes. Switch over to the curve corresponding to the new parameters ϕ_1, l_1, and r_1. Continue until the line $t_l\,t_r$ is reached.

Theorem 1. *By using strategy WCA we can search a goal in an unknown street with a competitive factor of $\sqrt{2}$ at the most. This is optimal.*

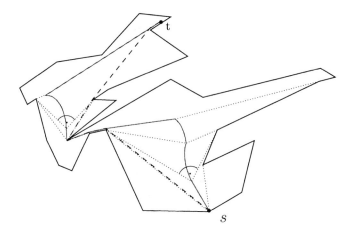

Fig. 3. A street and the path generated by WCA.

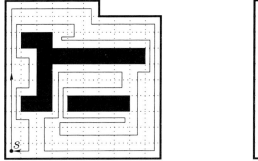
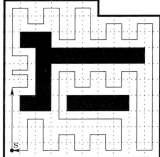

Fig. 4. (i) An exploration tour with two obstacles. (ii) A shortest TSP tour for the same scene.

3   Exploring an Unknown Cellular Environment

Here we study the model of a rather short-sighted robot. We assume that the environment is given by a polygon, P, which consists of square cells on an integer grid and which may contain some impenetrable areas, i. e. obstacles.

The robot starts from a cell, s, adjacent to P's boundary. From there it can enter one of the neighboring cells, and so on. Once inside a cell, the robot knows which of its 4 neighbors exist and which are boundary edges. The robot's task is to visit each cell inside P and to return to the start; see Fig. 4 (i) for an example. This example shows a tour that visits each cell at least once, and we are interested in producing an exploration tour as short as possible.

Even though our robot does not know its environment in advance it is interesting to ask how short a tour can be found in the offline situation, i. e. when the environment is already known. This amounts to constructing a shortest traveling salesperson (TSP) tour on the cells.

If the polygonal environment contains obstacles, the problem of finding such a minimum length tour is known to be NP-hard [25] and there are some approximation schemes [2, 3, 18, 36].

In a simple polygon without obstacles, the complexity of constructing offline a minimum length tour seems to be open. There are, however, some results concerning the related Hamiltonian cycle and path problems [14, 44] and approximations [2, 38].

3.1   The Competitive Complexity

The following result holds true no matter if the robot must return to the start or not.

Theorem 2. *The competitive complexity of exploring an unknown cellular environment with obstacles equals 2.*

Proof. Even if we do not know the environment we can apply depth first search (DFS) to the cell graph[7]. This results in a complete traversal of a tree on C nodes, where C is the number of cells. Such a traversal takes $2C - 2$ steps. Since the shortest tour needs at least C steps to visit all cells and to return to s, DFS turns out to be competitive with a factor of 2.

On the other hand, 2 is also a lower bound for the competitive factor of any strategy. To prove this, we construct a special cell graph which depends on the behavior of the strategy. The start position s is situated in a long corridor of width 1. We fix a large number Q and observe how the strategy explores the corridor. Two cases are distinguished.

Case 1: The robot comes back to s some time after having made at least Q and at most $2Q$ steps. At this time, we close the corridor such that there are only two unvisited cells, one at each end, see Fig. 5 (i). Let R be the number of cells visited at this point. The robot has already made at least $2R - 2$ steps and needs another $2R$ steps to visit the two remaining cells and to return to s while the shortest tour needs only $2R$ steps to accomplish this task.

Case 2: In the remaining case the robot, more or less, concentrates on one end of the corridor. Let R be the number of cells visited after $2Q$ steps. At that time, we add a fork at a cell b just behind the farthest visited cell on the corridor, see Fig. 5 (ii). Two paths arise which turn back and run parallel to the long corridor. If the robots comes back to s before exploring one of the two paths, then an argument analogous to Case 1 applies. Otherwise, one of the two paths will eventually be explored till the end at a cell e where it turns out that it is connected to the other end of the first corridor. At this time, the other path is fixed to be a dead end of length R' which closes just one cell after the last visited cell, e'.

---

[7] The nodes of the graph are the cells, two nodes are joined by an edge if their cells lie side by side.

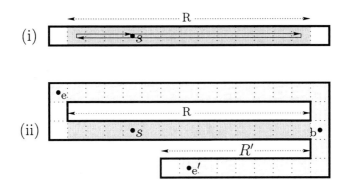

Fig. 5. Proving the lower bound of 2 for the competitive factor.

From e, the robot must still walk to the other end of the corridor and visit the dead end, and then return to s. Eventually, it will have walked at least four times the length of the corridor, R, plus four times the length of the dead end, R'. The optimal path only needs $2R + 2R'$, apart from a constant number of steps.

In any case the competitive factor tends to 2 while Q goes to infinity.

Now consider the case where the robot is not forced to return to the startpoint after the exploration is finished. This problem has also a competitive complexity of 2.

Again DFS provides the upper bound. Now the lower bound is even easier to prove. We let the robot explore the lenghty corridor of Fig. 5 (i). If it passes the start cell, s, every so often, we proceed as before and close the corridor, leaving one unexplored cell at either end. Then the robot still has to spend the same amount of work the optimum solution needs. Otherwise, the robot concentrates on—say—the right part of the corridor. Then we let it run for a while, and eventually close it, having one unexplored cell at the left end. To get there, the robot must backtrack along the very long right part, which is about as costly as the optimum solution.

3.2   A More Efficient Exploration Strategy

In the previous section we have seen that each exploration strategy can be forced to make about twice as many cell visits as necessary. On the other hand, the standard depth first search (DFS) technique makes exactly $2C - 2$ steps in any environment over C cells, while at least C visits are necessary for any strategy.

Yet, we hesitate to call DFS an optimal strategy since the lower bound was achieved in very thin corridors. For example, an optimum exploration tour for a rectangle of 3 by m cells makes at most $3m + 1$ many steps[8], while DFS needs

---

[8] An optimum tour could follow one of the long edges and meander back to the start. If m is odd, one extra step at the end is unavoidable since each closed tour is of even length.

about $6m$ many steps. In this case the pure competitive analysis leads to the observation that we have to take the shape of the environment into account.

A simple way of improving DFS would be as follows. When there is no unexplored cell adjacent to the robot's current position, it has to backtrack to the latest unexplored cell encountered on its path, and to resume from there. By the original DFS algorithm, the robot uses the unique path in the exploration tree for backtracking. Instead, it could use the *shortest path in the environment* it knows of.

In the $3 \times m$-rectangle, the resulting strategy DFSO would need an exploration tour of length only $4m$, which is quite an improvement. Unfortunately, this advantage is lost in more general environments. Figure 6 shows an example of a corridor of width 3 and length m that connects two small rooms. In (i), the exploration tour generated by DFSO is shown. First, the robot makes a complete round trip along the wall, cutting off the right part of the right room. It returns to the right room, and runs out of unexplored cells after clearing its left part. Now it returns to the left room because there are the unexplored cells discovered last. Then the robot has to move to the right room again and finish its right part. Even if the robot were not required to return to the start, a path of length $> 5m$ would result; if returning is necessary, we are back to the poor performance of DFS, for large values of m.

To overcome this difficulty, we present a new exploration strategy called URC. While proceeding with the exploration of new cells, the robot reserves adjacent cells for its return path[9]. In its *forward* mode, URC visits unexplored and unreserved cells by the left-hand rule, preferring a left turn over a straight step over a right turn. If no unexplored and unreserved cells are adjacent, the robot enters the *backward* mode. It follows the shortest known path to the cell reserved last—in general this cell will be adjacent to the robot's current position—and uses the reserved cells for returning. If an unexplored cell is passed on the way, the robot switches into forward mode, and continues. At each time, the reserved cells are kept in a stack, sorted by their order of reservation. Figure 6, (ii), shows the path generated by URC in the corridor example. As by DFSO, the robot first travels around the wall. When it returns to the left room, the corridor's center lane is reserved, in left-to-right order. Now, all cells of the left room that were not previously reserved, are visited. Only afterwards moves the robot back to the right room, to get to the remaining cells that were reserved last. Both parts of the right room are completely finished before the robot follows the reserved path back to the start. So, it passes the corridor only 4 times.

3.3 The Conjectured Performance of Strategy URC

To distinguish between fleshy and skinny environments we introduce the number E of boundary edges and the number H of obstacles, in addition to the number of cells, C. In a fleshy environment, E can be of order $\sqrt{C}$, while it can be as

---

[9] Planning the return path is not only helpful for the final path segment back to the start; it also helps in getting out of parts of the environment that have been finished.

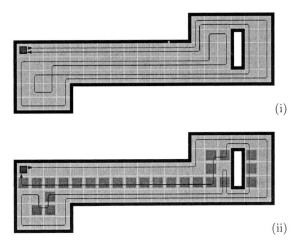

(i)

(ii)

Fig. 6. DFSO versus URC.

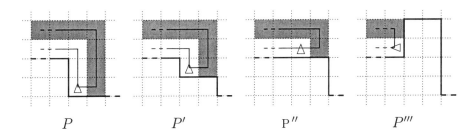

P P' P'' P'''

Fig. 7. Example: decomposing a polygon.

large as $2C$ in a corridor. We believe that strategy URC can be implemented to explore any environment by a closed tour of length S such that

$$S \le C + \frac{E}{2} + 3H - 2$$

holds. We are working towards a proof, using the following approach.

Let P denote the original environment. Now let P' denote the environment that results from P by removing the start cell, s, and all cells that are reserved for the return path during the first step in P. The start cell, s', in P' is the cell the robot enters by its first step in P; see Fig. 7. Now we observe the differences in the number of steps, cells, edges, and holes, between P and P', and check how the conjectured expression is affected. It turns out that in most cases the balance is zero. Those situations incurring a negative balance must be counted, and compared with the situations where the balance is positive. However, this approach requires the local analysis of quite a few cases, as well as some global arguments.

4 Conclusions

Apart from finally settling the exploration problem for cellular environments, there are some interesting open questions we would like to mention. Both are addressing a robot with precise and unbounded vision that operates in a polygonal environment containing polygonal obstacles.

It was shown by Albers et al. [1] that the exploration problem in this model is of competitive complexity at least $\sqrt{H}$, where H denotes the number of obstacles. However, the obstacles used in the proof were extremely long and thin. The first question is the following: Does the competitive complexity become constant if we assume that all obstacles are fat or have constant complexity[10]?

The other question is related to the problem of searching for a target whose position is unknown. How much does it help to know the environment?

References

1. S. Albers, K. Kursawe, and S. Schuierer. Exploring unknown environments with obstacles. In *Proc. 10th ACM-SIAM Sympos. Discrete Algorithms*, pages 842–843, 1999.
2. E. M. Arkin, S. P. Fekete, and J. S. B. Mitchell. Approximation algorithms for lawn mowing and milling. Technical report, Mathematisches Institut, Universität zu Köln, 1997.
3. S. Arora. Polynomial time approximation schemes for Euclidean TSP and other geometric problems. In *Proc. 37th Annu. IEEE Sympos. Found. Comput. Sci.*, pages 2–11, 1996.
4. R. Baeza-Yates, J. Culberson, and G. Rawlins. Searching in the plane. *Inform. Comput.*, 106:234–252, 1993.
5. R. Bellman. Problem 63-9*. *SIAM Review*, 5(2), 1963.
6. C. Bröcker and A. López-Ortiz. Position-independent street searching. In *Proc. 6th Workshop Algorithms Data Struct.*, volume 1663 of *Lecture Notes Comput. Sci.*, pages 241–252. Springer-Verlag, 1999.
7. C. Bröcker and S. Schuierer. Searching rectilinear streets completely. In *Proc. 6th Workshop Algorithms Data Struct.*, volume 1663 of *Lecture Notes Comput. Sci.*, pages 98–109. Springer-Verlag, 1999.
8. S. Carlsson and B. J. Nilsson. Computing vision points in polygons. *Algorithmica*, 24:50–75, 1999.
9. G. Das, P. Heffernan, and G. Narasimhan. LR-visibility in polygons. *Comput. Geom. Theory Appl.*, 7:37–57, 1997.
10. P. Dasgupta, P. P. Chakrabarti, and S. C. DeSarkar. A new competitive algorithm for agent searching in unknown streets. In *Proc. 16th Conf. Found. Softw. Tech. Theoret. Comput. Sci.*, volume 1180 of *Lecture Notes Comput. Sci.*, pages 147–155. Springer-Verlag, 1996.

---

[10] In geometry an object A is called α-fat with $0 < \alpha < 1$, if the following holds. Let a be a point of A, and let C be a circle centered at a that does not fully contain A. Then at least a fraction α of C's area belongs to A. This should hold for any point a and any corresponding circle C.

11. A. Datta, C. A. Hipke, and S. Schuierer. Competitive searching in polygons: Beyond generalised streets. In *Proc. 6th Annu. Internat. Sympos. Algorithms Comput.*, volume 1004 of *Lecture Notes Comput. Sci.*, pages 32–41. Springer-Verlag, 1995.

12. A. Datta and C. Icking. Competitive searching in a generalized street. In *Proc. 10th Annu. ACM Sympos. Comput. Geom.*, pages 175–182, 1994.

13. A. Datta and C. Icking. Competitive searching in a generalized street. *Comput. Geom. Theory Appl.*, 13:109–120, 1999.

14. H. Everett. Hamiltonian paths in non-rectangular grid graphs. Report 86-1, Dept. Comput. Sci., Univ. Toronto, Toronto, ON, 1986.

15. A. Fiat and G. Woeginger, editors. *On-line Algorithms: The State of the Art*, volume 1442 of *Lecture Notes Comput. Sci.* Springer-Verlag, 1998.

16. S. Gal. *Search Games*, volume 149 of *Mathematics in Science and Engeneering.* Academic Press, New York, 1980.

17. S. K. Ghosh and S. Saluja. Optimal on-line algorithms for walking with minimum number of turns in unknown streets. *Comput. Geom. Theory Appl.*, 8(5):241–266, Oct. 1997.

18. M. Grigni, E. Koutsoupias, and C. H. Papadimitriou. An approximation scheme for planar graph TSP. In *Proc. 36th Annu. IEEE Sympos. Found. Comput. Sci.*, pages 640–645, 1995.

19. C. Hipke, C. Icking, R. Klein, and E. Langetepe. How to find a point on a line within a fixed distance. *Discrete Appl. Math.*, 93:67–73, 1999.

20. F. Hoffmann, C. Icking, R. Klein, and K. Kriegel. The polygon exploration problem. *SIAM J. Comput.*, 2001. to appear.

21. C. Icking. *Motion and Visibility in Simple Polygons.* PhD thesis, Department of Computer Science, FernUniversität Hagen, 1994.

22. C. Icking, R. Klein, and E. Langetepe. An optimal competitive strategy for walking in streets. In *Proc. 16th Sympos. Theoret. Aspects Comput. Sci.*, volume 1563 of *Lecture Notes Comput. Sci.*, pages 110–120. Springer-Verlag, 1999.

23. C. Icking, R. Klein, and E. Langetepe. Searching a goal on m rays within a fixed distance. In *Abstracts 15th European Workshop Comput. Geom.*, pages 137–139. INRIA Sophia-Antipolis, 1999.

24. C. Icking, A. López-Ortiz, S. Schuierer, and I. Semrau. Going home through an unknown street. Technical Report 228, Department of Computer Science, FernUniversität Hagen, Germany, 1998.

25. A. Itai, C. H. Papadimitriou, and J. L. Szwarcfiter. Hamilton paths in grid graphs. *SIAM J. Comput.*, 11:676–686, 1982.

26. R. Klein. Walking an unknown street with bounded detour. In *Proc. 32nd Annu. IEEE Sympos. Found. Comput. Sci.*, pages 304–313, 1991.

27. R. Klein. Walking an unknown street with bounded detour. *Comput. Geom. Theory Appl.*, 1:325–351, 1992.

28. J. M. Kleinberg. On-line search in a simple polygon. In *Proc. 5th ACM-SIAM Sympos. Discrete Algorithms*, pages 8–15, 1994.

29. E. Kranakis and A. Spatharis. Almost optimal on-line search in unknown streets. In *Proc. 9th Canad. Conf. Comput. Geom.*, pages 93–99, 1997.

30. E. Langetepe. *Design and Analysis of Strategies for Autonomous Systems in Motion Planning.* PhD thesis, Department of Computer Science, FernUniversität Hagen, 2000.

31. A. López-Ortiz and S. Schuierer. Going home through an unknown street. In *Proc. 4th Workshop Algorithms Data Struct.*, volume 955 of *Lecture Notes Comput. Sci.*, pages 135–146. Springer-Verlag, 1995.

32. A. López-Ortiz and S. Schuierer. Generalized streets revisited. In *Proc. 4th Annu. European Sympos. Algorithms*, volume 1136 of *Lecture Notes Comput. Sci.*, pages 546–558. Springer-Verlag, 1996.

33. A. López-Ortiz and S. Schuierer. Walking streets faster. In *Proc. 5th Scand. Workshop Algorithm Theory*, volume 1097 of *Lecture Notes Comput. Sci.*, pages 345–356. Springer-Verlag, 1996.

34. A. López-Ortiz and S. Schuierer. The exact cost of exploring streets with CAB. Technical report, Institut für Informatik, Universität Freiburg, 1998.

35. A. López-Ortiz and S. Schuierer. Lower bounds for searching on generalized streets. University of New Brunswick, Canada, 1998.

36. J. S. B. Mitchell. Guillotine subdivisions approximate polygonal subdivisions: A simple new method for the geometric k-MST problem. In *Proc. 7th ACM-SIAM Sympos. Discrete Algorithms*, pages 402–408, 1996.

37. J. S. B. Mitchell. Geometric shortest paths and network optimization. In J.-R. Sack and J. Urrutia, editors, *Handbook of Computational Geometry*, pages 633–701. Elsevier Science Publishers B.V. North-Holland, Amsterdam, 2000.

38. S. Ntafos. Watchman routes under limited visibility. *Comput. Geom. Theory Appl.*, 1(3):149–170, 1992.

39. S. Schuierer. Lower bounds in on-line geometric searching. In *11th International Symposium on Fundamentals of Computation Theory*, volume 1279 of *Lecture Notes Comput. Sci.*, pages 429–440. Springer-Verlag, 1997.

40. S. Schuierer and I. Semrau. An optimal strategy for searching in unknown streets. In *Proc. 16th Sympos. Theoret. Aspects Comput. Sci.*, volume 1563 of *Lecture Notes Comput. Sci.*, pages 121–131. Springer-Verlag, 1999.

41. I. Semrau. Analyse und experimentelle Untersuchung von Strategien zum Finden eines Ziels in Straßenpolygonen. Diploma thesis, FernUniversität Hagen, 1996.

42. D. D. Sleator and R. E. Tarjan. Amortized efficiency of list update and paging rules. *Commun. ACM*, 28:202–208, 1985.

43. L. H. Tseng, P. Heffernan, and D. T. Lee. Two-guard walkability of simple polygons. *Internat. J. Comput. Geom. Appl.*, 8(1):85–116, 1998.

44. C. Umans and W. Lenhart. Hamiltonian cycles in solid grid graphs. In *Proc. 38th Annu. IEEE Sympos. Found. Comput. Sci.*, pages 496–507, 1997.

Geometry and Part Feeding

A. Frank van der Stappen[1], Robert-Paul Berretty[1,2],
Ken Goldberg[3], and Mark H. Overmars[1]

[1] Institute of Information and Computing Sciences,
Utrecht University,
P.O.Box 80089, 3508 TB Utrecht, The Netherlands
[2] *Current address:* Department of Computer Science,
University of North Carolina,
Campus Box 3175, Sitterson Hall, Chapel Hill, NC 27599-3175, USA
[3] Department of Industrial Engineering and Operations Research,
University of California at Berkeley,
Berkeley, CA 94720, USA

Abstract. Many automated manufacturing processes require parts to be oriented prior to assembly. A part feeder takes in a stream of identical parts in arbitrary orientations and outputs them in uniform orientation. We consider part feeders that do not use sensing information to accomplish the task of orienting a part; these feeders include vibratory bowls, parallel jaw grippers, and conveyor belts and tilted plates with so-called fences. The input of the problem of sensorless manipulation is a description of the part shape and the output is a sequence of actions that moves the part from its unknown initial pose into a unique final pose. For each part feeder we consider, we determine classes of orientable parts, give algorithms for synthesizing sequences of actions, and derive upper bounds on the length of these sequences.

1 Introduction

Manipulation tasks such as part feeding generally take place in structured factory environments; parts typically arrive at a more-or-less regular rate along for example a conveyer belt. The structure of the environment removes the need for intricate sensing capabilities. In fact, Canny and Goldberg [22] advocate a RISC (Reduced Intricacy in Sensing and Control) approach to designing manipulation systems for factory environments. Inspired by Whitney's recommendation that industrial robots have simple sensors and actuators [38], they argue that automated planning may be more practical for robot systems with fewer degrees of freedom (parallel-jaw grippers instead of multi-fingered hands) and simple, fast sensors (light beams rather than cameras). To be cost-effective industrial robots should emphasize efficiency and reliability over the potential flexibility of anthropomorphic designs. In addition to these advantages of RISC hardware, RISC systems also lead to positive effects in software: manipulation algorithms that are efficient, robust, and subject to guarantees.

G.D. Hager et al. (Eds.): Sensor Based Intelligent Robots, LNCS 2238, pp. 259–281, 2002.

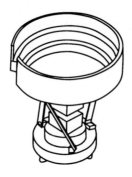

Fig. 1. A bowl feeder [19].

We consider part feeders in the line of thought of the RISC approach. More specifically, we shall focus on the problem of sensorless orientation of parts in which *no* sensory information at all is used to move the part from an unknown initial pose into a unique – and known – final pose. In sensorless orientation or part feeding parts are oriented using passive mechanical compliance. The input of the problem of sensorless orientation is a description of the shape of the part and the ouput is a sequence of open-loop actions that moves the part from an unknown initial pose into a unique final pose. Among the sensorless part feeders considered in the literature are the traditional bowl feeder [18, 19], the parallel-jaw gripper [23, 26], the single pushing jaw [3, 29, 31, 34], the conveyor belt with a sequence of fences rigidly attached to both its sides [20, 35, 39], the conveyor belt with a single rotational fence [2], the tilting tray [25, 33], and vibratory plates and programmable vector fields [16, 17].

Traditionally, sensorless part feeding is accomplished by the *vibratory bowl feeder*, which is a bowl that is surrounded by a helical metal track and filled with parts [18, 19], see Figure 1. The bowl and track undergo an asymmetric helical vibration that causes parts to move up the track, where they encounter a sequence of mechanical devices such as wiper blades, grooves and traps. The majority of these mechanical devices act as filters that serve to reject (force back to the bottom of the bowl) parts in all orientations except for the desired one. Eventually, a stream of parts in a uniform orientation emerges at the top after successfully running the gauntlet. The design of bowl feeders is, in practice, a task of trial and error. It typically takes one month to design a bowl feeder for a specific part [30]. We will see in Section 5 that it is possible to compute whether a given part in a given orientation will safely move across a given trap. More importantly, we will see that it is possible to use the knowledge of the shape of the part to synthesize traps that allow the part to pass in only one orientation [9, 12, 13].

The first feeders to which thorough theoretical studies have been devoted were the parallel-jaw gripper and pushing jaw. Goldberg [26] showed that these devices can be used for sensorless part feeding or orienting of two-dimensional parts. He gave an algorithm for finding the shortest sequence of pushing or

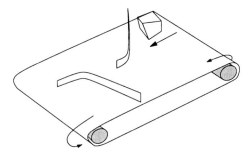

Fig. 2. Rigid fences over a conveyor.

squeezing actions that will move the part from an unkown initial orientation to a known final orientation. Chen and Ierardi [23] showed that the length of this sequence is $O(n)$ for polygonal parts with n vertices. In Section 2 we shall provide theoretical foundation to the fact that the sequence length often stays well below this bound [37]. As the act of pushing is common to most feeders that we consider in this paper we will first study the pushing of parts in some detail.

The next feeder we consider consists of a sequence of fences which are mounted across a conveyor belt [20, 35, 39]. The fences brush the part as it travels down the belt thus reorienting it (see Figure 2). The motion of the belt effectively turns the slide along a fence into a push action by the fence. It has long been open whether a sequence of fences can be designed for any given part such that this sequence will move that part from any initial pose into a known final pose. We report an affirmative answer in Section 3. In addition we give an $O(n^3)$ algorithm (improving an earlier exponential algorithm by Wiegley et al. [39]) for computing the shortest sequence of fences for a given part along with several extensions [8, 10, 11].

A drawback of most of the achievements in the field of sensorless orientation of parts is that they only apply to planar parts, or to parts that are known to rest on a certain face. In Section 4 we present a generalization of conveyor belts and fences that attempts to bridge the gap to truly three-dimensional parts [15]. The feeder consists of a sequence tilted plates with curved tips; each of the plates contains a sequence of fences (see Figure 3). The feeder essentially tries to orient the part by a sequence of push actions by two orthogonal planes. We analyze these actions and use the results to show that it is possible to compute a set-up of plates and fences for any given asymmetric polyhedral part such that the part gets oriented on its descent along plates and fences.

This paper reports on parts of our research in the field of sensorless manipulation of the last few years. The emphasis will be on the transformation of various sensorless part feeder problems into geometric problems, a sketch of the algorithms that solve these problems, and on determining classes of orientable parts. For proofs and detailed descriptions of the algorithms and their extensions the reader is in general referred to other sources [8–15, 37].

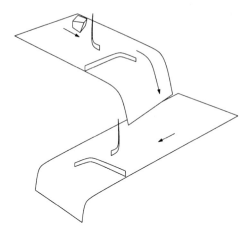

Fig. 3. Feeding three-dimensional parts with a sequence of plates and fences.

2   Pushing Planar Parts

2.1   Push Functions

Throughout the entire paper, we assume zero friction – unless stated otherwise – between the part and the orienting device. Let c be the center-of-mass and P be the convex hull of the planar part. As a pushing device always touches the part at its convex hull, we can only orient a part up to rotational symmetries in its convex hull. Without loss of generality, our problem is now to orient the convex part P with given center-of-mass c.

We assume that a fixed coordinate frame is attached to P. Directions are expressed relative to this frame. The *contact direction* of a supporting line (or tangent) l of a part P is uniquely defined as the direction of the vector perpendicular to l and pointing into P (see Figure 4 for a supporting line with contact direction π.). As in Mason [31], we define the *radius function* $\rho : [0, 2\pi) \rightarrow \{x \in \mathbb{R} | x \geqslant 0\}$ of a part P with a center-of-mass c; ρ maps a direction ϕ onto the distance from the center-of-mass c to the supporting line of P with contact direction ϕ. Recall that the direction ϕ is measured with respect to the frame attached to P. The (continuous) radius functions determines the push function, which, in turn, determines the final orientation of a part that is being pushed.

Throughout this paper, parts are assumed to be pushed in a direction perpendicular to the pushing device. The *push direction* of a single jaw is determined by the direction of its motion. The push direction of a jaw pushing a part equals the contact direction of the jaw. In most cases, parts will start to rotate when pushed. If pushing in a certain direction does *not* cause the part to rotate, then we refer to the corresponding direction as an *equilibrium* (push) *direction* or *orientation*. These equilibrium orientations play a key role throughout this paper. If pushing does change the orientation, then this rotation changes the orientation of the pushing gripper relative to the part. We assume that pushing continues until the part stops rotating and settles in a (stable) equilibrium pose.

The *push function* $p : [0, 2\pi) \to [0, 2\pi)$ links every orientation ϕ to the orientation $p(\phi)$ in which the part P settles after being pushed by a jaw with push direction ϕ (relative to the frame attached to P). The rotation of the part due to pushing causes the contact direction of the jaw to change. The final orientation $p(\phi)$ of the part is the contact direction of the jaw after the part has settled. The equilibrium push directions are the fixed points of the push function p.

The push function p consists of *steps*, which are intervals $I \subset [0, 2\pi)$ for which $p(\phi) = v$ for all $\phi \in I$ and some constant $v \in I$, and *ramps*, which are intervals $I \subset [0, 2\pi)$ for which $p(\phi) = \phi$ for all $\phi \in I$. Note that the ramps are intervals of equilibrium orientations. The steps and ramps of the push function are easily constructed [26, 36] from the radius function ρ, using its points of horizontal tangency; these orientations of horizontal tangency are the equilibrium push orientations. Angular intervals of constant radius turn up as ramps of the push function. Notice that such intervals only exist if the boundary of the part contains certain specific circular arcs. Thus, ramps cannot occur in the case of polygonal parts. If the part is pushed in a direction corresponding to a point of non-horizontal tangency of the radius function then the part will rotate in the direction in which the radius decreases. The part finally settles in an orientation corresponding to a local minimum of the radius function. As a result, all points in the open interval I bounded by two consecutive local maxima of the radius function ρ map onto the orientation $\phi \in I$ corresponding to the unique local minimum of ρ on I. (Note that ϕ itself maps onto ϕ because it is a point of horizontal tangency.) This results in the steps of the push function. Note that each half-step, i.e., a part of a step on a single side of the diagonal $p(\phi) = \phi$, is a (maximal) angular interval without equilibrium push orientation. An equilibrium orientation v is *stable* if it lies in the interior of an interval I for which $p(\phi) = v$ for all $\phi \in I$. Besides the steps and ramps there are isolated points satisfying $p(\phi) = \phi$ in the push function, corresponding to local maxima of the radius function. Figure 4 shows an example of a radius function and the corresponding push function.

Similar to the push function we can define a *squeeze function* that links every orientation ϕ to the orientation in which the part settles after being simultaneously pushed from the directions ϕ and $\phi + \pi$. The steps and ramps of the squeeze function can be computed from the part's *width function* (see [26, 36] for details).

Using the abbreviation $p_\alpha(\phi) = p((\phi + \alpha) \bmod 2\pi)$, we define a *push plan* to be a sequence $\alpha_1, \ldots, \alpha_k$ such that $p_{\alpha_k} \circ \ldots \circ p_{\alpha_1}(\phi) = \Phi$ for all $\phi \in [0, 2\pi)$ and a fixed Φ. In words, a push plan is an alternating sequence of jaw reorientations – by angles α_i – and push actions that will move the part from any initial orientation ϕ into the unique final orientation Φ. Observe that a single push action puts the part into one of a finite number of stable orientations. Most algorithms for computing push plans proceed by identifying reorientations that will cause a next push to reduce the number of possible orientations of the part.

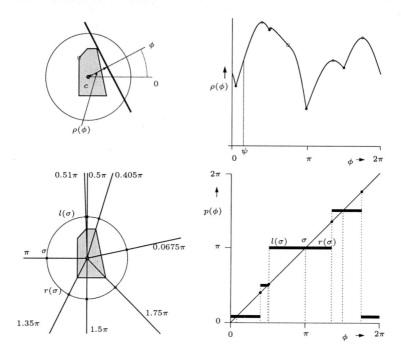

Fig. 4. A polygonal part and its radius and push function. The minima of the radius function correspond to normals to polygon edges that intersect the center-of-mass. The maxima correspond to tangents to polygon vertices whose normals intersect the center-of-mass. The horizontal steps of the push function are angular intervals between two successive maxima of the radius function.

2.2 Push Plan Length

Goldberg [26] considered the problem of orienting (feeding) polygonal parts using a parallel-jaw gripper. A parallel-jaw gripper consists of a pair of flat parallel jaws that can close in the direction orthogonal to the jaws, which can *push* and *squeeze* the part. When the initial orientation of the part is unknown, a sequence of gripper operations can be used to orient the part – relative to the gripper – without sensing. Let N be the number of gripper operations in the shortest sequence that will orient the part up to symmetry. Goldberg showed that N is $O(n^2)$ for polygonal parts with n vertices and gave an algorithm for finding the shortest squeeze plan. He also conjectured that N is $O(n)$.

Chen and Ierardi [23] proved Goldberg's conjecture by constructing simple push and squeeze plans of length $O(n)$. They also presented pathological polygons where N is $\Theta(n)$, showing that the $O(n)$ bound is tight in the worst case. Such pathological polygons are 'fat' (approximately circular), while N is almost always small for 'thin' parts. Consider the two parts shown in Figure 5. Imagine grasping part A. Regardless of the orientation of the gripper, we expect the part to be squeezed into an orientation in which its longest edge is aligned with a jaw of the gripper. Hence, the number of possible orientations of the part (relative

Fig. 5. Both polygonal parts have $n = 11$ vertices, but part A is thin, while B is fat. Part A is intuitively easier to orient than part B.

to the gripper) after a single application of the gripper is very small. Part B can end up with any of its n edges against a gripper jaw; the number of possible orientations (again relative to the gripper) after a single application of the gripper is considerably higher than in the case of the thin part. In general, we observe that thin parts are easier to orient than fat ones.

A theoretical analysis confirms this intuition. To formalize our intuition about fatness, we define the *geometric eccentricity* of a planar part based on the length-to-width ratio of a distinguished type of bounding box. We deduce an upper bound on the number of actions required to orient a part that depends only on the eccentricity of the part. The bound shows that a constant number of actions suffices to orient a large class of parts. The analysis also applies to curved parts and provides the first complexity bound for non-polygonal parts.

The inspiration for our thinness measure comes from ellipses. The eccentricity of an ellipse equals $\sqrt{1 - (b/a)^2}$, where a and b are the lengths of the major and minor axes respectively. Our (similar) definition of eccentricity for a convex object relies on the maximum of all aspect ratios of bounding boxes of the object.

Definition 1 *The* eccentricity ϵ *of a convex object* $P \subset \mathbb{R}^2$ *is defined by* $\epsilon = r - 1$, *where* r *equals the maximum of all aspect ratios of bounding boxes of* P.

Note that the minimum eccentricity of 0 is in both our definition and in the definition for ellipses obtained for circles.

Chen and Ierardi [23] proposed a class of plans for orienting polygonal parts based on repeating a unique push-and-reorient operation. The length of the longest angular interval without equilibrium orientation, or, in other words, of the longest half-step of the push function, determines the angle of reorientation. Assume that this half-step is uniquely defined and has length α. A reorientation by $\alpha - \mu$ for some very small positive μ in the proper direction followed by a push action will cause the part to rotate to the next equilibrium orientation unless it is in the orientation ϕ corresponding to the height – in the push function – of the longest half-step. Since the number of steps is bounded by n, it will take at most n of these combined actions to make the part end up in orientation ϕ. The case where the longest half-step is not uniquely defined requires additional techniques but again a plan of linear length can be obtained.

We use ideas similar to those of Chen and Ierardi to establish a relation between the length of the longest half-step and the number of push actions required to orient the part. The bound applies to arbitrary parts and is given in the following lemma.

Lemma 1. *A part can be oriented by* $N = 2\lceil 2\pi/\alpha \rceil + 1$ *applications of the gripper, where* α *is the longest-half-step of the push function.*

Eccentricity imposes a lower bound on the length of the longest half-step. Intuitively it is clear that a part can only be eccentric when its radius is allowed to increase over a relatively long angular interval (about its center-of-mass). A thorough analysis [37] confirms this intuition. The result of the analyis is given below.

Lemma 2. *The eccentricity* ϵ *of a part with a push function with a longest half-step of length* α *is bounded by*

$$\epsilon \leqslant \frac{\cos^{k-1}\alpha \cdot \sin{(k+1)\alpha}}{\cos^k 2\alpha} - 1,$$

where $k = \lceil \pi/(2\alpha) \rceil$.

Lemmas 1 and 2 yield the following theorem.

Theorem 1. *Let* P *be a part with eccentricity*

$$\epsilon > \frac{\cos^{k-1}\alpha \cdot \sin{(k+1)\alpha}}{\cos^k 2\alpha} - 1$$

$(k = \lceil \pi/(2\alpha) \rceil)$, *for some* $\alpha \in (0, \pi/4)$. *Then,* P *can be oriented by a push plan of length*

$$N \leqslant 2\lceil \frac{2\pi}{\alpha} \rceil + 1.$$

Theorem 1 shows that the number of push actions needed to orient a part is a function of its eccentricity. It provides the first upper bound on the length of a push plan for non-polygonal parts. Sample values show that the upper bound provided by Theorem 1 is relatively low even for smaller values of ϵ; $N \approx 75$ for $\epsilon = 0.5$, N is below 50 for $\epsilon = 1$ and below 30 for $\epsilon = 2.5$. Similar bounds can be obtained for squeeze plans [37].

2.3   Pulling Parts

We have recently studied sensorless orientation of planar parts with elevated edges by inside-out pull actions [14]. In a pull action a finger is moved (from the inside of the part) towards the boundary. As it reaches the boundary it continues to pull in the same direction until the part is certain to have stopped rotating. Subsequently, the direction of motion of the finger is altered and the action is repeated. The problem os sensorless orientation by a pulling finger is to find a sequence of motion directions that will cause the finger to move the part from any initial pose into a unique final pose.

Although intuitively similar to pushing it turns out that sensorless orienting by pull actions is considerably harder than pushing [14]. As the finger touches

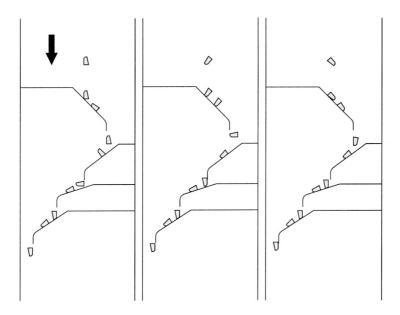

Fig. 6. Three overhead views of the same conveyor belt and fence design. The traversals for three different initial orientations of the same part are displayed. The traversals show that the part ends up in the same orientation in each of the three cases.

the part from the inside it does no longer make sense to assume that the part is convex. In fact, it can be shown that certain non-convex parts cannot be oriented by a sequence of pull actions. Most convex parts are orientable by $O(n)$ pull actions, and the shortest pull plan is computable in $O(n^3)$ time.

3   Fence Design

The problem of *fence design* is to determine a sequence of fence orientations, such that fences with these orientations align a part as it moves down a conveyor belt and slides along these fences [20, 35, 39]. Figure 6 shows a fence design that orients the given part regardless of its initial orientation. We shall see below that fence design can be regarded as finding a constrained sequence of push directions. The additional constraints make fence design considerably more difficult than sensorless orientation by a pushing jaw.

Wiegley *et al.* [39] gave an exponential algorithm for computing the shortest sequence of fences for a given part, if such a sequence exists. They conjectured that a fence design exists for any polygonal part. We prove the conjecture that a fence design exists for any polygonal part. In addition, we give an $O(n^3)$ algorithm for computing a fence design of minimal length (in terms of the number of fences used), and discuss extensions and possible improvements.

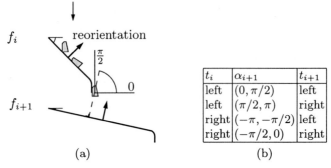

t_i	α_{i+1}	t_{i+1}
left	$(0, \pi/2)$	left
left	$(\pi/2, \pi)$	right
right	$(-\pi, -\pi/2)$	left
right	$(-\pi/2, 0)$	right

(a)   (b)

Fig. 7. (a) For two successive left fences, the reorientation of the push direction lies in the range $(0, \pi/2)$. (b) The ranges op possible reorientations of the push direction for all pairs of successive fence types.

We address the problem of designing a shortest possible sequence of fences $f_1, \ldots, f_k$ that will orient P when it moves down a conveyor belt and slides along these fences. Let us assume that the conveyor belt moves vertically from top to bottom, as indicated in the overhead view in Figure 6. We distinguish between left fences, which are placed along the left belt side, and right fences, which are placed along the right side. The angle or orientation of a fence f_i denotes the angle between the upward pointing vector opposing the motion of the belt and the normal to the fence with a positive component in upward direction. The motion of the belt turns the sliding of the part along a fence into a push by the fence. The direction of the push is – by the zero friction assumption – orthogonal to the fence with a positive component in the direction opposing the motion of the belt. Thus, the motion of the belt causes any push direction to have a positive component in the direction opposing the belt motion. We now transform this constraint on the push direction relative to the belt into a constraint on successive push directions relative to the part.

Sliding along a fence f_i causes one of P's edges, say e, to align with the fence. The carefully designed [20] curved tip of the fence guarantees that e is aligned with the belt sides as P leaves the fence. If f_i is a left (right) fence then e faces the left (right) belt side (see Figure 7). Assume f_i is a left fence. The reorientation of the push direction is the difference between the final contact direction of f_i and the initial contact direction of f_{i+1}. At the moment of leaving f_i, the contact direction of f_i is perpendicular to the belt direction and towards the right belt side. So, the reorientation of the push direction is expressed relative to this direction.

Figure 7(a) shows that the reorientation α_{i+1} is in the range $(0, \pi/2)$ if we choose f_{i+1} to be a left fence. If we take a right fence f_{i+1} then the reorientation is in the range $(\pi/2, \pi)$. A similar analysis can be done when P leaves a right fence and e faces the left belt side. The results are given in Figure 7(b).

The table shows that the type t_i of fence f_i imposes a bound on the reorientation α_{i+1}. Application of the same analysis to fences f_{i-1} and f_i and reorientation α_i leads to the following definition of a valid fence design [39].

Definition 2 [39] *A fence design is a push plan* $\alpha_1, \ldots, \alpha_k$ *satisfying for all* $1 \leqslant i < k$:

$$\alpha_i \in (0, \pi/2) \cup (-\pi, -\pi/2) \Rightarrow \alpha_{i+1} \in (0, \pi/2) \cup (\pi/2, \pi)$$
$$\wedge \ \alpha_i \in (-\pi/2, 0) \cup (\pi/2, \pi) \Rightarrow \alpha_{i+1} \in (-\pi/2, 0) \cup (-\pi, -\pi/2).$$

Definition 2 immediately shows that the linear-length push plans by Chen and Ierardi are valid fence designs for parts with a push function with a uniquely defined longest half-step of length $\alpha < \pi/2$. In other words, such parts can be oriented by a sequence of equivalent fences along one side of the belt of length $O(n)$. It is much harder to prove that all other parts can also be oriented by a sequence of fences [8].

Theorem 2. *Any polygonal part with n vertices can be oriented up to symmetry by a fence design.*

The results from the preceding section indicate that eccentric parts can be oriented by a constant number of fences.

3.1   A Simple Graph-Based Algorithm

We now turn our attention to the computation of the shortest fence design that will orient a given part. We denote the sequence of *stable* equilibrium orientations of P by Σ. As every fence puts the part in a stable equilibrium orientation, the part is in one of these $|\Sigma| = O(n)$ orientations as it travels from one fence to another. Let us label these stable equilibria $\sigma_1, \ldots, \sigma_{|\Sigma|}$. The problem is to reduce the set of possible orientations of P to one stable equilibrium $\sigma_i \in \Sigma$ by a sequence of fences. We build a directed graph on all possible *states* of the part as it travels from one fence to a next fence. A state consists of a set of possible orientations of the part plus the type (left or right) of the last fence, as the latter imposes a restriction on the reorientation of the push direction. Although there are $2^{|\Sigma|}$ subsets of Σ, it turns out that we can restrict ourselves to subsets consisting of sequences of adjacent stable equilibria. Any such sequence can be represented by a closed interval I of the form $[\sigma_i, \sigma_j]$ with $\sigma_i, \sigma_j \in \Sigma$. The resulting graph has $|\Sigma|^2$ nodes.

Consider two graph nodes (I, t) and (I', t'), where $I = [\sigma_i, \sigma_j]$ and I' are intervals of stable equilibria and t and t' are fence types. Let $A_{t,t'}$ be the open interval of reorientations admitted by the successive fences of types t and t' according to Figure 7(b). There is a directed edge from (I, t) to (I', t') if there is an angle $\alpha \in A_{t,t'}$ such that a reorientation of the push direction by α followed by a push moves any stable equilibrium in I into a stable orientation in I'. To check this condition, we determine the preimage $(\phi, \psi) \supseteq I'$ of I' under the push function. Observe that if $|I| = \sigma_j - \sigma_i < \psi - \phi$, any reorientation in the open interval $(\phi - \sigma_i, \psi - \sigma_j)$ followed by a push will map I into I'. We add an edge from (I, t) to (I', t') if $(\phi - \sigma_i, \psi - \sigma_j) \cap A_{t,t'} \neq \emptyset$, and label this edge with this non-empty intersection. For convenience, we add a source and a sink to the graph. We connect the source to every node $(I = [\sigma_i, \sigma_{i-1}], t)$, and we connect every node $(I = [\sigma_i, \sigma_i], t)$ to the sink. The graph has $O(n^4)$ edges. Every path from the source to the sink now represents a fence design. A fence design of minimum length corresponds to a shortest such path.

An important observation is that some graph edges are redundant if we are just interested in a fence design of minimum length. Consider a node (I, t) and all its outgoing edges to nodes $(I' = [\sigma_i, \sigma_j], t')$ for a fixed σ_I and t'. Lemma 3 [11] shows that only the edge to the node corresponding to the shortest such I' is required.

Lemma 3. *Let (I, t), (I', t'), and (I'', t') be nodes such that I' and I'' have a common left endpoint, and $I' \subset I''$. If there are edges from (I, t) to both (I', t') and (I'', t') then the edge from (I, t) to (I'', t') can be deleted without affecting the length of the shortest path.*

Informally, the lemma says that we can afford to be greedy in our wish to reduce the length of the interval of possible orientations. It reduces the number of graph edges to $O(n^3)$.

The computation of the reduced graph for fence design is easy. In the reduced graph, each node with interval $[\sigma_i, \sigma_j]$, has just one outgoing edge to the set of nodes with intervals with a common left endpoint $\sigma_{i'}$ and a common fence type t'. The shortest interval with left endpoint $\sigma_{i'}$ is obtained by a push direction which maps σ_i onto $\sigma_{i'} - \ell$, where ℓ is the length of the half-step left of $\sigma_{i'}$. The construction of the graph follows directly from this observation. We align the interval with the left environment of the reachable orientations for a valid reorientation of the push direction, and compute the resulting interval after application of the push function. If it is not possible to align σ_i with $\sigma_{i'} - \ell$, then we take the reorientation of the jaw that gets us as close as possible to $\sigma_{i'} - \ell$.

The computation of the outgoing edges for one node can be accomplished in linear time, by shifting $[\sigma_i, \sigma_j]$ along the possible reorientations of the push direction. As a result, the total time required to compute the graph edges is $O(n^3)$. A breadth-first search on the graph takes $O(n^3)$ time, and results in the shortest fence design.

Theorem 3. *Let P be a polygonal part with n vertices. The shortest fence design that orients p up to symmetry can be computed in $O(n^3)$ time. The resulting design consists of $O(n^2)$ fences in the worst case.*

Theorem 3 immediately provides an upper bound of $O(n^2)$ on the length of the shortest fence design. We expect, however, that the true bound is $O(n)$.

3.2   An Output-Sensitive Algorithm

The running time of the preceding algorithm could be considered quite high when realizing that fence designs will often (or maybe even always) have linear (in the case of parts push functions with a unique longest half-step) or even constant (in the case of eccentric parts) length. This suggests that an algorithm whose running time is sensitive to the length of the fence design is to be preferred.

The main idea of the output-sensitive algorithm is to maintain the shortest interval of possible orientations after k fences, instead of precomputing the whole graph of all possible intervals of orientations. This is basically the same technique as used by Goldberg's algorithm to compute push plans [26]. Goldberg maintains the interval of possible orientations, and greedily shrinks this interval per application of the pushing jaw. We, however, must take into account the constraints of fence design. It is not sufficient to maintain a single shortest interval of possible orientations. Lemma 3 indicates that it is sufficient to maintain for each pair of a fence type and a stable orientation the shortest interval after leaving a fence of the given type starting with the given stable orientation. The algorithm should terminate as soon as one of the $2|\Sigma|$ intervals has shrunk to a single orientation. Updating the candidate intervals can be accomplished in $(\log n)$ time per interval using a range tree data structure (see [10] for details).

Theorem 4. *Let P be a polygonal part with n vertices. A shortest fence design that orients P up to symmetry can be computed in $O(kn \log n)$ time, where k is the length of the resulting fence design.*

The output-sensitive algorithm will in most cases be more efficient than the simpler graph-based approach; in fact, the former will only have a chance to be outperformed by the latter if parts exist that require a quadratic-length fence design.

Both algorithms can be modified to deal with situations in which there is friction between the part and the fences. This modification has no impact on the running time. On the other hand we lose the guarantee that a fence design always exists, so that the algorithm may have to report failure. The output-sensitive algorithm will be able to do so in $(n^3 \log n)$ time. See [10, 8] for other extensions.

4   Pushing Three-Dimensional Parts

A drawback of most achievements in the field of sensorless orientation of parts is that they only apply to planar parts, or to parts that are known to rest on a certain face. The generalization of conveyor belts and fences that we describe here attempts to bridge the gap to truly three-dimensional parts. The device we use is a cylinder with plates tilted toward the interior of the cylinder attached to the side. Across the plates there are fences. The part cascades down from plate to plate, and slides along the fences as it travels down a plate (see Figure 8(a)). The plate on which the part slides discretizes the first two degrees of freedom of rotation of the part. A part in alignment with a plate retains one undiscretized rotational degree of freedom. The orientation of the part is determined up to its roll, i.e. the rotation about the axis perpendicular to the plate. The fences, which are mounted across the plates, push the part from the side, and discretize the roll. We assume that P first settles on the plate before it reaches the fences which are mounted across the plate. Moreover, we assume that the fences do not topple the part but only cause it to rotate about the roll axis.

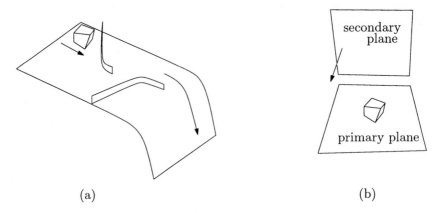

(a)   (b)

Fig. 8. (a) A part sliding down a plate with fences. (b) The same part on the jaw.

The objective of this section is to compute a set-up of plates and fences that is guaranteed to move a given asymmetric polyhedral part towards a unique final orientation. Such a set-up, or design, consists of a sequence of plate slopes, and for each plate a sequence of fence orientations.

When a part moves along a fence on a plate, it is essentially pushed from two orthogonal directions. This motivates us to first study the fundamental (but artificial) problem of pushing in three-dimensional space. Here, the part is assumed to float in the air while we push it from two orthogonal directions.

We show that a three-dimensional polyhedral part P can be oriented up to symmetry by a (particular) sequence of push actions, a *push plan*, of length $O(n^2)$, where n is the number of vertices of P. Furthermore, we give an $O(n^3 \log n)$ time algorithm to compute such a push plan. We show how to transform this three-dimensional push plan to a three-dimensional design for the plates and fences. The resulting design consists of $O(n^3)$ plates and fences, and can be computed in $O(n^4 \log n)$ time.

A polyhedral part in three-dimensional space has three rotational degrees of freedom. We assume that a fixed reference frame is attached to P and denote the orientation of P relative to this reference frame by (ϕ, ψ, θ), where (ϕ, ψ) denotes a point on the sphere of directions, and θ is the roll about the ray emanating from the origin through (ϕ, ψ).

4.1   Push Plan

We study the push actions of the plates and the fences in a more general setting by replacing a plate and a fence by two orthogonal planes. We call the planes the primary and secondary plane, respectively. A picture of the resulting jaw is given in Figure 8(b). Since the planes can only touch P at its convex hull, we assume without loss of generality that P is convex. We assume that the center-of-mass of P, denoted by c, is in the interior of P. Analogously to the cylindrical feeder, we assume that only after P has aligned with the primary plane, we apply the secondary plane. As the part rests on the primary plane, the secondary plane pushes P at its orthogonal projection onto the primary plane. We assume that the feature on which P rests retains contact with the primary plane as the secondary plane touches P. We assume that for any equilibrium orientation, which is an orientation for which P rests on the jaw, the projection of P onto the primary plane has no symmetry. We refer to a part with this property as being *asymmetric*.

In order to be able to approach the part from any direction, we make the (obviously unrealistic) assumption that the part floats in the air, and assume that we can control some kind of gravitational field which attracts the part in a direction towards the jaw. Also, we assume that the part quasi-statically aligns with the jaw, meaning that we ignore inertia.

A basic action of the jaw consists of directing and applying the jaw. The result of a basic action for a part in its reference orientation is given by the *push function*. The push function $p : [0, 2\pi) \times [-\pi/2, \pi/2] \times [0, 2\pi) \to [0, 2\pi) \times [-\pi/2, \pi/2] \times [0, 2\pi)$ maps a push direction of the jaw relative to P in its reference orientation onto the orientation of P after alignment with the jaw. The orientation of P after a basic action for a different initial orientation than its reference orientation is equal to the push function for the push direction plus the offset between the reference and the actual initial orientation of P.

In our approach to finding a push plan we do not explicitly compute the push function. Instead we occasionally query some data structure for the reorientation of the part when being pushed from a certain direction. Without going into the details, which are far from easy, we claim that this query takes $O(n \log n)$ time. We now use this fact to show that any asymmetric polyhedral part P can be oriented by a push plan of length $O(n^2)$. The part P has at most $O(n)$ equilibria with respect to the primary plane, and any projection of P onto the primary plane has $O(n)$ vertices. Hence, the total number of orientations of P compliant to the jaw is $O(n^2)$, and this bound turns out to be tight.

Let us, for a moment, assume that the part lies in a stable orientation on the primary plane. We can now reorient the jaw in such a way that the contact direction of the primary plane remains unchanged while the direction of the secondary plane is altered. A subsequent push by the jaw will cause the part to rotate about the normal to the pimary plane – keeping the same face of P in contact with the primary plane. The application of the jaw in this manner can therefore be regarded as a push operation on the 2D orthogonal projection of P. In Section 2 we have seen that an asymmetric 2D part with m vertices can be oriented up by means of planar push plan of length $O(m)$. Consequently, we can orient P in stable contact with the primary plane by $O(n)$ applications of the secondary plane.

Lemma 4. *Let P be an asymmetric polyhedral part with n vertices. There exists a plan of length $O(n)$ that puts P into a given orientation (ϕ, ψ, θ) from any initial orientation (ϕ, ψ, θ')*

We call the operation that orients P for a single stable equilibrium contact direction (ϕ, ψ) of the primary plane COLLIDEROLLSSEQUENCE(ϕ, ψ). It allows us to eliminate the uncertainty in the roll for any stable contact direction of the primary plane. In an initialization phase we reduce the number of possible orientations of P to $O(n)$ by executing COLLIDEROLLSSEQUENCE(ϕ, ψ) for all equilibrium contact directions (ϕ, ψ) of the primary plane. We let Σ be the set of the resulting possible orientations. Lemma 5 (see [15] for a proof) provides us with push operations to further reduce the number of possible orientations.

Lemma 5. *There exist two antipodal reorientations of the primary plane that map any pair of orientations (ϕ, ψ, θ), and (ϕ', ψ', θ') of a polyhedral part onto orientations $(\tilde{\phi}, \tilde{\psi}, \tilde{\theta})$ and $(\tilde{\phi}', \tilde{\psi}', \tilde{\theta}')$ that satisfy $\tilde{\phi} = \tilde{\phi}'$ and $\tilde{\psi} = \tilde{\psi}'$.*

We call the basic operation that collides two orientations onto the same equilibrium for the primary plane COLLIDEPRIMARYACTION. Combining Lemma 4 and 5 leads to a construction of a push plan for a polyhedral part. The following algorithm orients a polyhedral part without symmetry in the planar projections onto supporting planes of its stable faces.

ORIENTPOLYHEDRON(P):
▷   After initialization $|\Sigma| = O(n)$
1. **while** $|\Sigma| > 1$ **do**
    2.1   pick (ϕ, ψ, θ), $(\phi', \psi', \theta') \in \Sigma$
    2.2   plan $\leftarrow$ COLLIDEPRIMARYACTION$((\phi, \psi, \theta), (\phi', \psi', \theta'))$
        ▷ Lemma 5;
        ▷ plan$(\phi, \psi, \theta) = (\phi'', \psi'', \theta'')$, and plan$(\phi', \psi', \theta') = (\phi'', \psi'', \theta''')$
    2.3   **for all** $(\tilde{\phi}, \tilde{\psi}, \tilde{\theta}) \in \Sigma$
        2.3.1 $(\tilde{\phi}, \tilde{\psi}, \tilde{\theta}) \leftarrow$ plan$(\tilde{\phi}, \tilde{\psi}, \tilde{\theta})$.
    2.4   plan $\leftarrow$ COLLIDEROLLSSEQUENCE(ϕ'', ψ'')
        ▷ Lemma 4
    2.5   **for all** $(\tilde{\phi}, \tilde{\psi}, \tilde{\theta}) \in \Sigma$
        2.5.1 $(\tilde{\phi}, \tilde{\psi}, \tilde{\theta}) \leftarrow$ plan$(\tilde{\phi}, \tilde{\psi}, \tilde{\theta})$.

The algorithm repeatedly takes two of the remaining possible orientations of the part and computes a reorientation that maps these two orientations onto two different orientations whose representations share the first two coordinates. Step 2.3 maps all currently possible orientations onto the orientations result from applying the appropriately reoriented jaw. We recall that this step takes $O(n \log n)$ for each of the at most $O(n)$ remaining orientations. At this stage, the number of faces of P that can be aligned with the primary plane is reduced by one. The remaining steps map the two orientations that share the first two coordinates onto a single orientation – essentially by means of a planar push plan of $O(n)$ length for the projection of P. Since the number of iterations of ORIENTPOLYHEDRON(P) is $O(n)$ the algorithm runs in $O(n^3 \log n)$ time and results in a push plan of length $O(n^2)$.

Theorem 5. *A push plan of length $O(n^2)$ for an asymmetric polyhedral part with n vertices can be computed in $O(n^3 \log n)$ time.*

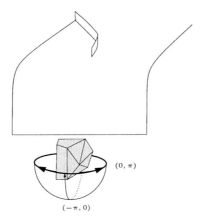

(0, π)

(−π, 0)

Fig. 9. The next plate can only touch the lower half of the part.

4.2   Plates and Fences

We use the results from the preceding subsection to determine a design for the feeder consisting of tilted plates with curved tips, each carrying a sequence of fences. The motion of the part effectively turns the role of the plates into the role of the primary pushing plane, and the role of the fences into the role of the secondary pushing plane. We assume that the part quasi-statically aligns to the next plate, similar to the alignment with the primary plane of the generic jaw. Also, we assume that the contact direction of the plate does not change as the fences brush the part, i.e. the part does not tumble over.

The fact that the direction of the push, i.e., the normal at the fence, must have a non-zero component in the direction opposite to the motion of the part, which slides downward under the influence of gravity, imposes a restriction on successive push directions of the secondary plane. The restriction is equivalent to that in planar fence design. Theorem 3 shows that it is possible to orient a planar polygonal part (hence a polyhedral part resting on a fixed face) using $O(n^2)$ fences. The optimal fence design can be computed in $O(n^3)$ time.

As the part moves towards the end of a plate, the curved end of the plate causes the feature on which the part rests to align with the vertical axis, while retaining the roll of the part. When the part leaves the plate, the next plate can only push the part from below. This draws restrictions on the possible reorientations of the primary plane, in the model with the generic three-dimensional jaw (see Figure 9). Careful analysis shows that the reorientation of the primary plane is within $(-\pi, 0) \times (0, \pi)$ when the last fence of the last plate was a left fence. Similarly, for a last right fence, the reorientation of the primary plane is within $(0, \pi) \times (0, \pi)$.

The gravitational force restricts our possible orientations of the primary plane in the general framework. Fortunately, Lemma 5 gives us two antipodal possible reorientations of the primary plane. It is not hard to see that one of these reorientations is in the reachable hemisphere of reorientations of the push direction

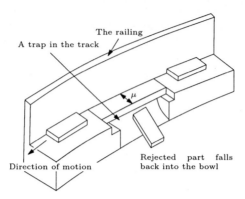

Fig. 10. Vibratory bowl feeder track [19].

of the primary plane for two succesive plates. This implies we can still find a set-up of plates and fences of $O(n^3)$ complexity.

Theorem 6. *An asymmetric polyhedral part can be oriented using $O(n^3)$ fences and plates. We can compute the design in $O(n^4 \log n)$ time.*

5   Trap Design

The oldest and still most common part feeder is the *vibratory bowl feeder*. It consists of a bowl filled with parts surrounded by a helical metal track [18, 19]. The bowl and track undergo an asymmetric helical vibration that causes parts to move up the track, where they encounter a sequence of mechanical devices such as wiper blades, grooves and traps. Most of these devices are filters that serve to reject (force back to the bottom of the bowl) parts in all orientations except for the desired one. In this section, we consider the use of traps to filter polygonal parts on a track. A trap is a (partial) interruption of the track. We focus on polygonal traps. Figure 10 shows a section of track with a rectangular trap. Parts in undesired orientations fall back into the bowl, other orientations remain supported.

Specific to vibratory bowls, researchers have used simulation [7, 27, 32], heuristics [28], and genetic algorithms [24] to design traps. Perhaps closest in spirit to our work is Caine's PhD thesis [21] which develops geometric analysis tools to help designers by rendering the configuration-space for a given combination of part, trap, and obstacle. Caine also gives some heuristics to design feeder track features.

This section reports on the analysis and design of traps that allow a part to pass in only one orientation. To the extent of our knowledge, no research in the systematic algorithmic design of vibratory bowl feeders has previously been conducted. As the techniques and analyses used in trap design differ largely from those used for the feeders in the preceding sections – which are all based on pushing – we do not provide extensive coverage of all our algorithms for traps. Instead we confine ourselves to a brief characterization of when a part falls into a trap and to reporting our algorithmic results for trap analysis and design. We focus on two-dimensional parts, or, in other words on three-dimensional parts that are known to rest on a certain stable face.

5.1   Modeling and Analysis

The track in a bowl feeder is slightly tilted to keep the part in contact with the railing of the track as it moves. Although the vibration of the bowl causes the part to hop along the track we simplify our analysis by assuming that is slides. The radius function of the part P determines the at most $O(n)$ stable orientations in which the part can move; these correspond to local minima of the radius function.

Let T be a polygonal trap, and assume it has m vertices. In reality, the part P (which is assumed to be in a fixed orientation) slides across the trap. Since is it convenient to have a stationary center-of-mass c of the part in our analysis, we choose to assume that the trap moves underneath the part (which is clearly equivalent). We denote the trap in a configuration q by $T(q)$. Note that a configuration is – because of the simple sliding motion of the part in reality – nothing more than a horizontal displacement, and as such representable by a single value. The supported area $S(q)$ of a part P with the trap placed underneath in configuration q is defined by $S(q) = P - \text{int}(T(q))$. We denote the convex hull of a shape X by $CH(X)$. Lemma 6 says when a part is safe when placed over a trap, i.e., when it does not fall into the trap.

Lemma 6. *The part P is safe above T in a configuration q if and only if $c \in CH(S(q))$, or, in other words, if and only if there is no line through c that has $CH(S(q))$ entirely on one side.*

It is clear that a part will safely move across a trap if P is safe in every configuration q in the motion of T.

The above characterization is the key to our algorithm for computing whether a part P in a given orientation will safely move across a trap. The crucial convex hull $CH(S(q))$ is determined by vertices of T and P, and by intersections of edges of T and P. Our algorithm plots the directions of the rays emenating from c towards each of the aforementioned vertices and intersections as a function of q. A simple sweep (see e.g. [6]) suffices to detect whether the interval of all these directions remains longer than π at all q. A somewhat more efficient algorithm exists for the case where both the trap and the part are convex.

Theorem 7. *Let P be a polygonal part with n vertices and T be a polygonal trap with m vertices. We can report whether P will move safely across T in $O(n^2 m \log n)$ time, or in $O((n + m) \log n)$ time if both P and T are convex.*

The result for a general part and trap has recently been improved by Agarwal et al. [1].

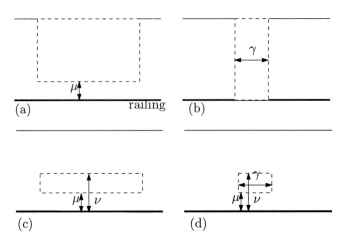

Fig. 11. The four rectilinear traps of this section: (a) a balcony, (b) a gap, (c) a canyon, and (d) a slot. The bold lines at the bottom of the pictures depict the railing. The line at the top depicts the edge of the track at the inside of the bowl. The traps are dashed.

5.2   Design of Traps

In this subsection we report results on the design of traps that allow a given part to pass in only one of its stable orientations. A trap with this property is said to have the feeding property. We consider four different specific rectangular traps and arbitrary polygonal traps. Figure 11 shows the four rectilinear traps along with the parameters that specify their measures.

A balcony is a long interruption of the upper part of the supporting area of the track. Let us consider the radii of the part P, or, in other words, the distances of the center-of-mass c to the railing in all of its stable orientations. Assume that there is a uniquely defined minimum, corresponding to an orientation ϕ. If we choose the height μ of the balcony slightly larger than this minimum, then it is immediately clear that P will only be able to pass T when in orientation ϕ. As the minimum radius is computable in $O(n)$ time we obtain the following result.

Theorem 8. *In $O(n)$ time we can design a balcony with the feeding property for a polygonal part with n vertices, or report that no such balcony exists.*

Unfortunately, the design of the other feeders is considerably harder.

A gap is an interruption of the trach that spans the entire width of the track. Its shape is determined by a single parameter, the gap length γ. Our algorithm (see [9, 12, 13] for details) determines a choice for γ that allows P to pass in only one orientation.

Theorem 9. *In $O(n^2 \log n)$ time we can design a gap with the feeding property for a polygonal part with n vertices, or report that no such gap exists. The bound reduces to $O(n^2)$ if the part is convex.*

A canyon is a long rectangular interruption of the supporting area of the track. Its shape is defined by the distances μ and ν from the lower and upper boundary to the railing. Our algorithm [9, 12, 13] determines a suitable choice for μ and ν.

Theorem 10. *In $O(n^2\alpha(n)\log n)$ time we can design a canyon with the feeding property for a polygonal part with n vertices, or report that no such canyon exists; $\alpha(n)$ is the extremely slowly growing inverse Ackermann function.*

A slot is a true rectangular interruption of the supporting area of the track, and as such specified by three parameters μ, ν, and γ. Our algorithm [9, 12, 13] finds a slot with the feeding property if one exists.

Theorem 11. *In $O(n^8)$ time we can design a slot with the feeding property for a polygonal part with n vertices, or report that no such slot exists.*

Finally, we consider arbitrary polygonal traps with k vertices. Such a trap can be represented by $2k$ parameters. Our approach to computing a k-vertex trap that allows a given part to pass in only one orientation uses high-dimensional arrangements and quantifier elimination. Using recent results by Basu et al. [4, 5], we obtain our final result [12], which is given below.

Theorem 12. *In $O((nk)^{O(k^2)})$ time we can design a polygonal trap with k vertices with the feeding property for a polygonal part with n vertices, or report that no such trap exists.*

References

1. P.K. Agarwal, A.D. Collins, and J.L. Harer. Minimal trap design. 2000.
2. S. Akella, W. Huang, K. M. Lynch, and M. T. Mason. Parts feeding on a conveyor with a one joint robot. *Algorithmica*, 26:313–344, 2000.
3. S. Akella and M. T. Mason. Posing polygonal objects in the plane by pushing. In *IEEE International Conference on Robotics and Automation (ICRA)*, pages 2255–2262, 1992.
4. S. Basu. New results on quantifier elimination over real closed fields and applications to constraint databases. *Journal of the ACM*, 46(4):537–555, 1999.
5. S. Basu, R. Pollack, and M-F. Roy. On the combinatorial and algebraic complexity of quantifier elimination. *Journal of the ACM*, 43:1002–1045, 1996.
6. M. de Berg, M. van Kreveld, M. H. Overmars, and O. Schwarzkopf. *Computational Geometry: Algorithms and Applications*. Springer-Verlag, Berlin, 1997.
7. D. Berkowitz and J. Canny. Designing parts feeders using dynamic simulation. In *IEEE International Conference on Robotics and Automation (ICRA)*, pages 1127–1132, 1996.
8. R-P. Berretty. *Geometric design of part feeders*. PhD thesis, Institute of Information and Computing Sciences, Utrecht University, 2000.
9. R-P. Berretty, K. Y. Goldberg, L. Cheung, M. H. Overmars, G. Smith, and A. F. van der Stappen. Trap design for vibratory bowl feeders. In *IEEE International Conference on Robotics and Automation (ICRA)*, pages 2558–2563, 1999.

10. R-P. Berretty, K. Y. Goldberg, M. H. Overmars, and A. F. van der Stappen. Algorithms for fence design. In *Robotics, the algorithmic perspective*, pages 279–295. A.K. Peters, 1998.

11. R-P. Berretty, K. Y. Goldberg, M. H. Overmars, and A. F. van der Stappen. Computing fence designs for orienting parts. *Computational Geometry: Theory and Applications*, 10(4):249–262, 1998.

12. R-P. Berretty, K. Y. Goldberg, M. H. Overmars, and A. F. van der Stappen. Geometric techniques for trap design. In *Annual ACM Symposium on Computational Geometry*, pages 95–104, 1999.

13. R-P. Berretty, K. Y. Goldberg, M. H. Overmars, and A. F. van der Stappen. Geometric trap design for automatic part feeders. In *International Symposium on Robotics Research (ISRR)*, pages 139–144, 1999.

14. R-P. Berretty, K. Y. Goldberg, M. H. Overmars, and A. F. van der Stappen. Orienting parts by inside-out pulling. In *IEEE International Conference on Robotics and Automation (ICRA)*, 2001. To appear.

15. R-P. Berretty, M. H. Overmars, and A. F. van der Stappen. Orienting polyhedral parts by pushing. *Computational Geometry: Theory and Applications*, 2001. To appear.

16. K-F. Böhringer, V. Bhatt, B.R. Donald, and K. Y. Goldberg. Algorithms for sensorless manipulation using a vibrating surface. *Algorithmica*, 26:389–429, 2000.

17. K-F. Böhringer, B. R. Donald, and N.C. MacDonald. Upper and lower bounds for programmable vector fields with applications to mems and vibratory plate part feeders. *Algorithms for Robotic Motion and Manipulation*, J.-P. Laumond and M. Overmars (Eds.), A.K. Peters, pages 255–276, 1996.

18. G. Boothroyd and P. Dewhurst. *Design for Assembly – A Designers Handbook*. Department of Mechanical Engineering, University of Massachusetts, Amherst, Mass., 1983.

19. G. Boothroyd, C. Poli, and L. Murch. *Automatic Assembly*. Marcel Dekker, Inc., New York, 1982.

20. M. Brokowski, M. A. Peshkin, and K. Y. Goldberg. Optimal curved fences for part alignment on a belt. *ASME Transactions of Mechanical Design*, 117, 1995.

21. M. E. Caine. The design of shape interaction using motion constraints. In *IEEE International Conference on Robotics and Automation (ICRA)*, pages 366–371, 1994.

22. J. Canny and K. Y. Goldberg. Risc for industrial robotics: Recent results and open problems. In *IEEE International Conference on Robotics and Automation (ICRA)*, pages 1951–1958, 1994.

23. Y-B. Chen and D. J. Ierardi. The complexity of oblivious plans for orienting and distinguishing polygonal parts. *Algorithmica*, 14:367–397, 1995.

24. A. Christiansen, A. Edwards, and C. Coello. Automated design of parts feeders using a genetic algorithm. In *IEEE International Conference on Robotics and Automation (ICRA)*, pages 846–851, 1996.

25. M. A. Erdmann and M. T. Mason. An exploration of sensorless manipulation. *IEEE Journal of Robotics and Automation*, 4:367–379, 1988.

26. K. Y. Goldberg. Orienting polygonal parts without sensors. *Algorithmica*, 10(2):201–225, 1993.

27. M. Jakiela and J. Krishnasamy. Computer simulation of vibratory parts feeding and assembly. In *International Conference on Discrete Element Methods*, pages 403–411, 1993.

28. L. Lim, B. Ngoi, S. Lee, S. Lye, and P. Tan. A computer-aided framework for the selection and sequencing of orientating devices for the vibratory bowl feeder. *International Journal of Production Research*, 32(11):2513–2524, 1994.

29. K. M. Lynch and M. T. Mason. Stable pushing: Mechanics, controllability, and planning. *International Journal of Robotics Research*, 15(6):533–556, 1996.

30. M. T. Mason. Mechanics of robotic manipulation. Unpublished book.

31. M. T. Mason. *Manipulator grasping and pushing operations*. PhD thesis, MIT, 1982. published in *Robot Hands and the Mechanics of Manipulation*, MIT Press, Cambridge, 1985.

32. G. Maul and M. Thomas. A systems model and simulation of the vibratory bowl feeder. *Journal of Manufacturing Systems*, 16(5):309–314, 1997.

33. B. K. Natarajan. Some paradigms for the automated design of parts feeders. *International Journal of Robotics Research*, 8(6):89–109, 1989.

34. M. A. Peshkin and A. C. Sanderson. The motion of a pushed sliding workpiece. *IEEE Journal of Robotics and Automation*, 4(6):569–598, 1988.

35. M. A. Peshkin and A. C. Sanderson. Planning robotic manipulation strategies for workpieces that slide. *IEEE Journal of Robotics and Automation*, pages 696–701, 1988.

36. A. Rao and K. Y. Goldberg. Manipulating algebraic parts in the plane. *IEEE Transactions on Robotics and Automation*, 11:589–602, 1995.

37. A. F. van der Stappen, K. Y. Goldberg, and M. H. Overmars. Geometric eccentricity and the complexity of manipulation plans. *Algorithmica*, 26:494–514, 2000.

38. D. E. Whitney. Real robots don't need jigs. In *IEEE International Conference on Robotics and Automation (ICRA)*, volume 1, pages 746–752, 1986.

39. J. A. Wiegley, K. Y. Goldberg, M. Peshkin, and M. Brokowski. A complete algorithm for designing passive fences to orient parts. *Assembly Automation*, 17(2):129–136, 1997.

CoolBOT: A Component-Oriented Programming Framework for Robotics

Jorge Cabrera-Gámez, Antonio Carlos Domínguez-Brito,
and Daniel Hernández-Sosa

Centro de Tecnología de los Sistemas y de la Inteligencia Artificial (CeTSIA)
University of Las Palmas de Gran Canaria (ULPGC)
Edificio de Informática y Matemáticas
Campus Universitario de Tafira
35017 Las Palmas, Spain
{jcabrera,acdbrito,dhernandez}@dis.ulpgc.es
http://mozart.dis.ulpgc.es

Abstract. This paper introduces at the specification level CoolBOT, a component-oriented programming framework for robotics designed to assist robotic system developers in obtaining more structured and reusable systems without imposing any specific architecture. Within this framework components are conceived as Port Automata (PA)[13] that interact through their ports and that can be composed to build up new components from existing ones. Components, no matter if they are atomic or compound, are internally modeled as Discrete Event Systems and controlled using the same state control graph. CoolBOT hides the programmer any aspects related to communications and provides standard mechanisms for different modes of data exchange between components, exception handling and support for distributed computing environments.

1 Introduction

During the last years we have known about a number of successful projects in robotics in very different fields, ranging from exploration in space and harsh environments on Earth to medical robotics and entertainment. These systems illustrate from different perspectives that there is actually a wealth of well developed solutions to many of the basic problems that need to be solved and integrated when designing a robotic system, even though many of them still remain as very active research fields. This situation is fostering the development of more ambitious systems of increasing complexity to face new challenges but also makes evident some difficulties.

One of them is the lack of a methodology to develop robotic systems in a principled way, a problem that has been identified by several authors, who have recognized that traditional programming and validation techniques are not adequate for intelligent robotic systems [10][4]. This methodology should help in designing systems that were more scalable, reusable in new scenarios, more robust and reliable, and easier to debug and profile.

G.D. Hager et al. (Eds.): Sensor Based Intelligent Robots, LNCS 2238, pp. 282–304, 2002.

These problems have been often tackled proposing new robot architectures and specification languages (see [4] for a good up-to-date review). Certainly, some architectures seem better suited than others to favor the goals stated above as proved by the fact that majority of current intelligent robot systems use some sort of hybrid architecture [9][2][1][11], effectively combining the advantages of reactive and deliberative architectures.

Other research has been focused in the design of specification languages for robot systems, with a large variety in objectives and scope. Relevant to the research presented here, are those languages designed for task-level control as RAP [6], ESL [8] or TDL [12]. Typically, these languages offer primitives for task coordination and control, task communication, and also basic primitives for exception handling. An advantage of ESL and TDL is that they are, respectively, extensions of Lisp and C++. This allows to use the same language for coding the whole system, that is, not only for task control, but also for the rest of computations.

Along this paper we will introduce CoolBOT, a component-oriented programming framework being developed at ULPGC, whose main goal is to bridge the gap between these two approaches for tackling the design of complex robot systems. That is, CoolBOT is aimed at providing the designer of such systems with a powerful environment where it is possible to synthesize different architectures using the same specification language. As it will be explained, CoolBOT's basic building blocks, i.e. components, are modeled as Port Automata [13] that share the same control and communication models.

Reactivity, adaptability, modularity, robustness and stability are some of the design goals for almost any system. All of them are related to or dependent on how the components of a system interact among themselves and with the environment. In CoolBOT, all components share the same interfaces and control scheme. In this way, it is possible to define a modular system that uses a set of common abstractions to carry-out communication, coordination and exception handling among modules.

The rest of this paper discusses first the design goals that are guiding the development of CoolBOT, followed by a description of the elements that make up CoolBOT's model of a component. Finally, a simulated programming example is explained in detail.

2   Objectives & Design Principles

Latest trends in Software Engineering are exploiting the idea of Components as the basic units of deployment when building complex software systems, specially if software reuse, modular composition and third-party software integration are important issues. CoolBOT should be understood as a component framework, in the sense defined in [15], as it offers "a collections of rules and interfaces that govern the interaction of components plugged into the framework. A component framework typically enforces some of the more vital rules of interaction by encapsulating the required interaction mechanisms".

The following are the most important considerations that have guided the design of CoolBOT:

- **Component-Oriented.** CoolBOT is conceived as a component-oriented programming that relies on a specification language to manipulate components as building blocks in order to functionally define a robotic system by integrating components. This approach not only enforces modularity and discipline but requires well defined rules for interfacing and interaction. We consider these as basic requirements for achieving the long term goal of interoperability with third-party developed components (i.e. developed by other research groups or companies) [10].
- **Component Uniformity.** A component-based approach clearly demands certain level of uniformity among components. Within CoolBOT this uniformity manifests itself in two important aspects. First a uniform interface is defined for all components based on the concept of port automata. Additionally, a uniform internal structure for components facilitates its observability and controllability, i.e. the possibility of monitoring and controlling the inner state of a component. We consider that these properties are key elements when defining robust systems, making its design and implementation less error-prone. At the same time, component uniformity sets the real basis for development of debugging and profiling tools.
- **Robustness & Controllability.** A component-oriented robot system will be robust and controllable because its components are also robust and controllable. A component will be considered robust when:
  1. It is able to monitor its own performance, adapting to changing operating conditions, and it also implements its own adaptation and recovery mechanisms to deal with all errors that are detectable internally.
  2. Any error detected by a component that cannot be recovered by its own means, should be notified using standard means through its interface, bringing the component to an idle state waiting for external intervention, that either will order the component to restart or to abort. Communications sent or received by a component when dealing with exceptions should be common to all components.

  Furthermore a component will be considered controllable when it can be brought with external supervision - by means of a controller or a supervisor - through its interface, along an established control path. In order to obtain such an external controllability, components will be modeled as automata whose states can be forced by an external supervisor, and where all components will share the same control automaton structure [7].
- **Modularity & Hierarchy.** The architecture of a robot system will be defined in CoolBOT using selected components as elementary functional units. As in almost any component-based framework, there will be atomic and compound units. An atomic component will be indivisible, i.e. one that is not made up of other components. A compound component will be a component which includes in its definition other components, atomic or not, and provides a supervisor for their monitoring and control. With this vision, a whole

system is nothing else but a large compound component including several components, which in turn include another components, and so on, until this chain of decompositions finishes when an atomic component is reached. Thus, a complete system can be envisioned as a hierarchy of components from the coordination and control point of view.

- **Distributed.** The distribution of components over a distributed computing environment is a basic need in many control systems. CoolBOT manages communications between components that are being executed in the same or different computers exactly in the same manner.
- **Reuse.** Components are units that keep their internals hidden behind a uniform interface. Once they have been defined, implemented and tested they can be used as components inside any other bigger component or system. Modern robot systems are becoming really complex systems and very few research groups have the human resources needed to build systems from scratch. Component-oriented designs represent a suitable way to alleviate this situation. We believe that research in robotics might enormously benefit from the possibility of exchanging components between labs as a mean for cross-validation of research results.
- **Completeness & Expressiveness.** The computing model underlying CoolBOT should prove valid to build very different architectures for robotic systems and expressive enough to deal with concurrence, parallelism, distributed an shared resources, real time responsiveness, multiple simultaneous control loops and multiple goals in a principled and stable manner.

3 Elements of the System: Components

In CoolBOT, components are modeled as Port Automata [13][14][5] which defines an active entity which carries out a specific task, and performs all external communication by means of its input and output ports. A component usually executes in parallel or concurrently with other components that interact and communicate between them through their input and output ports. Hence, a system programmed with CoolBOT could be seen as a network of components interacting among them to achieve the pretended system's behaviors.

Components act on their own initiative and are normally weakly coupled, that is, no acknowledgements are necessary when they communicate through their ports. Most components could be described as a data flow machine, producing output whenever it has data on its input. Otherwise, the component is idle waiting for new input data to process.

Components can be atomic, i.e. indivisible, or compound when it is made up of a composition or assemblage of other atomic and/or compound components. With independence of its type, components are externally equivalent, offering the same uniform external interface and internal control structure. These properties are extremely important in order to attain standard mechanisms that guarantee that any component can be externally monitored and controlled.

Once a component, atomic or not, has been designed, implemented and tested, it can be used wherever it should be necessary, it can be instantiated

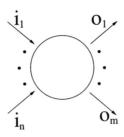

Fig. 1. The external view of a component.

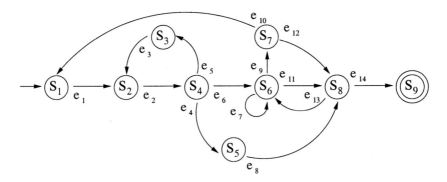

Fig. 2. The internal view of a component.

once or multiple times, locally or remotely in a computer network. Therefore components constitute in CoolBOT the functional building blocks to program robotic systems.

3.1   Components as Port Automata

Components have been modeled as port automata because this concept establishes a clear distinction between the internal functionality of an active entity, the automaton, and its external interface, the input and output ports. Figure 1 displays the external view of a component where the component itself is represented by the circle, input ports (i_i) by the arrows oriented towards the circle, and output ports (o_i) by arrows oriented outwards. As shown by the figure, the external interface keeps the component's internals hidden. Figure 2 depicts an example of the internal view of a component, concretely the automaton that models it, where the circles are the states of the automaton, and the arrows, the transitions between states. These transitions are triggered by events (e_i), caused either by incoming data through a port, by an internal condition, or by a combination of port incoming data and internal conditions. Doubled circles are automaton final states. Modeling the internal functionality as an automaton provides with a mean to make the component observable and controllable.

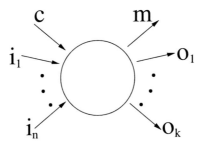

Fig. 3. The default ports. The control port **c** and the monitoring port **m**.

3.2 The Default Ports

In order to be able to build modular systems from reusable units, in CoolBOT all components must be observable and controllable at any time from outside the component itself. The approach followed is to impose a uniform interface and a common control structure on components. Figure 3 displays part of this external uniform interface, the **default ports**, present in every component:

- The control port **c**. A component can be externally controlled using the so called *controllable variables*.
- The monitoring port **m**. Variables exported by the component, termed *observables*, can be monitored through this port.

3.3 The Default Automaton

Figure 4 shows the automaton that represents the possible states and transition between them for every component. This **default automaton** contains all possible control paths for a component. In the figure some transitions are labeled as c_i's denoting that they are provoked by a command through the control port **c** displayed in figure 3. The default automaton is said to be "controllable" because it can be brought externally by means of its control port **c** to any of the controllable states of the automaton: **ready, running, suspended** and **dead**, in finite time. The rest of states are reachable only internally, and from them a transition to one of the controllable ones can be forced. The **running** state, the dashed state in figure 4, represents the state or set of states that structures the specific functionality of a certain component. This particular automaton, termed **user automaton**, varies among components and must be defined by the developer/user when the component is implemented.

When a component is instantiated, it is brought to the **starting** state, where the component captures resources needed for its operation and performs its initialization. Any error requesting resources may provoke a new attempt of asking for them, until a maximum number of attempts has been tried. In such a case, initialization is unsuccessful and the automaton transits to the **starting error** state, where it must await for external intervention through the control port,

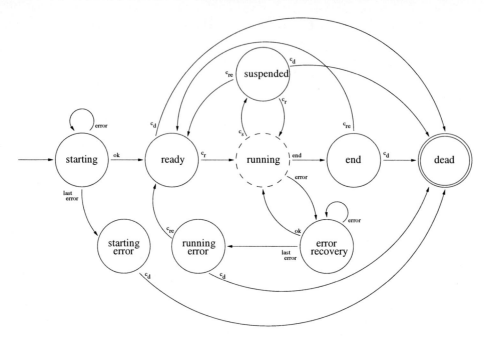

Fig. 4. The Default Automaton.

before jumping to the **dead** state for its destruction. Alternatively, if initialization is accomplished successfully, the component is brought into the **ready** state, there the component waits idle either for its first task execution getting into the user automaton by means of the **running** state, or for its destruction if it is driven to **dead** state. The **running** state is the part of the automaton – the user automaton – defined by the user/developer who implements the component and endows the component with its particular functionality. This automaton has its own states and transitions, but the default automaton imposes some requirements on it:

1. From any of the user states the component can be driven to the **suspended** state of the default automaton. This implies that the component should be externally interruptible at any user state, with a latency that will depend on its internal design. The component should save its internal status in case of continuing task execution, transition labeled as c_r in the default automaton. From **suspended** state the component can also be reset, i.e. driven to **ready** state, or destroyed, i.e. taken to **dead** state.
2. Some states, provided by the default automaton to indicate general states in all components, must be accessible from states in the user automaton:
   - **error recovery:** this state is reached when an error is detected during task execution, and it is conceived as a state for recovering from an error without canceling task execution. In this state, an error recovery procedure can be tried several times, until a maximum number of attempts

have been tried unsuccessfully. In that case, the automaton goes to the **running error** state (see below). If the error is fixed in any of the recovery attempts, the automaton continues where normal task execution was interrupted previously. That implies that the component internal status must be preserved during error recovery as well.

- **running error:** when a component has not been able to recover itself from a error, it transits into this state. Only external intervention can drive the component again to **ready** state to start a new task execution, or to **dead** state and instance destruction.
- **end:** if a component has finished its task, then it goes directly to the **end** state. From that state it can be brought to destruction, or it may start a new task execution by means of transiting to the **ready** state.

3.4 Observable and Controllable Variables

Additionally, components may define *observable* and *controllable* variables.

- **An Observable Variable.** Features of components that should be of interest from outside, should be declared as *observable variables*. Any change in an observable variable within a component is published for external observability through the component's monitoring port.
- **A Controllable Variable.** It is a variable in a component that represents an aspect of it which can be controlled, i.e. modified or updated, externally. Controllable variables are accessed through the component's control port.

As implied by figure 4, CoolBOT will endow all components with several default observable variables:

- **State.** The default automaton state where the component is at each instant.
- **User State.** The state where the automaton is when the component is in the user automaton, i.e. the actual state within the **running** pseudostate.
- **Result.** This variable indicates if a task has been finished by a component, and in that case, the results achieved. More precisely, this observable can take three values:
  - **null:** The component is not in the **end** state.
  - **success:** The component is in the **end** state, and task execution was completed successfully.
  - **fail:** The component is in the **end** state, and task execution failed.
  Aside from the value published through this observable, a component will probably have specific output through which it may communicate not only the results, but also a description of the situation in which it succeeded or failed.
- **Error Description.** When the component enters **starting error** or **running error** states, information about the situation could be provided through this variable.

and one default controllable variable:

– **New State.** It is used to bring the component to one of the controllable states of the default automaton, requesting its modification through the component's control port.

3.5 Port Compatibility and Cardinality

A pair output port/input port constitutes a **connection** between two components. Data is transmitted through port connections in discrete units called **port packets**, like in [5]. Port packets can carry information or not. In the last case they are used to signal the occurrence of an event and are called **event packets**; the other ones will be termed **data packets**. Each port can only transmit a type of data packet. To establish a port connection it is necessary that both ports, the output port and the input port, match the port packet type they can transmit. Currently, CoolBOT provides three typologies for output ports:

– **Poster.** A poster output port is a finite-length circular buffer that stores the data packets issued through it. When it is full, a new transmitted packet overwrites the oldest previously stored port packet. A poster output port is used by a component to publish data for multiple components, and is overall, used to uncouple consumers of data from the data producer [7].
– **Tick.** A tick output port is not buffered and just emits event packets.
– **Generic.** A generic output port may be associated to any kind of port packet, usually data packets. It is not buffered.

The typologies for input ports are:

– **Poster.** A poster input port is the counterpart of a poster output port.
– **Tick.** It is the counterpart of a tick output port. They will be normally used to associate timers to components.
– **Circular or LIFO.** It is one of the counterparts of generic output port packets. It is a finite-length buffered input port storing received port packets that always delivers the most recent data first (Last In First Out). When the buffer gets full, any new received port packet overwrites the oldest port packet previously stored.
– **FIFO.** Another counterpart for generic output port packets. The input port is a FIFO - First In First Out - buffer with two variants, depending on if its depth is fixed or it can grow until a limit is reached. The policy followed when the buffer gets full and/or it cannot grow further is to overwrite the oldest data with the most recent.
– **Unbounded FIFO or UFIFO.** It is a FIFO whose depth is not limited. When the buffer gets full its depth is increased, so that new port packets do not overwrite previously stored ones.

Table 1 resumes the compatibility and cardinality of connections among the different kinds of input and output ports. Remember that, to establish a connection between an input port and an output port, besides of the restrictions depicted on the table, it is also necessary that both ports match the type of port packet they transport.

Table 1. Compatibility and Cardinality (n, $m \in N$; n, $m \geq 1$).

Ports Compatibility		Input Ports			
& Cardinality	Poster	Tick	Circular	FIFO	UFIFO
Output Poster	$1 \rightarrow n$	-	-	-	-
Ports Tick	-	$n \rightarrow m$	-	-	-
Generic	-	-	$n \rightarrow m$	$n \rightarrow m$	$n \rightarrow m$

3.6   Port Communication Models

There are two basic communication models for port connections:

- **Push Model.** In a push connection the initiative for sending a port packet relies on the output port part, that is, the data producer sends port packets on its own, completely uncoupled from its consumers.
- **Pull Model.** A pull connection implies that packets are emitted when the input part of the communication – the consumer – demands new data to process. In this model the consumer keeps the initiative, and it supposes that it is necessary to send a request whenever a port packet is demanded. Only "pulled" pairs of output/input ports can be connected together. The pull model only can be used with Circular, FIFO and unbounded FIFO input ports.

As experience demonstrates communications is one of the most fragile aspects of distributed systems. In CoolBOT, the rationale for defining standard methods for data communications between components, is to ease interoperation among components developed independently offering optimized and reliable communication abstractions.

3.7   Atomic Components

An atomic component is embodied as a thread and models a port automaton [5]. It is atomic in the sense that it is indivisible and can not be decomposed in other components. Atomic components have been devised to abstract hardware like sensors and effectors, and/or other software libraries like third party software.

As any component in CoolBOT, an atomic component is implemented as a port automaton following the model of the default automaton shown in figure 4. For each component it is necessary to define the part of the component automaton that is specific to its internal control. That part of the automaton, the user automaton, is represented by the dashed **running** state in figure 4.

Once the developer has completed the automaton defining the user automaton, he/she will have to complete the component coding filling in the transitions between automaton states and the states themselves where several possible sections are provided for each state:

- **entry point:** this is the starting code section and it is executed each time the automaton gets into the state.

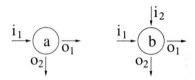

Fig. 5. Two atomic components: **a** and **b**.

- **exit point:** this is the ending code section and it is executed each time the automaton is about to leave the state.
- **period:** this is a periodic code section that is executed periodically when the automaton remains in the state. The period can be established statically or dynamically.

It is not necessary for the developer to fill in all these sections and transitions for all automaton states, default or not. The developer will fill in certain sections and transitions depending on the functionality of the component that is being coded. Unused ones will remain by default empty or idle.

Atomic components are implemented as simple DES, according to the definition given in [5]. Their definition declares observable, controllable and local variables, input and output ports, the states of the user automaton and transitions between them, and sections of code that were required.

The transitions and the states of the default automaton for each component are empty by default. If any of these transitions or sections of one of these states is also necessary, they must be also included in the definition of the component. At the same time, the description code should include information about what third party libraries - hardware drivers or other software libraries - must be linked with the component to achieve an executable component. This component description will be then compiled generating a C++ class embodying the component where all transitions and necessary state sections will be codified as class function members that should be filled in by the developer. Also the necessary makefiles will be created to compile the component with the specified third party libraries.

3.8   Compound Components

A compound component, is a composition of instances of another components which, in turn, can be either atomic or compound. Figure 6 graphically illustrates this concept, where the compound component **c** is a composition of two instances, one of an atomic component **a**, a_i, and one of another atomic component **b**, b_i, both shown in figure 5. Figure 7, depicts a compound component **d** made of an instance of the compound component **c**, c_i, and an instance of the atomic component **b**, b_i, evidencing that instances of compound components are functionally equivalents to atomic components in terms of composition and instantiation. Control and monitoring ports are not shown.

A compound component is a component that uses the functionality of instances of another atomic or compound components to implement its own func-

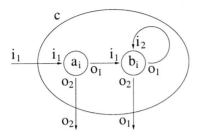

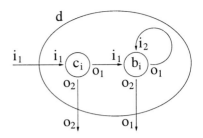

Fig. 6. The compound component **c**: a composition of atomic components **a** and **b**.

Fig. 7. The compound component **d**: a composition of compound component **c** and atomic component **b**.

tionality, and similarly to atomic components, it is modeled as a port automaton. Components whose instances are used inside a compound component will be its **local components**, thus, **b** and **c** are local components of **d** in figure 7.

The Supervisor. The automaton that coordinates and controls the functionality of a compound component is called its **supervisor**, and like atomic components it follows the control graph defined for the default automaton (see figure 4). It must be completed by the developer/user specifying the part of it that has been previously termed the user automaton.

Similarly to atomic components, compound components will be specified by means of a description code. In it, the developer/user completes the rest of the automaton describing the states and state transitions that constitute the user automaton, and also what sections – entry point, exit point and period – will be filled in on each state. Alike atomic components, when this description code is compiled, a C++ class is generated implementing the compound component. The description code of the compound component must specify all its interface. This includes how its local components' input and output ports of are mapped as input and output ports of the compound component; its observable and controllable variables, where some of them can be either completely new or mappings of observable and controllable variables of instances of its local components. Additionally local variables, internal to the compound component can be defined too.

It is at the level of compound components where we can talk about component oriented programming, as new components can be assembled from existing ones and their supervisors deal directly with components, through instructions that manipulate different components using the same abstract model. In the scope of the automaton of a compound component, new instances of any component can be created dynamically, then they can be observed and controlled through its control and monitoring ports, these instances can reside locally in the same machine, or remotely in another machine reachable through a computer network. Once an instance is created it can be run for a task execution, suspended, run again, ... and finally killed.

Rules. Automaton transitions between states in the supervisor of a compound component are triggered by **rules** which are conditions involving observable and controllable variables of its local components and its own local variables.

3.9 Exception Handling

CoolBOT's exception handling mechanisms exploits two basic ideas. First, all components, no matter if they are atomic or compound share the same exception handling, communication and control schemes. Second, CoolBOT capitalizes on the idea presented in 2 to build up a reliable system from reliable components. These ideas dictate the following design guidelines:

- A component should incorporate the capability to measure its own performance. For example, in case of performing a periodic task, it should verify that it is respecting its frequency of operation. Also timers can be associated with input ports, so that if another component that should be sending data is not keeping the pace or not working at all, it can be identified easily.
- The component's definition includes a list of the exceptions that the component can detect, along with specific "continuity plans" whenever they are available. Currently, in CoolBOT exceptions are declared using the following simple pattern:

```
On Error: <error_id>
          <description>
          <handler> [<retries> <lap>]
          <on_success_handler>
          <on_failure_handler>
```

  where first the error number and description are declared; then it is declared the error handler for this error, normally a function internal to the component; optionally it can be specified how many times and how frequently it should be triggered. The last two fields contain, respectively, the handlers that should be invoked if the error recovery procedure has succeeded or if it failed.
- The evolution of the component's state when an exception has occurred is the same for all components. The possible transitions are evident from the control graph of the default automaton (see figure 4 in section 3.3).

As it was explained previously, when a component detects an error that it cannot deal with, either because there is not any possible recovery mechanism at this level or because the error recovery plan has failed, it communicates the error to its supervisor, and goes into a running error state where it waits for external intervention to restart or die.

Errors arriving to a supervisor from included components must be managed first by this supervisor. They can be either ignored, propagated to higher levels in the hierarchy or handled as explained above. However, when handling exceptions within compound components some standard recovery mechanisms are possible, aside from the obvious re-instantiation of the faulty component. Let's suppose, for example, that we have several components that constitute equivalent alternatives for developing the same task, possibly using different resources, but offering the same external interface. Such components could be used alternatively to carry out a specific task and hence, a general strategy to cope with components in running error might be just **substitution** of one component with another one providing an equivalent interface and functionality. A complementary strategy may also be useful to avoid suspending a compound component whenever a member of the composition gets into running error. Equivalent components can be declared as **redundant** and executed concurrently or in parallel (i.e. if redundant components execute on different processors), so that if one of then fails, the others will keep the whole component running.

When a local component instance gets into running error, if a substitute exists, the supervisor will create an instance of it to carry out a substitution and keep the compound component working, and the erroneous instance will be put in a queue of instances to be recovered. Instances in that recovery queue are restarted periodically to check out if the running error persists. There is a deadline for each instance in this recovery queue, if the deadline expires the instance is deleted from the queue and destroyed. Otherwise, if any of them is recovered, the previous situation before its substitution is restored.

If a local instance in a running error can not be substituted, it will be added to the recovery queue previously mentioned. If its deadline in the queue is reached then the instance is retired from the queue and destroyed. This may provoke the whole compound component to go to running error, or not, depending on its functionality.

An error that needs a special treatment is when a component hangs during execution. In such a situation, it can not attend its control and monitoring port, turning it uncontrollable. When this exception is detected the component is destroyed, this time using an operating system call like *kill()*, and obviously it is not added to the recovery queue.

4   Example

To illustrate the concepts explained previously along this document, we will present as a qualitative example how a **goto** behavior for a mobile robot could be defined using CoolBOT. Figures 8 and 9 depict the kind of mobile robot conceived for this example compounded by a mobile platform and a visual stereo system mounted on a robotic head. The pictures show the robot ELDI [3], a Museum Robot developed for the Science and Technology Museum of Las Palmas de Gran Canaria, which has been in daily operation since December 1999.

The **goto** behavior we are going to devise is depicted as a component in fig. 10. In this figure and the following ones components are represented as rounded-corner rectangles. Input ports are the arrows oriented inwards the component, and output ports are the ones oriented outwards. In this example, the **goto** component provides several controllable and observable variables, and one local

Fig. 9. The robot Eldi interacts with visitors engaged in some demos and games that are carried out on a backlighted board.

Fig. 8. The robot Eldi.

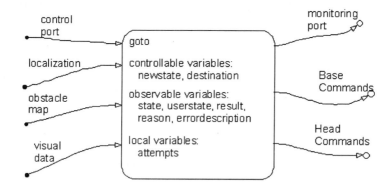

Fig. 10. The **goto** component.

variable. On figure 11 we can see the portion of the description code which corresponds to the graphic shown previously on fig. 10.

Apart from default variables and ports transparently added by CoolBOT, the component has been endowed with one extra controllable variable, one local variable, an several input and output ports which will be briefly explained:

- The controllable variable *destination* that, on each task execution, indicates where the **goto** behavior should drive the robot.
- The local variable *attempts* tracks the number of failed attempts of any of the behaviors that the component uses internally: a door crossing behavior, a corridor navigation behavior, and a freeway behavior (to be presented later).
- Three input ports, namely *localization*, *obstaclemap* and *visualdata*, permit the component to receive respectively, the position of the robot in the world coordinate system, and obstacle map and visual data. This information is supplied by other component/s in the system not shown in the example. The visual data could be used by the component, for example, to localize certain objects in the scene that were relevant to the navigation purposes, e.g. to locate the door's knob.

```
component goto {
    ...

    input ports
    {
        /*
         * NOTE: The control port is transparently defined
         * by CoolBOT.
         */
        localization ... ;
        obstaclemap ... ;
        visualdata ... ;
    };

    output ports
    {
        /*
         * NOTE: The control port is also transparently
         * defined by the software framework.
         */
        basecommands ... ;
        headcommands ... ;
    };

    controllable variables
    {
        /*
         * NOTE: The framework adds implicitly the default
         * one: newstate.
         */
        location destination;
    };

    observable variables
    {
        /*
         * NOTE: The default ones are: state, userstate,
         * result, reason and errordescription.
         *
         * No other observable variable is needed.
         */
    };

    local variables { integer attempts; };
    ...
};
```

Fig. 11. Skeleton definition of **goto** component: input and output ports and variables.

- The **goto** component's two output ports connect with the controller of the head system and with the motion controller of the mobile platform or base.

Once the **goto** component were completely designed and tested, the procedure to use it at runtime would be:

1. Instantiate the component.
2. Connect its inputs and output ports to other interacting components.
3. Once connected and ready, on each execution, it should receive a *destination* and then set the *newstate* controllable variable to running.

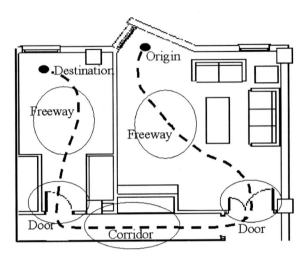

Fig. 12. The **goto** component: a typical situation.

4. Execution finishes when the *state* gets to end with the observable variable *result* indicating if execution was successful or not. In case of failure, the observable variable *reason* will hold the identified cause.
5. At any time during execution the component may be suspended or aborted by means of the controllable variable *newstate*, setting it to suspended in order to suspend execution, and then either setting running again to resume, or ready to restart execution.
6. To destroy the component we have to drive it to **dead** using the controllable variable *newstate*.

It is worthy to note that updating the controllable variable *newstate* is equivalent to send control port commands to the component to make it transit along the default automaton, see fig. 4.

Typically, the **goto** component is activated in a situation like that depicted in figure 12, where the robot in order to get to a destination point should navigate through different environments, namely corridors, doors and open spaces. The **goto** component incorporates specialized components to deal each situation, see figure 13, that will be described below.

The **planner** component, shown in figure 13 performs path planning along an ideal map known a priori. Such a planning is carried out in terms of behaviors that should be used to go from place to place along a trajectory inside this map. Through its input port *localization* it knows the robot's position inside the map and computes a trajectory to reach the destination point. This trajectory is divided into steps with its own sub destinations inside the whole trajectory, each one of these steps is classified depending on what behavior should be used to complete the step, i.e. door crossing, corridor traversing and free space traversing. To use this component we must set its controllable variable *destination* to the the final location where we want the robot to move. The **planner** will

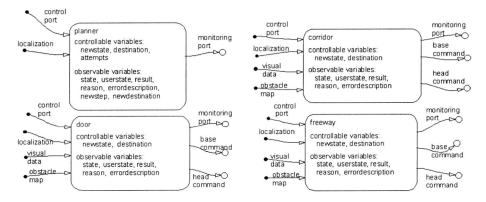

Fig. 13. The **goto** component: internal components.

finish current execution successfully when localization data match this controllable variable. During execution, each time its controllable variable *attempts* is updated, the planner carries out path planning in the following terms: if the variable is 0, then it has to plan the next step inside the trajectory to the final destination. If the variable is greater than 0, it means that the behavior launched to complete the previous step failed because maybe either a door was closed or a corridor was blocked by an obstacle. Situation that the component tries to solve looking for an alternative trajectory. Besides, it uses *attempts* to track how many unsuccessful attempts have been launched to complete a specific trajectory step. When the number of failed attempts exceeds a maximum value, it considers that current task has failed and finishes itself updating its observable variable *result* accordingly.

The remaining components of figure 13 implement the different behaviors that will drive the robot to reach a goal position in three different situations: door crossing – component **door** –, corridor traversing (or wall following) – the component **corridor** –, and free space navigation, the component **freeway**. All of them present the same external interface, they have a controllable variable, *destination*, where the goal position to complete the behavior should be set before running the component. During task execution, as input, they get sensory information from the different sources available in the system, as output of these behaviors, the robot actuators' are commanded.

On figure 14 internal components and internal connections for the **goto** component are depicted. An excerpt of the code describing this component appears on figure 15. Note how the supervisor, the automaton supervising the component, controls and monitors all internal components through control and monitoring ports. The rest of connections must be explicitly described by the component designer as shown in figure 15. The software framework makes a lot of work behind the scenes as mentioned in the commented text in figure 15.

The **Supervisor** in figure 14 embodies the automaton which monitors and controls the **goto** component. Figure 16 depicts part of the automaton defining

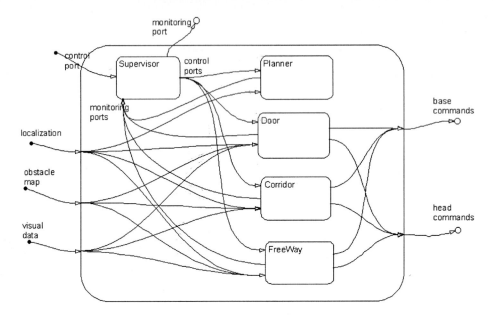

Fig. 14. The **goto** component: internal connections. Rounded-corner rectangles represent automaton states and oriented arrows state transitions.

the **Supervisor** of the **goto** component. The condition provoking a transition is put in square brackets and the action to carry out along the transition appears after a slash following the condition. Actions to be executed when a state is reached, the entry point, are expressed following a slash after the state name. This figure shows how the component launches a door crossing behavior depending on what the planner decides on each step. Although only two states basically are depicted, **planning** and **door**, there are currently two additional states, **corridor** and **freeway**, that do not appear because its related transitions and actions will be equivalent to the corresponding ones of **door** state. Here also the software framework makes work behind the scenes and the designer does not have to code instantiation of internal components and internal mapping, this is included transparently as part of the default automaton, figure 4, and the same holds for destruction of instances when the whole component goes to **dead** state, or when the component is suspended. For further clarification figure 17 presents the code corresponding the automaton representation of figure 16.

5   Final Discussion

Along this paper we have presented the underlying concepts and ideas that are guiding the implementation of CoolBOT, an undergoing research initiative being developed at ULPGC.

Systems developed with CoolBOT will share the same communication abstractions and inner control organization for the making up components, so that

```
component goto {
    components instances
    {
        /*
         * NOTE: The supervisor is generated implicitly by the software framework.
         */
        planner Planner;
        door Door;
        corridor Corridor;
        freeway FreeWay;
    };
    input ports
    {
        /*
         * NOTE: The control port is transparently defined by the framework.
         */
        localization to Planner.localization,
                        Door.localization,
                        Corridor.localization,
                        FreeWay.localization;
        obstaclemap to Door.obstaclemap,
                        Corridor.obstaclemap,
                        FreeWay.obstaclemap;
        visualdata to Door.visualdata,
                        Corridor.visualdata,
                        FreeWay.visualdata;
    };
    output ports
    {
        /*
         * NOTE: The monitoring port is also transparently defined
         * by the software framework.
         */
        basecommands to Door.basecommands,
                        Corridor.basecommands,
                        FreeWay.basecommands;
        headcommands to Door.headcommands,
                        Corridor.headcommands,
                        FreeWay.headcommands;
    };
    internal mapping
    {
        /*
         * NOTE: The framewok establishes transparently internal
         * connections between control and monitoring ports of
         * each instance and the supervisor.
         */
        /*
         * No internal mapping is necessary in this component.
         */
    };
    ...
};
```

Fig. 15. The **goto** component: description code for internal connections.

it will be possible to monitor and debug any of these components using a standard set of tools. We expect that theses features will reveal as essentials to achieve reliable, modular and easy to extend systems.

CoolBOT shouldn't be understood as a new architecture for perception-action systems but as an alternative design methodology and its associated set of development tools, that should assist the robotics researcher in the process of conceiving and validating different architecture proposals.

In our opinion, it is just in this aspect in which CoolBOT differences itself from many other architecture proposals for perception-action systems that populate the robotics literature. We think it belongs to a group of recent proposals

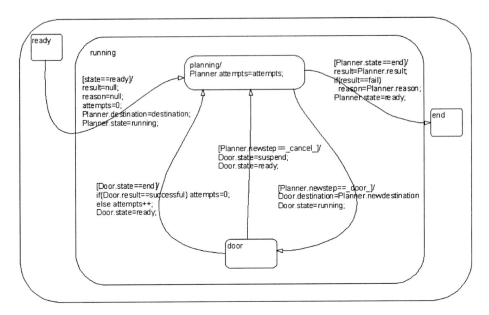

Fig. 16. The **goto** component: partial view of the user automaton.

that are aimed at defining new languages (e.g. ESL[8] or TDL [12]) not new
"architectures".

Obviously, it is too premature to make any claims about the superiority of
the approach described in this paper over others addressing the same or similar
goals. Only a posteriori, that is, through extensive experimentation and cross-
validation it will be possible to validate the CoolBOT's approach if the systems
so built proved to be more reliable, extensible and easier to maintain or adapt.
With this aim we will be willing to release CoolBOT to other interested research
groups that would like to apply ColBOT in the development of their systems.

Acknowledgements

The authors want to thank the anonymous reviewers for their proof-checking
reading of this manuscript and for their comments and suggestions. This work
has been partially funded by the Government of Canary Islands through project
contract PI1999/153.

References

1. R. Alami, R. Chatila, S. Fleury, M. Ghallab, and F. Ingrand. An architecture for
   autonomy. *International Journal of Robotics Research*, 17(4):315–337, April 1998.
2. R.P. Bonasso, R.J. Firby, E. Gat, D. Kortenkamp, D. Miller, and M. Slack. A
   proven three-tiered architecture for programming autonomous robots. *Journal of
   Experimental and Theoretical Artificial Intelligence*, 9(2):237–256, 1997.

```
component goto {
    ...
    automaton
    {
        /*
         * NOTE: Next definition of state ready adds actions to
         * default behavior already provided by the software
         * framework.
         */
        state ready
        {
            transition if (newstate==running)
            {
                result=null;
                reason=null;
                attempts=0;
                Planner.destination=destination;
                Planner.newstate=running;
            } goto planning;
        };
        /*
         * NOTE: Implicitly for all user states, several transitions
         * are included transparently by the framework, for example,
         * the one which is provoked externally to suspend the the
         * component by means of setting the controllable variable
         * newstate to suspended, in this case, all current running
         * components are suspended in the transition.
         */
        state planning
        {
            entry point
            {
                Planner.attempts=attempts;
            };
            transition if (Planner.newstep==_door_)
            {
                Door.destination=Planner.newdestination;
                Door.newstate=running;
            } goto door;
            transition if (Planner.newstep==_corridor_)
            {
                Corridor.destination=Planner.newdestination;
                Corridor.newstate=running;
            } goto corridor;
            transition if (Planner.newstep==_freeway_)
            {
                FreeWay.destination=Planner.newdestination;
                FreeWay.newstate=running;
            } goto freeway;
            transition if (Planner.state==end)
            {
                result=Planner.result;
                if(result==fail)
                    reason=Planner.reason;
                Planner.newstate=ready;
            } goto end;
        };

        state door
        {
            transition if (Planner.newstep==_cancel_)
            {
                Door.newstate=suspended;
                Door.newstate=ready;
            } goto planning;

            transition if (Door.state==end)
            {
                if (Door.result==successful) attempts=0;
                else attempts++;
                Door.newstate=ready;
            } goto planning;
        };
        ...
    };
    ...
};
```

Fig. 17. The **goto** component: code for the automaton.

3. J. Cabrera, D. Hernández, A.C.Domínguez, M. Castrillón, J. Lorenzo, J. Isern, C. Guerra, I. Pérez, A. Falcón, and J. Méndez M. Hernández. Experiences with a museum robot. Workshop on Edutainment Robots 2000, Institute for Autonomous Intelligent Systems, German National Research Center for Information Technology, Bonn, 27-28 September, Germany, 2000. Also available through URL ftp://mozart.dis.ulpgc.es/pub/Publications/eldi5p.ps.gz.

4. E. Coste-Maniere and R. Simmons. Architecture, the backbone of robotic systems. Proc. IEEE International Conference on Robotics and Automation (ICRA'00), San Francisco, 2000.

5. Antonio C. Domínguez-Brito, Magnus Andersson, and Henrik I. Christensen. A software architecture for programming robotic systems based on the discrete event system paradigm. Technical Report CVAP 244, Centre for Autonomous Systems, KTH - Royal Institute of Technology), S-100 44 Stockholm, Sweden, September 2000.

6. R.J. Firby. *Adaptive Execution in Dynamic Domains*. PhD thesis, Departament of Computer Science, Yale University, 1989.

7. S. Fleury, M. Herrb, and R. Chatila. G^{en}oM: A tool for the specification and the implementation of operating modules in a distributed robot architecture. IROS 97, Grenoble, France. LAAS Report 97244, 1997.

8. E. Gat. ESL: A language for supporting robust plan execution in embedded autonomous agents. Proc. of the AAAI Fall Symposium on Plan Execution, AAAI Press, 1996.

9. D. Kortenkamp, R. Peter Bonasso, and R. Murphy (Eds). *Artificial Intelligence and Mobile Robots: Case Studies of Successful Robot Systems*. MIT Press, 1998.

10. D. Kortenkamp and A.C. Schultz. Integrating robotics research. *Autonomous Robots*, 6:243–245, 1999.

11. B. Pell, D.E. Bernard, S.A. Chien, E. Gat, N. Muscettola, P. Nayak, M.D. Wagner, and B.C. Williams. An autonomous spacecraft agent prototype. *Autonomous Robots*, 5:29–52, 1998.

12. R. Simmons and D. Apfelbaum. A task description language for robot control. Proc. International Conference on Intelligent Robotics and Systems, Vancouver, Canada, October 1998.

13. M. Steenstrup, M. A. Arbib, and E. G. Manes. Port automata and the algebra of concurrent processes. *Journal of Computer and System Sciences*, 27:29–50, 1983.

14. D.B. Stewart, R.A. Volpe, and P.K. Khosla. Design of dynamically reconfigurable real-time software using port-based objects. *IEEE Transaction on Software Engineering*, 23(12):759–776, 1997.

15. C. Szyperski. *Component Software: Beyond Object-Oriented Programming*. Addison-Wesley, 1999.

Particle Filtering with Evidential Reasoning

Christopher K. Eveland

University of Rochester, Rochester NY 14627
eveland@cs.rochester.edu

Abstract. Particle filtering has come into favor in the computer vision community with the CONDENSATION algorithm. Perhaps the main reason for this is that it relaxes many of the assumptions made with other tracking algorithms, such as the Kalman filter. It still places a strong requirement on the ability to model the observations and dynamics of the systems with conditional probabilities. In practice these may be hard to measure precisely, especially in situations where multiple sensors are used.

Here, a particle filtering algorithm which uses evidential reasoning is presented, which relaxes the need to be able to precisely model observations, and also provides an explicit model of ignorance.

1   Introduction

This paper is concerned with the task of state estimation for tracking, also known as "filtering". The problem is generally formulated as the task of estimating the unknown internal state of some system that can be observed only though noisy observations.

This problem has many useful applications in computer vision and robotics, such as tracking a person in a video sequence, or estimating position through reading noisy GPS data.

A popular attack for solving this problem has been through a Bayesian approach, where a PDF modeling the probability of an observation given a state is assumed to be known, which is combined with a prior distribution to produce a PDF estimating the state given the observations.

This approach has produced many impressive results, but the requirement to model the observations with an explicit probability can sometimes prove difficult, and an incorrect model will produce incorrect results. Furthermore, it is useful to have a measure of confidence, so that a higher level process using the results of the tracker may make intelligent decisions.

It is these two aspects of the tracking problem that will be addressed here. The tool used to solve them is evidential reasoning, which will be introduced below, after a short review of the Bayesian method.

1.1   Bayesian Tracking

There is a nice introduction to filtering, and in particular the Bayesian approach to filtering in Chapter 2 of MacCormick's thesis [7]. A summary is given here to motivate further discussion below.

G.D. Hager et al. (Eds.): Sensor Based Intelligent Robots, LNCS 2238, pp. 305–316, 2002.
© Springer-Verlag Berlin Heidelberg 2002

Let $\mathcal{X}$ be a state space of possible configurations of some process that is to be estimated over time. This space can include dimensions describing the static position of the object to be tracked, such as 3-dimensional Cartesian coordinates, the dynamics of the object, such as its velocity or acceleration, the internal configuration of the object, such as joint angles of an articulated object, or anything else that may be desired to track through time. At each point in time t, we wish to reason about possible configurations of the tracked system, $\mathbf{x} \in \mathcal{X}$.

The only source of information on which to base this reasoning on is a series of observations up to time t, $\mathcal{Z}_t = \{\mathbf{Z}_1, \mathbf{Z}_2, \ldots, \mathbf{Z}_t\}$. These observations are frequently in a different space than the states, as the state can not be measured directly (otherwise filtering would not be a hard problem). For instance, it may be the case that the task is to track an object in a video stream. The state space may be the Cartesian coordinates of the object, but the measurements may be limited to image features such as gray values or edges. In order to relate $\mathcal{Z}_t$ to some configuration $\mathbf{x}_t$, it is assumed that there is an observation model, $p(\mathbf{Z}_t|\mathbf{x}_t)$. The assumption is made that observations are independent, so $p(\mathbf{Z}_t|\mathcal{Z}_{t-1}, \mathbf{x}_t) = p(\mathbf{Z}_t|\mathbf{x}_t)$.

Furthermore, it is assumed that there is some rule governing the dynamics of the system, such that $\mathbf{x}_t = f(\mathbf{x}_{t-1})$. This expresses the physics that govern the system. For instance it might relate velocity to position, or it might describe how uncertainty increases over time. Although this rule might not be known explicitly, it is assumed that there is some model of it that is known, which is expressed as $p(\mathbf{x}_t|\mathbf{x}_{t-1})$.

Armed with these models, and the observations $\mathcal{Z}_t$, it is possible to determine $p_t(\mathbf{x}_t|\mathcal{Z}_t)$. This is done by an application of Bayes' rule combined with an application of the dynamics equation to obtain:

$$p_t(\mathbf{x}_t|\mathcal{Z}_t) = \frac{p_t(\mathbf{Z}_t|\mathbf{x}_t) \int_{\mathbf{x}_{t-1}} p_t(\mathbf{x}_t|\mathbf{x}_{t-1}) p_{t-1}(\mathbf{x}_{t-1}|\mathcal{Z}_{t-1}) d\mathbf{x}_{t-1}}{p_t(\mathbf{Z}_t)} \tag{1}$$

This is not something that is likely to be directly implemented in real time for the general model described above. One solution is to limit the observation model to be Gaussian, and to limit the dynamics model to be linear, which produces the Kalman filter [5]. A more general solution is particle filtering, known in the computer vision community as the CONDENSATION algorithm [3, 4]. CONDENSATION represents the PDFs in equation 1 with weighted sets of "particles". These allow the representation and efficient manipulation of multi-modal PDFs, which can be a great improvement over the Kalman filter in cases such as clutter.

1.2   Evidential Reasoning

Suppose we wish to do particle filtering on a sensor that can give only positive information. An example of such a sensor is an IR proximity sensor, commonly used on mobile robots. These are binary sensors, that in the ideal case, have a sensing distance $\mathbf{d}_s$, such that if there is an obstacle along their axis at distance

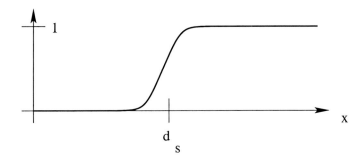

Fig. 1. A representation of a simple model $P(\mathbf{z}_o|\mathbf{d}_o = x)$ for a near-IR-proximity sensor. This does not model the false negatives that can occur when the obstacle does not reflect in the near-IR wavelengths.

$\mathbf{d}_o$, the sensor will return true (we will call this observation $\mathbf{z}_o$ for the obstacle observation), otherwise the sensor returns false (which we will label $\mathbf{z}_{\neg o}$). Of course, there will be some uncertainty about the exact position at which it turns on or off. Examining how $P(\mathbf{z}_o|\mathbf{d}_o)$ changes with $\mathbf{d}_o$, might reveal a Gaussian convolved with a step function, both centered about $\mathbf{d}_s$. This is shown in figure 1.

This may seem like a good model so far, except that these sensors tend to be highly dependent on how the obstacle interacts with light in the near-IR wavelengths. For instance, black trash cans are not detected by the sensors, nor are black jeans. This clearly leads to problems when using the above model in environments that have not been white-washed ahead of time.

To simplify the following argument, consider only the cases when $\mathbf{d}_o \ll \mathbf{d}_s$ (called O, the obstacle predicate), or $\mathbf{d}_o \gg \mathbf{d}_s$ (called $\neg O$). Furthermore, simply assume that the tails of the Gaussian describe above are truncated off at this point, so they are 0 or 1. For notation, let W be the "obstacle is white" predicate, and $B \equiv \neg W$.

Now consider the conditional distribution for $\mathbf{z}_O$. The following conditions hold:

$$
\begin{array}{ll}
P(\mathbf{z}_o|O, W) = 1 & P(\mathbf{z}_o|O, B) = 0 \\
P(\mathbf{z}_o|\neg O, W) = 0 & P(\mathbf{z}_o|\neg O, B) = 0 \\
P(\mathbf{z}_{\neg o}|O, W) = 0 & P(\mathbf{z}_{\neg o}|O, B) = 1 \\
P(\mathbf{z}_{\neg o}|\neg O, W) = 1 & P(\mathbf{z}_{\neg o}|\neg O, B) = 1
\end{array}
$$

In order for this information to be useful when B and W are not part of the tracked state, we have to have a prior model for $P(B)$ to obtain $P(\mathbf{Z}_o|O)$ and the others. There are several ways to approach this. The simplest is to simply go out and measure what fraction of the world is white, and what fraction is black. This has clear drawbacks. First, it is difficult to get an accounting of the fraction of the world that is different colors. Second, you may not be able to train on the same environment as you test on. Third, the regions of black and white are not evenly distributed. For instance, taking time into account for locality, it may be the case that $P(B_t|B_{t-1}) \gg P(B_t|W_{t-1})$. Clearly if an accurate model is to be made, it will get very complicated very quickly.

One way to handle these problems is to fold the model into the state to be estimated by the filtering process. So you could add one dimension to the state representing "black obstacle or "white obstacle", and have the tracker flip back and forth between models as appropriate. This has in fact been done by [2]. In the case when multiple sensors are being used, the number of additional dimensions needed may grow quickly, rapidly increasing the complexity of the filtering task, so this is not always a desirable solution.

Rather than trying to model every aspect of the world explicitly, it can be advantageous to admit ignorance in certain situations. This can be done by generalizing probabilities into belief, by relaxing the constraint that $P(A) + P(\neg A) = 1$ into $Bel(A) + Bel(\neg A) \leqslant 1$. [9] So a modified model for the above sensor might be:

$$Bel(\mathbf{z}_o|O) = w \quad Bel(\mathbf{z}_o|\neg O) = 0$$
$$Bel(\mathbf{z}_{\neg o}|O) = b \quad Bel(\mathbf{z}_{\neg o}|\neg O) = 1$$

where w is a lower bound on the expectation of seeing white, and b is a lower bound on the expectation of seeing black. This might not seem very different than what would be obtained by using probability theory, in which case $w = P(W)$ and $b = P(B)$, but the key is that here $w + b \leqslant 1$, so if absolutely nothing is known about $P(W)$, then w and b may both be set to 0. The result is that the model does not need to be as detailed, since by explicitly modeling ignorance, things that are not modeled can not come back to cause problems later when assumptions are violated.

While all of the above has been discussed in the context of one sensor, it should be noted that there are many sensors that have such characteristics. One example is stereo range sensors, where many stereo algorithms produce either a range value of a "don't know" token (i.e., [6]). In addition, consider the case of tracking edges in both visible and long-wave IR cameras. (See figure 2.) When entering a shadow region, the visible camera will return only black, which could be considered a "don't know" as well. There are similar occurrences for the IR sensor, such as when an object passes behind glass. This is an interesting case where the two sensors can provide more information than just one. A tracker for this camera is under development.

A model of ignorance can be useful on another level as well. There are cases when the additional semantics of ignorance *vs.* uncertainty can be useful. Take, for example, a tracker whose job it is to track people through a cluttered scene, and take snapshots of the people entering an area. Here it is important that the precise moment in time at which the shutter release is pressed, there is a high degree of confidence in the tracked state, or the camera may "miss" the target.

2   Filtering with Evidential Reasoning

To evaluate how evidential reasoning can be used in a particle filtering framework, a test system has been built. The key questions in building the system are first how to represent the particles, and second, how to propagate them with time. The solutions provided here are not definitive answers to these questions,

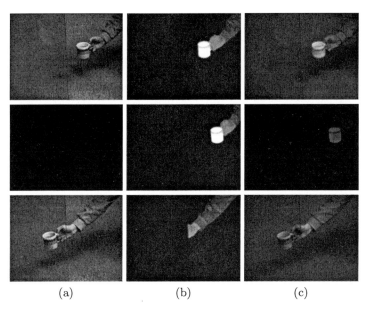

(a)   (b)   (c)

Fig. 2. Output under various conditions of a coregistered visible to long-wave IR (8-12μm) sensor. The column (a) shows the visible channel, the column (b) shows the long-wave IR channel, and column (c) shows a color visualization with the visible on the green channel and the IR on the red channel. Three scenes are shown, illustrating the trade-offs between the channels. The first scene shows a hot cup being held. In the second scene, the lights are turned off, while in the final scene, the cup passes behind a glass window.

but do provide a working algorithm. A more formal framework is the subject of ongoing research.

2.1   The Particles

To describe how the particles are defined in the evidential framework, some more background on belief functions is needed. Belief functions are defined over a frame of discernment Θ, which is an exhaustive set of mutually exclusive possibilities. An example from above would be "there is an obstacle present" or "there is not an obstacle present". In this case the frame of discernment is $\Theta = \{O, \neg O\}$. A mass function is then defined over subsets of the frame of discernment. The two constraints on a mass function are that it sums to unity:

$$\sum_{X \in 2^\Theta} m(X) = 1 \tag{2}$$

and no mass is assigned to nothing: $m(\{\}) = 0$. This mass function is then used to define both belief and plausibility. Belief is the sum of all mass that *must* be assigned to a set,

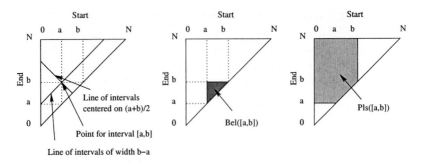

Fig. 3. Strat's continuous frame of discernment.

$$Bel(X) = \sum_{Y \subseteq X} m(Y) \qquad (3)$$

Plausibility is the sum of all the mass that *could* be assigned to a set,

$$Pls(X) = \sum_{Y \cap X \neq \{\}} m(Y) \qquad (4)$$

A key question that should arise at this point is what happens when the belief to be represented involves continuous values. For instance, suppose the belief in a sonar range sensor's value given a distance to obstacle needs to be represented in a mass function. What should the frame of discernment be? One possibility is to split the range into several discrete quanta, such as between 1 and 2 feet, between 2 and 3 feet, etc. To have these approach a continuous spectrum, there must be an increasingly large frame of discernment, which in turns means that the number of subsets that the mass function must be defined over grows exponentially. To get around the problem, mass functions can be limited to convex subsets. [10] In this manner the exponential growth is limited to quadratic growth, while still retaining most of the likely subsets to have mass assigned to. Under most situations, belief is not assigned to non-convex intervals, so this is a minor limitation.

To help visualize this type of mass function, it can be seen as a right triangle where one side is the start of an interval, and the other side is the end of an interval. The interval $[a, b]$ is simply represented by the Cartesian coordinate (a, b). The belief for an interval $[a, b]$ is simply found by integrating the region of the triangle that $[a, b]$ contains, and the plausibility is simply the region that covers intervals intersecting $[a, b]$. This can be seen visually in figure 3.

The particles used in the filtering test system are simply points in the interval space. If the state space is more than one dimension, the interval space will have twice as many dimensions. To date, testing has been done with only one dimension.

2.2   Combining Beliefs

There are many ways in which multiple sources of belief may be combined. One such way is the Dempster-Shafer rule of combination. [9] Given two mass functions, m_a and m_b, these can be combined to produce $m_c = m_a \oplus m_b$ as follows:

$$m_c(Z) = \frac{\sum_{X \cap Y = Z} m_a(X) m_b(Y)}{1 - \sum_{X \cap Y = \{\}} m_a(X) m_b(Y)} \tag{5}$$

Dempster-Shafer combination is not well suited to particle filtering applications due to several problems. Murphy [8] points out several problems and invalid assumptions, some of which are:

1. Observations are assumed to be independent.
2. Combination is commutative.
3. Combination with total ignorance is the identity.

None of these assumptions and properties apply in the filtering application.

First, observations are not independent; an observation at time t can be highly dependent on the observation at time $t-1$ [1]. Intuitively, taking the same observation twice should not increase your belief in the result more than the portion of your uncertainty due to sensor noise. If however the same observation is combined with itself repeatedly using Dempster-Shafer combination, the combined belief will quickly converge to absolute certainty.

Furthermore, combination should not be commutative. The sequence of observations A, B, A, B, A, B should intuitively produce a high degree of uncertainty, while the sequence A, A, A, B, B, B should first converge to A, and then rapidly converge to B. In essence, events further back in history should not be as heavily weighted as current events.

The third flaw is related to the second one. If an agent observes ignorance, it should reduce its total belief, since memory alone will not prove useful over a long time. The Demspter-Shafer rule of combination only allows for the accumulation of evidence, so uncertainty cannot increase with time.

Of these, the most serious flaw is the first one. The second two could just as well be covered by the process model. The process model should produce more uncertainty when it is applied, since it includes a model of noise. If the process model relaxes the belief in the prior model, the the second two problems are avoided. This doesn't entirely fix the problem of the rule of combination over-committing belief, but it does help to mitigate the problem, by relaxing belief at each stage. It also leaves open the question of how exactly to apply the process model, since clearly it shouldn't be done with Dempster-Shafer combination, or the relaxation would not occur. While alternatives exists (ie. [8]), here we stick to Dempster-Shafer for simplicity.

---

[1] Note that this is not the same assumption that the CONDENSATION algorithm makes. Here we are considering direct evidence for a state: the belief in a state given the observation. CONDENSATION examines the probability of an observation given a state. This conditional probability is independent of past observations.

2.3   Filtering

The evidential filtering algorithm works in the following stages, where m_t is the mass function representing belief at time t:

1. Predict m'_t from m_{t-1}.
2. Relax m'_t.
3. Observe to produce m''_t.
4. Combine observation m''_t with knowledge m'_t to produce the updated knowledge $m_t = m'_t \oplus m''_t$, and prune.

To elaborate, the first two steps correspond to the process model. The first step is the deterministic part of the model. For each $m_{t-1}(X \in 2^\Theta)$, $m'_t(F(X)) = m_{t-1}(X)$, where $F : 2^\Theta \longrightarrow 2^\Theta$ applies the dynamics f to each element of the set it operates on:

$$F(X) = \{f(x)|x \in X\} \tag{6}$$

This is essentially convolution by the predictive part of the process model.

The second step is relaxation, where the belief in the model from the previous stage is reduced. It is possible to again simply convolve by a noise model, as is done in CONDENSATION, to handle the non-deterministic part of the process model, but this is not quite desirable. The reason is essentially the same as the reason why re-sampling must be performed before the convolution step in the CONDENSATION algorithm. Consider what happens to an impulse mass, $m(x \in \Theta)$, when uncertainty is added. The noise model will be some distribution, and if you move this one mass by a random value from the distribution, you loose the rest of the distribution. CONDENSATION gets around this by re-sampling, so the single mass will (if it has sufficient mass) get split into multiple masses each of which can undergo a different random perturbation. Since masses in the belief framework cover sets of states, not just a single state, a different approach can be used to relax belief and apply this noise model. It is best described pictorially, so see figure 4. The single mass is split into several masses that model the distribution. Given a PDF, there are several ways that the masses can be assigned. See, for example [11].

The observation step (step 3) of the filtering is straightforward, the sensor used should provide some belief function m''_t. This is then combined with the result of prediction to produce the final result in step 4. Simple Dempster-Shafer combination is used. This is not ideal, as has been mentioned above, but the relaxation stage serves to discount the belief in memory, and helps to avoid amassing belief too quickly.

Since combination can significantly increase the number of particles, a pruning is done as part of the combination step. After combination is done, if there are more than M particles, M particles are chosen from the set of particles with probability of the particle's mass. This is a rather crude way of doing the pruning, but it seems to work in practice. Several other methods of making the combination more computationally feasible are reviewed in [1]. Using one of these methods is a matter for future work.

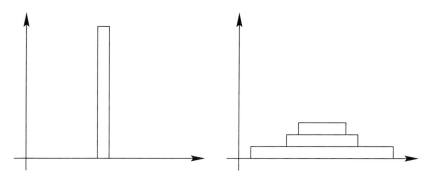

Fig. 4. Applying a noise model to an impulse interval. On the left is the original mass, on the right is has been split into several masses to model the noise and uncertainty.

Fig. 5. The map used in egolocation trials. The map was created by blending together a sequence of ceiling snapshots to create a mosaic. This particular mosaic was collected in a hallway at the University of Rochester. The white grid is composed of bars suspended from the ceiling above which, wires, smoke detectors, air-conditioning conduits, and other building internals are attached.

3   Results

This filtering method has been tested in an egolocation framework. The "state" to be tracked is the position of the observer. Observations are ceiling images taken by the observer (a mobile robot) and compared to a ceiling mosaic image. In addition, the odometer of the robot is made available for rough local distance measurements. For this test, the position is limited to only one dimension. So the state space is $\mathcal{X} = [0, N]$, where N is the number of pixels in the map. The initial position is unknown.

In general, given any single image of the ceiling to match with the map, there are many possible matches, since the ceiling has a regular structure. However, as more images are collected, it is possible to rule out certain possibilities, and more evidence will be collected for the true position.

The map is shown in figure 5. Since observations must assign mass to intervals $[a, b] \in \mathcal{X}$, a multi-scalar approach is used. A sum of absolute difference metric is used. Thus if $I_t(x, y, s)$ is the intensity of the pixel at position (x, y) at scale s and time t, and M is the same function for them map, then

$$m_t''([a, b])^{-1} \propto \max_y \sum_{i=0}^{w} \sum_{j=0}^{h} \left| I_t \left(i, j, \frac{1}{b-a} \right) - M \left(\frac{a+b}{2} + i, y + j, \frac{1}{b-a} \right) \right| \quad (7)$$

Note that no account is taken for the depth of the object being matched to the map; the map is simply assumed to be planar. This holds well for most indoor office environment scenes, where there is a regular grid of ceiling tiles suspended from the ceiling. In our case, this does not entirely hold, since the architect of the building chose to just use the frames for ceiling tiles, omit the tiles themselves, and paint everything above the frame black. Close examination of the hallway image sequence reveals that there are features effected by parallax, but in general the effect is not significant enough to prevent decent matching.

The observation is guided by the knowledge. Rather than evaluate the belief at some fixed set of intervals, belief is evaluated at locations determined by the knowledge from previous iterations. For each interval in the knowledge model, there are four possibilities as to how it can direct the observation. First, it may be chosen with probability proportional to its weight to be split into several smaller intervals. The motivation for this is that a large interval represents ignorance, but if a large mass is applied to this large area, then effort should be put into disambiguating the interval, so it is split into equal length intervals spanning the same total interval, and each of these sub intervals will be evaluated. Of the intervals that are left after splitting, intervals that significantly overlap are joined into one interval. The rationale behind this is that if one interval can explain several intervals without significantly increasing its size (and therefore increasing uncertainty), the representation is more efficient. Third, if an interval spans too many pixels for the evaluation to make sense, it is simply ignored. For instance, this is the case when an interval is wider than the map is tall. In the sub-sampling stage, nothing is left of the image, so no comparison can take place. Finally, all the remaining intervals are left as-is and evaluated as such. This has the effect of dynamically adapting the size of the sample set. In this experiment, the sample size fluctuated from about 80 to 250 samples. The algorithm runs significantly faster when it can get by with a smaller sample size.

The dynamics used for the experiment simply used the odometry to translate masses. The odometry used was not actually from wheel encoders but estimated by hand to be a fixed 100 centimeters per frame. The noise model used splits each interval mass into a pyramid distribution. If a granularity of n is used, then the single mass $m([a, b])$ will be split into n masses:

$$m_i\left(\left[a - \frac{i\sigma}{2n}, b + \frac{i\sigma}{2n}\right]\right) = \frac{m([a, b])}{n} \tag{8}$$

for $1 \leqslant i \leqslant n$.

At each iteration, belief and plausibility histograms were produced. Selected results are shown in figure 6. The upper left plot is for the third frame in the sequence. At this point, there are still several plausible hypotheses for the correct state. The upper right plot shows a few frames later when there is only one hypothesis. In the lower left panel, some contradictory evidence has been received: the odometry was significantly off, causing it to contradict the observation. The process model put all of the mass from the previous time at a place where there was little evidence to support it with the image observation. This lowers the belief in the current hypothesis relative to the plausibility. A few frames later, in the lower right panel, belief is recovered as more information comes in.

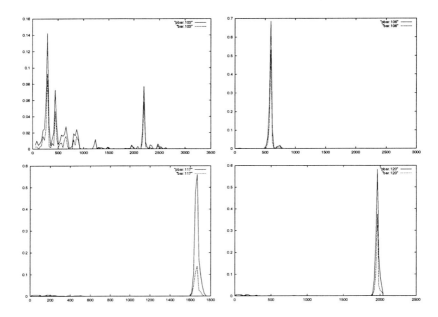

Fig. 6. Results for the egolocation trial at several time instances. The upper line represents the plausibility, while the lower line represents the belief.

4   Conclusion

In the previous sections, a method of particle filtering with belief functions has been demonstrated to work. The performance is similar to what can be obtained using other techniques such as CONDENSATION, but there are two advantages to this system.

First, the task of modeling observations and the process as simplified by using a looser framework. This also makes it easier to combine sources of information. Second, there is an explicit model of ignorance, which is part of the output of the tracker. This extra information can be used to decide when to perform critical tasks, such as take a picture of a target.

Not only is the explicit model of ignorance useful to the user, but it is used by the algorithm to guide the observations, and to automatically adjust the number of particles to use in tracking. This dynamic re-sizing leads to dramatic speed improvements when it is able to get by with fewer particles.

These advantages do come at a price. The computational cost of using this method are roughly $O(n^2)$, which is significantly more expensive than the $O(n \log n)$ cost for CONDENSATION. Another drawback, is that since some of the problems with Dempster-Shafer combination are overcome with heuristics, the formalism of CONDENSATION and the Bayesian approach are lost.

Given this, we feel that evidential reasoning is an important method to consider in future filtering applications and research.

References

1. Mathias Bauer. Approximations for decision making in the dempster-shafer theory of evidence. In *Proceedings of the 12th Conference on Uncertainty in Artificial Intelligence*, pages 73–80, 1996.
2. Michael J. Black and David J. Fleet. Probabilistic detection and tracking of motion discontinuities. In *IEEE 7th International Conference on Computer Vision*, volume 1, pages 551–558, 1999.
3. M. Isard and A. Blake. Contour tracking by stochastic propagation of conditional density. In *Proceedings ECCV*, 1996.
4. M. Isard and A Blake. Condensation - conditional density propagation for visual tracking. *International Journal of Computer Vision*, 29(1):5–28, 1998.
5. R. E. Kalman. A new approach to linear filtering and prediction problems. *Transactions of the ASME Journal of Basic Engineering*, pages 35–45, March 1960.
6. Kurt Konolige. Small vision systems: Hardware and implementation. In *The Eights International Symposium of Robotics Research*, October 1997.
7. John MacCormick. *Probabilistic Modeling and Stochastic Algorithms for Visual Localisation and Tracking*. PhD thesis, University of Oxford, January 2000.
8. Robin R. Murphy. Adaptive rule of combination for observations over time. In *Multisensor Fusion and Integration for Intelligent Systems*, 1996.
9. Glenn Shafer. *A Mathematical Theory of Evidence*. Princeton University Press, 1976.
10. Thomas M. Strat. Continuous belief functions for evidential reasoning. In *AAAI*, pages 308–313, August 1984.
11. Doug Y. Suh. Transformation of mass function and joint mass function for evidence theory in the continuous domain. *Journal of Mathematical Analysis and Applications*, 176:521–544, 1993.

Structure and Process: Learning of Visual Models and Construction Plans for Complex Objects*

G. Sagerer, C. Bauckhage, E. Braun, J. Fritsch, F. Kummert,
F. Lömker, and S. Wachsmuth

Faculty of Technology, Bielefeld University, Germany
`sagerer@techfak.uni-bielefeld.de`

Abstract. Supervising robotic assembly of multi-functional objects by means of a computer vision system requires components to identify assembly operations and to recognize feasible assemblies of single objects. Thus, the structure of complex objects as well as their construction processes are of interest. If the results of both components should be consistent there have to be common models providing knowledge about the intended application. However, if the assembly system should handle not only exactly specified tasks it is rather impossible to model every possible assembly or action explicitly. The fusion of a flexible dynamic model for assemblies and a monitor for the construction process enables reliable and efficient learning and supervision. As an example, the construction of objects by aggregating wooden toy pieces is used. The system also integrates a natural speech dialog module, which provides the overall communication strategy and additionally supports decisions in the case of ambiguities and uncertainty.

1 Introduction

Natural instruction of robots and natural communication with robots are major challenges if we want robots to be used as service or flexible construction tools. In both cases, the machine must be able to acquire knowledge about the environment it is working in and about the tasks it has to provide. The acquisition process should at least be supported by sensory input. Programming should not be involved or should be reduced to minimal effort. But even if no programming is necessary by the end-user there is a certain amount of *a priori* skills, tasks, and dialog capabilities that must be implemented for the system. In the following we will call these the *baseline competence* of the system. Communication between system and user enables the acquisition of further competencies.

In order to establish this *communication* as naturally as possible, instructions based on speech dialog, on gestures, and on presentation must be supported by

---

* This work has been supported by the German Research Foundation within the Collaborative Research Center 'Situated Artificial Communicators' and the Graduate Program 'Task-oriented Communication'.

G.D. Hager et al. (Eds.): Sensor Based Intelligent Robots, LNCS 2238, pp. 317–344, 2002.

the baseline competence of the system. Learning of new words, new objects, new skills, new plans and strategies relies on such communication processes. Incrementally, the robot enriches its *acquired competence*. Already a simple instruction like "Bring me a cup of coffee" can illustrate different aspects of *a priori* necessary capabilities and possible learning procedures. Assume this would be the first command to the system in a so-far unknown environment. Speech recognition with a sufficient lexicon is expected by the user. Even knowledge about the meaning of almost all words in this sentence is assumed. The problems start with the linkage between environment and words. According to the fixed *service* task, the system knows something about apartments or offices. A map can be partially or completely acquired by some automatic tools or by communication. This map may also include information about some special objects and their locations. The word "me" indicates the goal location of the "bring"-process, but this location is not known at the moment the process starts. The person could move. Therefore, features characterizing the person must be stored, and strategies to find a person must be implemented as a baseline competence. "a cup of coffee" refers to both a place where coffee is available and a location where cups are. The cup to bring can be specialized according to "me", which may select one unique cup out of a certain set. But it can also denote one arbitrary object of type "cup". In both cases either some information about the object class "cup" must be covered by the baseline competence or a communication is necessary to explain the object "my cup" or the visual concept for "cup". If knowledge that is necessary to perform the task is not available according to the baseline and already-acquired competence, the system must initiate a dialog to acquire and learn the necessary information. Also in cases of ambiguities, uncertainty and errors the dialog capability can improve robustness and support decisions.

Of course, we can not present a system that is able to cover all the capabilities and behaviors described so far. It is even debatable which competences are required for a baseline system and which ones are due to communication-based acquisition processes. Additionally, the acquisition processes can be done in separated learning phases or just be part of task-oriented communication. The goal of the research work presented in this contribution is to achieve natural man-machine communication that integrates speech, gestures, object recognition as well as learning capabilities and to go some steps towards system capabilities as described above. As an example, the *cooperative construction* of objects by aggregating wooden toy pieces is used. Figure 1 gives an impression of the environment, called the baufix scenario. According to the toy pieces, the set of baseline objects and possible construction rules is fixed. The set of aggregated objects is not known and not limited. Instructing by showing in task oriented communication is natural due to this environment. Learning the meaning of words as well as visual features and concepts according to real-world objects, events, and actions can be clearly described as tasks that enrich step by step the competence of the system.

Supervising and learning assembly processes of multi-functional objects by means of a computer vision system requires components to identify assembly op-

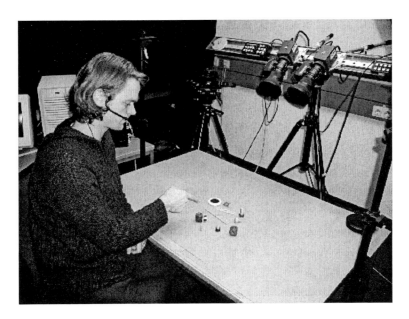

Fig. 1. The baufix scenario: a human instructor advises the system in an construction task. The scene on the table is received by a stereo head consisting of two cameras. A third camera in the background captures hand actions such as connecting parts of the scene.

erations and to recognize feasible assemblies of single objects. Thus, the structure of complex objects as well as their construction processes are of interest. If the results of both components should be consistent, there have to be common models providing knowledge about the intended application. However, if the assembly system should handle not only exactly specified tasks it is rather impossible to model every possible assembly or action explicitly. The fusion of a flexible dynamic model for assemblies and a monitor for the construction process enables a reliable and efficient acquisition and improvement of competences.

The remaining part of this introduction will give a short outline of the system competences and along the way an outline of the presentation. One important baseline competence is the recognition and location of the elementary objects of the baufix set. In cases of overlapping and already aggregated pieces, knowledge about spatial or temporal context, the construction process, or the construction rules is necessary to solve this task. In a cooperative construction environment, speech recognition and, as a consequence, the construction of interrelations between objects in the scenes and verbal descriptions is of great importance. In section 3 a probabilistic model for multi-modal scene understanding using visual and verbal object and spatial descriptions is presented. This baseline competence enables the natural communication about objects in the scene, their features and locations. It therefore also provides algorithms for the construction of a visual semantics of words. Learning to construct or recognize aggregated objects becomes feasible if the basic construction rules and actions are available. The baseline competence of the system performs both a pure symbolic and a visual based module for action detection and classification. The first one takes into account that for every construction action by a human two hands are used, objects disappear, and a new aggregated object occurs in the scene. If the construction rules are known, this simple model allows an automatic plan construction for complex objects and can provide context information for the module recognizing elementary objects. But of course there are some limitations of this approach. The results depend on the quality of the elementary recognition and the detection of object changes. Some actions cannot be distinguished. Therefore, a visual action recognition module is added. It detects hands and is able to classify their actions according to a set of elementary and complex actions. For both modules, an initial set of actions is integrated into the baseline competence. Complex actions and plans can be acquired by observing construction processes and combining already known categories. Complex objects are results of construction processes but are also restricted to structural rules giving information about geometry and topology. *Structural and process information* together allows one to build up models that support object recognition and enables learning by watching capabilities. In the summary a qualitative result of the overall system will be discussed. Recognition rates and other quantitative evaluations are presented for the different modules in the corresponding sections.

2   Recognition of Elementary Objects

The recognition system is characterized by a combination of multiple cues from segmentation and classification algorithms (Fig. 2). The cues deliver different kinds of results, that differ in their information content and also in their reliability and robustness. In order to exploit these results, they are integrated into a unified representation for segmentation and classification. This is analyzed to get hypotheses for location and type of elementary objects. In the following the modules are described briefly, before the integration mechanism is outlined.

Following the mean-shift algorithm [Com97] the *color region segmentation* is done using problem-independent color properties. The algorithm reduces the number of colors occurring in the image by clustering within the color space. Afterwards pixels are associated to color classes and regions of equal color classes in space are built. Fig. 3(a) shows an example of the segmentation results. *Contour-based perceptual grouping* is done on the basis of Gestalt laws independent of the concrete scenario. The edge elements resulting from a standard Sobel operator are approximated with straight line segments and elliptical arcs. In order to overcome fragmentation and obtain more abstract image primitives, grouping hypotheses are built up using various Gestalt laws [Maß97]. Besides others, hypotheses for closed contours are extracted based on collinearity, curvilinearity and proximity and are used as hypotheses for object surfaces (Fig. 3(b)).

The *hybrid classification module* [Hei96] provides hypotheses for Baufix elements based on regions with Baufix color. The center of gravity of each region

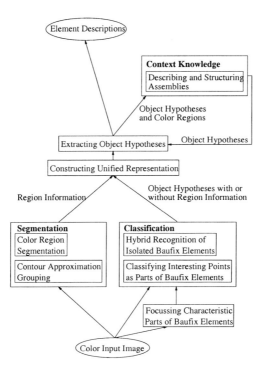

Fig. 2. Outline of the recognition module.

serves as a *point of interest*, where features are calculated in an edge enhanced intensity image and classification is done by an artificial neural network. This result initializes the instantiation process of a semantic network exploiting knowledge about the set of Baufix elements and the features of the color regions (Fig. 3(c)).

The goal of a second classification module is to recognize *parts of Baufix elements* in order to be independent of perspective occlusions. For this purpose windows around interesting points in the image, called *focus points*, are classified using a suitable set of classes of subparts of Baufix elements [Hei98]. Detecting these focus points is done either by searching for points with high entropy and symmetry of the gray value gradients or by taking the centers of gravity of homogeneous color regions. The classification of a window around a focus point as part of a Baufix element is done by a neural classifier (Fig. 3(d)). Additional object hypotheses are generated from *knowledge about complex objects*, which are assemblies built from elementary objects (see Sec. 6). Given a set of hypotheses for elementary objects, the assembly module is able to generate hypotheses for previously unlabeled color regions from assemblage knowledge [Sag01].

In our system object hypotheses are generated by identifying image regions, which are probably object regions, and labeling them. To find object regions in spite of structures on the surfaces, shadows *etc.*, the different segmentation and classification results are integrated into a unified representation, a hierarchy of

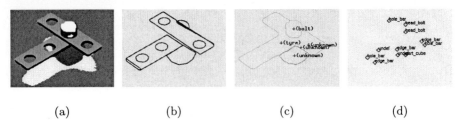

Fig. 3. Results for the single modules: (a) Color-based region segmentation (mean-shift algorithm, pseudo colors); (b) Closed contours from perceptual grouping; (c) Hybrid object recognition; (d) Holistic detection of object parts

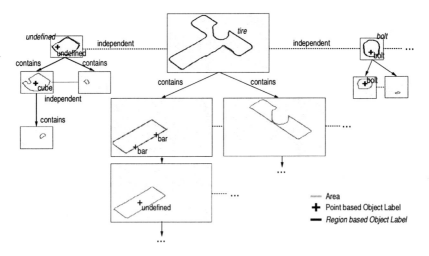

Fig. 4. Hierarchy of segmented areas. Available object labels are attached.

segmentation results with attached object labels. In a first step, all detected segments are represented as areas obtained by approximating the outer border of the segments with polygons. In doing so a hierarchical structure of areas containing other areas can be established. Areas from perceptual grouping may overlap several segments partially (e.g. see the area for the bar in Fig. 3(b) overlapping the bolt region of Fig. 3(c)). Those areas are excepted from the hierarchy and are considered later on. Object labels resulting from the classification modules are attached to the segmentation results within the hierarchy (Fig. 4). Region-based object hypotheses are associated with the corresponding area while point-based hypotheses are attached to the innermost area that contains the corresponding point (for details see [Sag01]). Neighbored areas on the highest level are clustered. As we expect one assembly or one isolated element per cluster, the clusters are processed independently from each other.

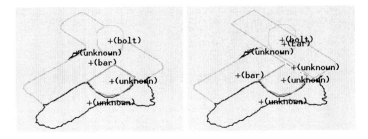

Fig. 5. Examples for competing results from data driven modules.

The interpreting procedure examines the independent areas sequentially, for each starting with its innermost parts. Object hypotheses for an area are generated according to the attached object labels by applying the plurality voting strategy [Par94]. The weight vector thereby used is obtained in two different ways. For a label from holistic object recognition, a classification error matrix is applied, which was generated from a test set labeled by hand. For the other labels, each weight vector contains a single nonzero element provided by the recognizer. Labels from the different sources are weighted equally against each other. For an area of a higher hierarchy level the labels directly attached to it and those attached to contained areas (which are assumed to belong to the same object) participate in the voting process for this area. Applying this voting strategy results for our example in a correct label for the bolt and the label 'unknown' for the cube, although there is one single focus point, which is classified as 'part of a cube'.

Besides classification errors, which are considered in applying the voting strategy, contradicting object labels within the hierarchy for one area possibly indicate that several objects are covered by this one area (think of a bolt area contained in a bar region). The existence of more than one object is also indicated by a higher level area containing several areas, which describe together the same image region. Within our example data, this situation occurs for the crossed bars (see Fig.4). There are two bars in the scene and the highest level area containing both occurs due to undersegmentation. During the interpretation of each level of the hierarchy these indicators, formulated as set of rules, are applied resulting in the generation of competing interpretations. The decision for the final interpretation is postponed in order to take additional information into account.

Fig. 5 shows the competing interpretations resulting from the uncertainty for the bar region. To complete the exploitation of the segmentation results the former excluded areas from perceptual grouping are incorporated. Additional competing results are generated by these areas replacing fully contained segments (Fig. 5, right).

The competing results obtained from data driven modules represent the context for the assembly module. This module hypothesizes elementary objects from

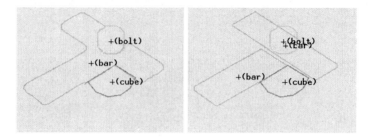

Fig. 6. Competing results for integrated recognition of elementary objects.

assembly knowledge for the competing interpretations independent from each other. For our example it generates 'cube' labels in both cases. Repeating the voting procedure with *all* object labels attached to the area results in 'cube' labels (Fig. 6).

Because subsequential modules expect a unique final result, the competing results for each cluster of image areas are evaluated. This is done according to the state of assemblage and the evaluation for the elementary objects, that results from the voting procedure of the single hypotheses. For a detailed description of the hierarchy and its analysis see [Sag01].

3   Verbal Descriptions

A vision system does not stand on its own. In our scenario it enables a human instructor to refer to objects in the scene in a natural way. He or she might advise the system to take some parts, tie them together, add another part, and put the constructed assembly back into the scene. Therefore, verbal object descriptions have to be related with observable objects. In order to realize this task in a *natural way*, the user should not adapt to the system, but the system has to be designed for the user. In our case the system can be instructed by speech, it is independent of the speaker, and the instructor needs to know neither a special command syntax nor the exact terms or identifiers of the objects. Consequently, the system has to face a high degree of *referential uncertainty* from vague meanings, speech recognition errors, unmodeled language structures, and, on the visual input channel, erroneous object classification results.

In this section we will briefly describe the speech understanding and speech and image integration parts of the system. More detailed information can be found in various articles (see e.g. [Fin99, BP99, Wac99, Wac00]). An essential aspect of the approach is that verbal descriptions play an active part in the interpretation of the visually observable scene instead of being a pure querying language. In order to cope with the different kinds of uncertainties Bayesian networks are employed. They provide a structured way of modeling probabilistic coherences and relationships.

3.1 Speech Recognition and Understanding

In the context of a multi-modal scene understanding system, speech is interpreted as a second source of information describing the visually perceived scene. Therefore, special attention is given to verbal object descriptions. The principle idea of the approach is to use a vertical organization of knowledge representation and integrated processing, in order to overcome the drawbacks of the traditional horizontal architecture: instead of organizing syntactic and semantic knowledge in isolated interpretation layers it is distributed in the understanding and recognition component of the system [BP99]. The integrated architecture makes use of an enhanced statistical speech recognizer as a baseline module [Fin99]. The recognition process is directly influenced by a partial parser which provides linguistic and domain-specific restrictions on word sequences [Wac98]. Thereby, instead of simple word sequences, partial semantic structures such as object descriptions are generated, e.g.

(*OBJECT:* the (*shape-adj:* thick) (*object-noun:* ring))),

or specifications of reference objects, e.g.

(*REF_OBJECT:*(*rel:* left) to the (*sizeadj:* short) (*object-noun:* bolt))).

These can easily be combined to form linguistic interpretations. Though there has been some progress recently, the detection of *out-of-vocabulary* words can still not be performed robustly on the level of acoustic recognition. Therefore, we employ a recognition lexicon that exceeds the one used by the understanding component but covers all lexical items frequently found in our corpus of human-human and human-machine dialogs. If such an additional word is used in the grammatical context of an OBJECT or REF_OBJECT description, it is interpreted as an unknown object noun.

3.2 Establishing Cross-Modal Interrelations

In order to calculate a common interpretation of the multi-modal input, we have to relate the visual representation of the scene and the partial scene description extracted from an instruction. Therefore, we propose a probabilistic model that integrates spatial information and evidences that indicate an object class. It is realized by using the Bayesian network formalism (cf. e.g. [Pea89]). An exemplary instantiation of a graphical model that is generated from a qualitative scene representation and the instruction is shown in Fig. 7.

The qualitative scene representation (Fig. 7d) comprises two complementary aspects. The nodes are associated with visual feature evidences *color* and *type* which indicate the class of an object and a *polygon* describing the boundary of the object. The first ones are used to instantiate the $F^V_{i\,type/color}$ random variables of the probabilistic model (Fig. 7b). The last one is instantiated as a positional feature from vision P^V_i. The edges represent possible object pairs that might be verbally related by a spatial proposition [Wac99]. They are labeled by a direction

"Please, take [the thick ring]₀ left to [the short bolt]₁."

(a) instruction

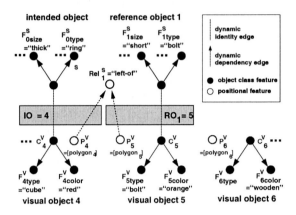

(b) probabilistic model

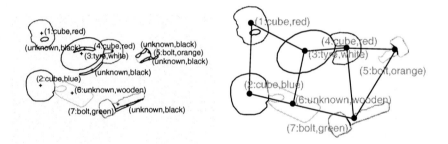

(c) recognition results   (d) qualitative scene representation

Fig. 7. Multi-modal scene understanding: An instruction (a), that includes the description of an intended object, a projective relation, and a reference object, is related to the qualitative visual representation of the scene (d) by a probabilistic model (b). The nodes in this model represent random variables where upper indices V and S refer to vision and speech, respectively. The edges are labeled by conditional probabilities. In the object recognition results (c), the bar (object 6) obtained no type label, and is interpreted as an unknown wooden object. The red felly (object 4) is wrongly classified as a cube. Unknown black regions are assumed to represent shadows and are subsequently ignored. By Bayesian inference, the most probable explanation consisting of the intended visual object $IO = 4$, the reference object $RO = 5$, and the intended object class $C_4^V = $ 'red-felly' can be calculated.

vector which is calculated from the positional features of the adjacent objects. In the following, we describe how we exploit the interpretation of the instruction in order to improve the reliability of the scene interpretation.

Psychologists distinguish the *what* system and the *where* system in the human brain [Ung82]. In this regard, Jackendoff [Jac92] has noted that there are many ways to describe *what* an object is, but few ways to describe *where* an object is. Nevertheless, the spatial meaning of linguistic expressions is not easy to capture (cf. e.g. [Muk97]). In our system spatial relations are represented by a computational model of space with a small finite vocabulary [Wac99]. We use the projective relations 'left', 'right', 'in-front-of', 'behind', 'above', 'below', and combinations of them. The projective relations are interpreted with regard to a speaker-centered reference frame which is assumed to be the default selection of the speaker. The spatial model calculates a *degree of applicability* of a projective relation to an object pair that is based on the graph structure of the qualitative scene representation and the directional information associated with it.

The meaning of attributes and nouns that verbally describe the visual appearance of an object is even harder to represent. Due to their huge number these cannot be completely modeled in a system – resulting in the *vocabulary problem* in human-system communication [Fur87].

We address this problem on different levels. First, we extract a list of frequently used words from a corpus of human-human and human-machine dialogs recorded in our domain context. In order to match these words to object classes used in the visual object recognition, we estimate conditional probabilities from data collected in two psycholinguistic experiments described in [Soc98]. In the first, type and color attributes are extracted from 453 utterances which name a marked object from a scene that is presented on a computer screen. In the second, size and shape attributes are collected from a multiple-choice questionnaire in the World Wide Web. All object classes are shown in eight different scene contexts. The subjects have to select all attributes from the multiple-choice list that correctly describe the marked object. A Total of 416 people completed the questionnaire. We estimate the conditional probabilities by counting the uttered types and colors, as well the selected sizes and shapes for each object class $C^V = c$:

$$P(F_{att}^S = f | C^V = c) = \frac{\#(f \text{ is selected for an object class } c)}{\#(\text{marked object has object class } c)}$$

$$att \in \{\text{color, type, shape, size}\}$$

The conditional probabilities are used for all verbal object specifications in the probabilistic model, e.g. $P(F_{1\,size}^S = short | C_5^V = c, RO_1 = 5) = P(F_{size}^S = short | C^V = c)$. These probabilities describe only elementary objects. Assembled objects are frequently denoted by metonymic names from various domains, like 'plane' for an assembled toy-plane. Such names are handled as an unknown object name in the understanding component that may match to an arbitrary elementary object or, with a somewhat higher probability, an assembled object.

The visual evidences used in the Bayesian network are 'type' and 'color'. Both may be affected by diverse sources of error, such as shadows, light reflections, or occlusions. Therefore, we estimated conditional probabilities for each object class using a hand labeled test set of 156 objects on 11 images:

$$P(F^V_{\text{type/color}} = f | C^V = c) = \frac{\#(\text{feature } f \text{ was classified for object class } c)}{\#(\text{object belongs to class } c)}$$

These probabilities are used for all visual objects in the probabilistic model.

The meaning of spatial relations is defined by the applicability function (App) of the computational model, which takes a projective relation r, a reference frame ref (currently, we assume only a speaker-centered reference frame), and the two polygons defining the object regions. It provides an applicability value which is based on the connectivity in the qualitative scene representation and the accordance with regard to the specified direction [Wac99]. Therefore, the corresponding conditional probabilities are defined as:

$$P(Rel = r | P^V_i = [\text{polygon}_i], P^V_j = [\text{polygon}_j], IO = i, RO = j)$$
$$= \alpha App(r, ref, [\text{polygon}_i], [\text{polygon}_j]) \qquad \text{where } \alpha \text{ is a normalizing constant.}$$

The Bayesian network is evaluated in the following way. The most probable explanation $\{(io^*, ro^*_1, \ldots, ro^*_m), c^*_{io}\}$ is calculated applying the following equations. In order to simplify the notation, variable names with fewer indices denote sets of random variables, e.g. $F^S_0 = \{F^S_{0j} | j = 1 \ldots k\}$:

$$(io^*, ro^*_1, \ldots, ro^*_m) = \underset{i, r_1, \ldots, r_m}{\text{argmax}} \; P(F^S, F^V, Rel^S, P^V | IO = i, RO_j = r_j, j = 1 \ldots m)$$
$$c^*_{io} = \underset{c}{\text{argmax}} \; P(F^S_0, F^V_{io^*} | c)$$

In the first identification step the configuration with the maximal probability for all evidences is selected. However, there may be other configurations that explain the observed evidence with a probability of similar magnitude. In such a case the assignment of the intended object io^* is said to be ambiguous, and all possible identification results are selected. Then the human instructor is asked to utter a more precise object description in the next dialog step.

If the verbal information is matched to a unique object, further inferences can be drawn and the most probable class c^*_{io} of the intended object io^* is calculated. The object class c^*_{io} that is derived from the visual and verbal evidences may be different than that hypothesized by the visual recognition process (Fig. 7). If we assume that the instructor definitively speaks about the perceived scene on the table, we can use this information in order to detect inconsistencies between the visual and the verbal interpretation and even to recover from erroneous classification results of the object recognizer by using this inferred object class.

4   Symbolic Actions and Plans

Besides analyzing image data to gain knowledge about a certain scene the information about previous actions leading to the actual situation can be used. In case of our Baufix domain the actions executed *during* the assembly construction process can be monitored to obtain information about the parts utilized for the assembly construction as well as a hypothesis for the resulting assembly. This information can serve as an additional cue for the visual recognition of elementary objects as well as for recognizing complex objects, i.e. assemblies.

To detect actions, a symbolic approach was developed based on the current state of the scene and the changes of parts in the scene. In the following, we will refer to parts as being either elementary objects (see Section 2) or complex objects (see Section 6) from the Baufix domain. The new and disappeared parts are extracted based on the results of the object recognition algorithm. Therefore this approach is computationally cheap since the object recognition results are needed by other modules, e.g. speech understanding, anyway.

As the actions are strongly related to the parts and their functions, we first turn to the description of the available parts before describing the symbolic action detection.

4.1   Assembly Model

To ensure flexibility, our system should be able to recognize any assembly that can be constructed using Baufix parts. As there are numerous different combinations of elementary parts, it is impossible to model every imaginable assembly. Thus, methods to represent assembled parts depending on specific knowledge of individual assemblies [Cha97, Hom90] will not solve our problem. Approaches that model feasible geometric transformations of elementary parts to cope with any possible assembly [Ana96, Tho96] are not applicable either. Our system must be able to recognize assembled parts in image data. However, qualifying the spatial relation of any two parts requires accurate geometric information, which is difficult and time-consuming to gain by means of vision. Therefore, we developed a syntactic method of modeling assemblies, which is based on the functional properties of their constituent parts. We understand an assembly to be composed of functional units, which are either represented by elementary parts or by subassemblies.

All the assemblies in our scenario are of the bolt-nut-type where miscellaneous parts like rings and bars can be put on bolts and are fastened using nut-type parts like cubes or rhomb-nuts. Hence, an assembled part consists at least of a bolt and a nut and optionally of several miscellaneous parts. Consequently, we model assemblies to be composed of a bolt part, a nut part and of an optional miscellaneous part. All parts either consist of single parts or of assembled parts. Using this compact model of bolted assemblies hierarchical structural descriptions can be derived for every possible assembly in our scenario.

Written as a grammar with terminals (Baufix parts in lowercase letters) and variables (uppercase letters) the model is:

```
ASSEMBLY:  (BOLT_PART MISC_PART*¹ NUT_PART)
BOLT_PART: ASSEMBLY² | round_head_bolt | hexagon_bolt
MISC_PART: ASSEMBLY² | 3_h_bar | 5_h_bar | 7_h_bar | felly | socket | ring
NUT_PART:  ASSEMBLY² | cube | rhomb_nut
```

Fig. 8. The recursive assembly model.

Note that the list of parts in an assembly is ordered, always starting with BOLT_PART followed by any number of MISC_PART and ending with NUT_PART.

It is important to note that there are several possible structural descriptions derivable from our model if an assembly contains more than one bolt. Generally, the number of different descriptions grows exponentially with an increasing number of assembled bolts.

4.2 Action Detection

For the detection of actions we use a two-hand model to hold the current state of the scene, which may not be completely observable since only the table scene is contained in the camera images used for object recognition. Two hand states are used to represent the parts or assemblies that are currently in the hands.

Possible hand states: empty | BOLT_PART | MISC_PART | NUT_PART
| (BOLT_PART MISC_PART⁺ ³) | ASSEMBLY

The state (BOLT_PART MISC_PART⁺) represents a partial assembly that is not yet tightened with a NUT_PART to form a complete assembly.

Using the hand states of our two-hand model together with information about new and disappeared parts extracted from object recognition results allows to apply a rule-based approach to infer actions occurring in the scene. Figure 9 shows the rules for the actions currently implemented; the presentation is similar to the notation for planning operators. To infer, for example, the *Take X* action two preconditions must be true: the part X must be on the table and one hand state must be empty. If the object recognition algorithms no longer detect part X on the table the action detection module is notified about its disappearance. Now the *Take X* action can be inferred since all preconditions are met. As a result the appropriate hand state is changed accordingly to capture the new state of the hand holding part X.

---

[1] The star operator indicates the possibility to put any number of MISC_PART on a BOLT_PART limited by the length of the thread of the BOLT_PART.

[2] A function can only be assigned to an assembly if there are free "ports" to use the assembly as a part (e.g. an assembly needs to have at least one free thread to be a NUT_PART).

[3] The plus operator indicates the possibility to put at least one MISC_PART on a BOLT_PART limited by its length.

New X
   Preconditions: ¬ (hand: X) ∧ X → new
   Effects: X on_table
Take X
   Preconditions: X on_table ∧ hand: empty ∧ X → disappeared
   Effects: ¬ (X on_table) ∧ hand: X
Put X Y
   Preconditions: hand_1: X (BOLT_PART) ∧ hand_2: Y (MISC_PART)
   Effects: hand_1: XY ∧ hand_2: empty
Screw X Y
   Preconditions: hand_1: X (BOLT_PART) ∧ hand_2: Y (NUT_PART)
   Effects: hand_1: XY ∧ hand_2: empty
Put down X
   Preconditions: X → new ∧ hand: X
   Effects: X on_table ∧ hand: empty

Fig. 9. The rules for inferring actions from the actual states of the two-hand-model and the changes in the scene. The notation hand_1 and hand_2 is only used to indicate the two different hand states, each of the two hand states can be hand_1 or hand_2.

In the introduced symbolic approach the *Put* and *Screw* actions are not observable because mounting parts together does not lead to new or disappeared parts. Without additional information from visually observing the hands of the constructor these actions can only be inferred if the next *Take* action has happened: If both hand states contain parts satisfying the preconditions of the *Put/Screw* action, it is inferred that the parts have been connected together if another part disappears. Inferring the *Put/Screw* action results in one hand state containing the (partial) assembly and the other hand state being empty. This empty hand state can now be used to hold the disappeared part.

Figure 10 shows an example assembly together with the action sequence derived during construction of the assembly.

Since our approach is based purely on symbolic information it is very important to reliably detect the scene changes. Besides this a few limitations are introduced by the fact that no direct visual observations of the hands are available: One important limitation is that the user may hold only one part or (partial) assembly as modeled by the hand states. This design of the hand states is motivated by the goal to monitor actions independently from whether a robot or a human is acting in the scene. Another limitation is that no parts may be put down outside the visual scene, as this cannot be detected without visual observation of the hands and therefore the two-hand-model would still contain the parts.

As all of these limitations are due to the lack of direct visual observations we developed a visual action recognition for our construction domain. This second approach is described in the next chapter and is intended to be integrated with

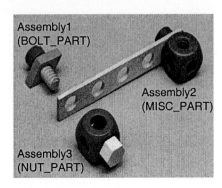

1. Take Assembly1 (BOLT_PART)

2. Take Assembly2 (MISC_PART)

3. Connect Assembly1 Assembly2

4. Take Assembly3 (NUT_PART)

5. Connect Assembly1 Assembly3

6. Put down Assembly1

Fig. 10. Example assembly with action sequence resulting from symbolic action detection.

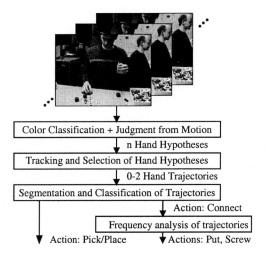

Fig. 11. Hierarchical processing strategy for the action classification.

the symbolic action detection to make it more robust and to allow for direct detection of the *Put* and *Screw* action instead of its delayed inference.

5 Visual Action Recognition

Figure 11 gives an overview of the system intended for the visual action recognition. The hierarchical architecture is domain independent on the first two levels. This enables the easy application to other scenarios with humans executing actions differing from the construction actions presented here. In the following subsections this system is described and an evaluation of its performance is given.

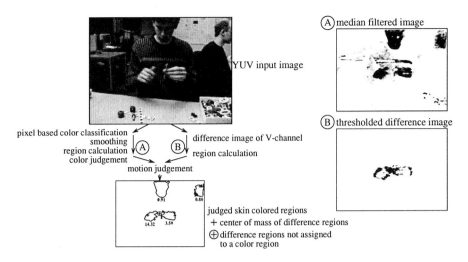

Fig. 12. Flowchart of the region calculation.

5.1 Fast Hand Segmentation

To segment skin-colored regions in images various methods have been proposed, see e.g. [Hun94], [Stö99], [Wei99], or [Boe98]. As the input image can contain many skin-colored regions, a reliable detection of hands based only on the color cue is not possible. Therefore we use motion as a second cue. Figure 12 shows a flowchart of the realized process, including the input image, the judged output regions and two intermediate images.

In the scenario, the lighting conditions are rather constant, making an approach with a static classifier with a large set of parameters for a detailed modeling of the input space feasible. Based on the classifier described in [Kum98], we chose a polynomial classifier of sixth degree using the YUV-values of every pixel of the input image as a three-dimensional feature vector. To speed up the classification, a lookup table containing the classification results for all possible input vectors is used. The output classes are skin color and background. To obtain the likelihood of a pixel belonging to the skin color class we applied a confidence mapping on the trained polynomial classifier as described in [Sch96].

After smoothing with a median filter, the image is thresholded and all regions contained in the binary image are approximated by polygons. During this step, different features like the size and the center of mass of the regions are calculated. Small regions are deleted. For each segmented region the average confidence value of all pixels inside the region is used as a judgment. This judgment is increased by using motion information from a difference image. This takes into account that hands often move while skin colored regions in the background or even faces typically move much less or not at all.

The difference image is calculated using the V-channel of two subsequent input images. After thresholding, the size and the center of mass of all regions contained in the image are calculated and small regions are deleted. For each difference region the distances between its center of mass and the polygons of the skin colored regions are calculated. The difference region is assigned to the closest skin colored region if it lies inside or close to the skin colored region and no other region lies in a comparable distance. The judgment of the skin colored regions is increased by adding a weighted size of the assigned regions.

5.2   Tracking the Hands

To determine the actions executed by the user the trajectories of segmented regions are analyzed. As the region segmentation yields several regions with judgments indicating their dynamics, it is not possible to determine which regions are representatives of the human hands based on a single image. During movements of objects in the foreground or due to shading, static objects can get high judgments in a single image. A second problem is the low judgment of hand regions while they are not moving.

To cope with these difficulties we track the center of mass of every skin colored region using standard Kalman filters. For every new input image all previously initialized Kalman filters are associated with the region that is closest to the predicted new position. We use a distance threshold to avoid assigning a region that is too far away. To restrict the tracking to dynamic regions for all untracked regions, a new Kalman filter is initialized only if the region reveals a high judgment. The two hands are selected based on the lifetime of the tracked trajectory and its distance from its initialization position.

5.3   Recognizing Complex Actions

After identifying the two trajectories generated by the hands of the human constructor we can now classify these trajectories. The actions to be classified consist of *Pick* (left hand/right hand (L/R)), *Place* (L/R), *Put*, and *Screw*. We use a hierarchical approach for the recognition. *Put* and *Screw* are discriminated from a *Connect* action by a post processing step. *Pick*, *Place*, and *Connect* are represented by the following elementary actions:

Pick/Place: $\{Hand\ down,\ Hand\ up\}^{L|R}$
Connect: $\{Approach,\ Move\ away\}$

The selected algorithm should be able to cope with the facts that the start and end points of the actions are not explicitly given and that the actions are not performed in the same way each time. The algorithm selected for the identification of the *Pick*, *Place*, and *Connect* actions is based on the *Conditional Density Propagation* (CONDENSATION) algorithm, which is a particle filtering algorithm introduced by Isard and Blake to track objects in noisy image sequences [Isa96]. Black and Jepson used the CONDENSATION algorithm in order to classify the trajectories of drawn commands at a blackboard [Bla98].

The actions are represented by parameterized models, which are consisting of 5-dimensional trajectories made up of the x- and y-velocities of both hands and the distance between them. The complex actions are concatenations of the

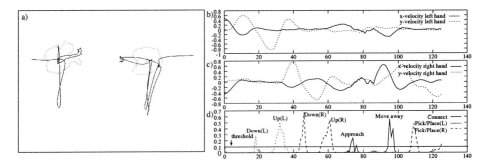

Fig. 13. The action sequence **Pick(L)**, **Pick(R)**, **Connect**, and **Place(R)**. a) Trajectories of the hands; b),c) Velocities of the hands; d) Probabilities for the model completion.

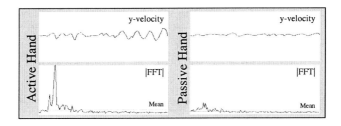

Fig. 14. Velocities and their Fourier transformations for a *Screw* action with one active hand regrasping the bolt while the passive hand just holds the threaded part.

elementary actions linked together by transition probabilities. For a detailed description see [Fri00]. Figure 13 shows an example trajectory of several complex construction actions and the corresponding hand velocities and classification results.

Due to the strongly varying length of time required by the *Screw* action, the recognition rates for this action performed with the CONDENSATION algorithm are poor [Fri00]. Therefore we perform the discrimination of *Put* and *Screw* in an additional post processing step performed after a *Connect* action is recognized. As the characteristic of a *Screw* action is the repeated movement of the hands to regrasp and turn the threaded bolt, we perform a Fourier transformation on the derivatives of the two Kalman trajectories, i.e. the hand velocities. In case of a *Screw* action the transformation exhibits high amplitudes in the low-frequency range ('Active Hand' in Figure 14). For a hand that is not moving, the resulting transformation exhibits no characteristic peak ('Passive Hand' in Figure 14). The difference in the FFT-magnitude between the maximum amplitude and the mean value of each hand is used together with the duration of the action as a three-dimensional feature vector for classifying the action with a k-nearest neighbor classifier.

Table 1. Results for 36 action sequences. The post processing step is only performed if the *Connect* model was recognized correctly.

Model	Elementary actions							Complex actions				PostPro.	
	Down L	Up L	Down R	Up R	Approach	Move away	Σ	Pick/Place L	Pick/Place R	Connect	Σ	Put	Screw
# Actions	57	57	87	87	72	72	468	57	87	72	216	36	36
# Recognized	51	54	78	86	64	70	430	50	77	64	191	32	32
Recognized (%)	89	95	90	99	89	97	92	88	88	89	88	89	89

5.4   Results

Using two Workstations (SPECInt95 13.9 and 27.3) the image segmentation and region tracking as well as the CONDENSATION algorithm are working in real-time. The segmentation is done at a frame rate of 25 Hz using an image size of 189x139 pixels. The CONDENSATION algorithm works with 3500 states.

Since the classification is done on trajectory data from a single camera we have only two-dimensional trajectories. This imposes the restriction on the acting humans to orient themselves towards the camera and execute the actions in an explicit way, i.e. not taking the parts and connecting them together directly on the table without any visible trajectory. The system was tested using 36 action sequences of 5 different people following these restrictions. Each action sequence consisted of taking a bolt and a bar, putting the bar onto the bolt and securing the parts with a cube screwed onto the bolt before placing the assembly back on the table. Each construction sequence therefore contained 6 complex actions (3 *Pick*, 2 *Connect*, 1 *Place*) with a total of 13 elementary actions. The results for the action classification with real-time constraints are shown in table 1.

6   Complex Objects

Techniques to recognize assemblies play a central role in vision-based construction cell analysis. As complex objects not only are the final product of an assembly process but usually appear at intermediate states, automatic observation has to cope with them. Concerning assembly recognition, however, little work has been reported yet. Moreover, most of these contributions deal with failure detection rather than with object recognition (cf. e.g. [AH97]). Known systems to recognize assemblies make use of highly specialized sensors like laser-range-finders [Ike94,Miu98] whereas pure vision systems merely deal with simple block world objects [Llo99].

This limitation to accurate sensors or simple objects is not surprising since in the cited works recognition relies on geometric object features. This, however, brings with it three major drawbacks with an impact on recognition: geometric features are sensitive to perspective occlusions, which occur naturally if objects are assembled; an accurate determination of geometric features by means of

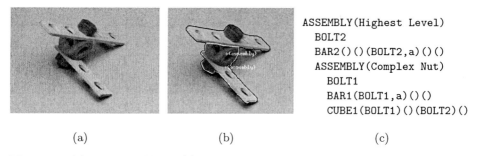

```
ASSEMBLY(Highest Level)
  BOLT2
  BAR2()()(BOLT2,a)()()
  ASSEMBLY(Complex Nut)
    BOLT1
    BAR1(BOLT1,a)()()
    CUBE1(BOLT1)()(BOLT2)()
```

(a)   (b)   (c)

Fig. 15. 15(a) An assembly, 15(b) the result of an assembly detection process, and 15(c) the corresponding high-level sequence plan.

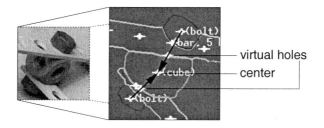

Fig. 16. A detail from Fig. 15(a) and the corresponding labeled image regions with 2D positions of mating features. As occupied holes of a cube are not visible, virtual positions are estimated. They are supposed to be the points where a line reaching from a bolt to the center of the cube intersects the border of the cube region.

vision is difficult and time consuming; it is rather impossible to provide a geometric model of every feasible assembly that might appear in a flexible and non-standardized construction environment.

6.1   From Hierarchical Models to Planar Graphs

In preliminary work we presented syntactic methods to detect assemblies in image data [Bau98,Bau00]. This kind of approach is well suited to detect assemblies since it can describe a large variety of feasible objects and does not rely on geometric features. Given the syntactic model described in section 4 and the results of elementary object recognition outlined in section 2, assembly detection is realized as a structural analysis of clusters of 2D objects yielding hierarchical descriptions (Fig. 15).

The topological appearance of a cluster of elementary objects yields mating relations among the objects, and our system automatically infers which holes of a bar or a cube are occupied by bolts. Therefore, syntactic descriptions enhanced with mating relations enable the generation of assembly plans [Bau99]. Note that these plans (Fig. 15(c)) do not contain information of spatial object positions but nevertheless are sufficient to generate *high-level assembly sequences* as defined in [Cha97].

However, syntactic methods cannot really *recognize* individual assemblies. Most complex objects have several syntactic structures and processing different views of an assembly usually results in different descriptions. But recognition requires matching a description against a prototypical model which is burdensome in case of hierarchical structures or plans.

Therefore we introduce another representation of assemblies which is based on the *semantics* of syntactically derived plans. All feasible plans of an assembly encode topological relations among its parts. They all describe how objects are attached to each other, i.e. they relate the *mating features* of elementary objects comprised in an assembly.

Mating features enable and characterize mechanical connections. Nut-type objects, for instance, have threaded holes to be rigidly connected to bolts. Due to their significance to mechanical connections, mating features are of course well known in assembly modeling. Even if by now their geometric properties were of primary concern (cf. e.g. [Rab96]) they also have topological significance. Relations among mating features uniquely characterize the structure of an assembly. Therefore, a graph describing the relations among mating features can be understood as the semantics of a set of high-level assembly plans. Such *mating feature graphs* extend the concept of liaison-graphs which qualify relations among objects and are well known from assembly planning literature (cf. e.g. [De 99]).

Figure 17(a) depicts the mating feature graph derived from the plan in Fig. 15(c). As mating feature graphs represent the semantics of a set of sequence plans, there is of course an abstract function calculating the semantics of any syntactically correct plan[4]. In this contribution we restrict ourselves to a short discussion of what is actually encoded in this graph: its vertices represent visible subparts of the objects constituting the assembly. They are labeled with their type and color. The edges relate the subparts, their direction indicates that objects were sequentially put onto bolts. Vertices connected by a pair of edges belong to the same object. Bolts connected to the same object are related by edges that are labeled with an angle specifying how these bolts are situated with respect to each other.

6.2   3D Reconstruction of Complex Objects

After transforming a sequence plan into a mating feature graph, assembly recognition becomes a problem of graph matching: an input graph obtained from image analysis has to be compared with a set of model graphs. Using graphs to encode knowledge for object recognition is a well established technique in computer vision (cf. e.g. [Son93, Cos95, Sid99]) In our system we make use of the *graph matching toolkit of the University of Berne* which implements a fast algorithm for calculating error-correcting subgraph isomorphisms developed by Messmer and Bunke [Mes98].

---

[4] An exemplary implementation using Haskell can be found under [Bau01].

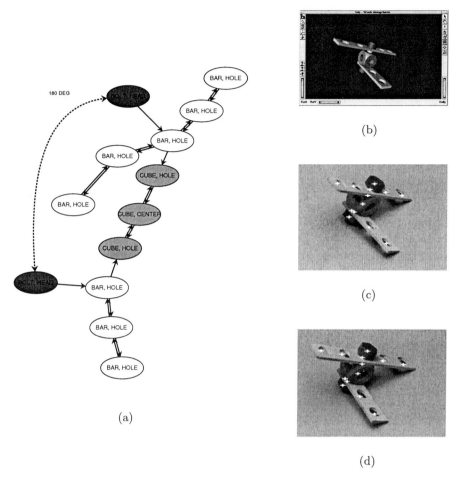

(b)

(c)

(d)

(a)

Fig. 17. 17(a) A mating feature graph describing the relations of functional subparts of the assembly in Fig.15. The virtual model in 17(b) was generated by means of a rough 3D reconstruction based on the calculated positions of mating features depicted in 17(c) and 17(d).

According to all previous considerations, our system realizes assembly recognition in a two-step procedure. After deriving a mating feature graph from an assembly plan, the graph is matched against a database of previously calculated graphs. If no match is found, the graph is assumed to represent an unknown assembly and is inserted into the database. This, in fact, realizes *a self learning vision system* which automatically extends its knowledge.

Another application of mating feature graphs lies in the field of 3D reconstruction. To this end vertices are labeled with image coordinates of the corresponding subpart which are calculated during plan generation. The position of a hole of a cube, for instance, is supposed to be the point where the line connecting the 2D centers of the cube and a bolt intersects the border of the cube region (Fig. 16). In this way only those holes can be labeled that are connected to a bolt. Thus, the center of a cube is regarded as a virtual mating feature. Consequently, cubes are represented by means of the maximum number vertices available (here: 3).

Matching two graphs derived from stereo images yields pairs of image coordinates that can be used to estimate the 3D spatial position of the corresponding features [Fau93]. Up to now, we used calibrated cameras in our experiments. Figure 17(c) and Fig. 17(d) show an example of a pair of stereo images with the calculated subpart positions cast into them. The corresponding coarse estimates of 3D positions are sufficient to create the virtual prototype of the assembly shown in Fig. 17(b). This is done using a CAD tool for virtual assembly developed within our research project [Jun98]. Given an assembly plan, this tool virtually interconnects the objects and automatically corrects faulty coordinates by means of detailed CAD models.

7   Summary

This contribution presents a system that performs learning of complex objects in a cooperative construction scenario. Visual models and construction plans are achieved during a task-oriented human-robot communication. The system covers already most components necessary for learning of words, skills, object models, and tasks. The baufix scenario — a wooden toolkit with multi-functional parts — allows a clear and obvious separation of baseline competences the system must provide *a priori* and competences to be acquired during communication. (i) Recognition and location of elementary objects, (ii) cross-modal interrelations of speech and visual data for objects, features, and spatial expressions, (iii) models for symbolic and visual action detection, and (iv) structural rules for complex objects form the baseline competences of the system. Complex objects and their verbal descriptions and naming are learned according to both their structure and their construction process. Fig. 18 shows an example. The object in 18(a) has been constructed and the objects in the scene were recognized. The structural description in 18(b) has been achieved and names that occurred during the process were assigned to the object and its substructures. The mating feature graph 18(c) provides a model for recognition of the object. For the construction of objects and their visualization the system Cody [Wac96] is used. The result of the corresponding virtual re-construction of the object is shown in 18(d).

Acknowledgment

The authors wish to thank H. Ritter and G. Heidemann for providing their subpart recognition tool, I. Wachsmuth and B. Jung for the Cody system, S. Posch and D. Schlüter for the grouping module, and G. Fink and H. Brandt-Pook for the speech recognition and dialog module, respectively. Furthermore, we acknowledge the members of the collaborative research center 'Situated Artificial Communicators' (SFB 360) for many fruitful discussions.

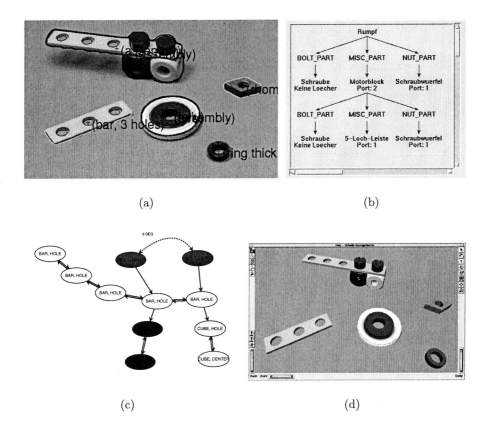

(a)   (b)

(c)   (d)

Fig. 18. 18(a) A scene containing a complex object that was constructed during an interactive session. 18(b) A structural description of this object where names and identifiers introduced during the construction are assigned to substructures. 18(c) The corresponding mating feature graph used for unique assembly recognition. 18(d) A virtual reconstruction of the scene.

References

[AH97]   M. Abu-Hamdan, A. Sherif-El-Gizawy: *Computer-aided monitoring system for flexible assembly operations, Computers In Industry*, Bd. 34, Nr. 1, 1997, S. 1–10.

[Ana96]   R. Anantha, G. Kramer, R. Crawford: *Assembly modelling by geometric constraint satisfaction, Computer-Aided Design*, Bd. 28, Nr. 9, 1996, S. 707–722.

[Bau98]   C. Bauckhage, F. Kummert, G. Sagerer: *Modeling and Recognition of Assembled Objects*, in *Proc. IECON'98*, 1998, S. 2051–2056.

[Bau99]   C. Bauckhage, F. Kummert, G. Sagerer: *Learning Assembly Sequence Plans Using Functional Models*, in *Proc. IEEE International Symposium on Assembly and Task Planning (ISATP'99)*, 1999, S. 1–7.

[Bau00]   C. Bauckhage, S. Kronenberg, F. Kummert, G. Sagerer: *Grammars and Discourse Theory to Describe and Recognize Mechanical Assemblies*, in *Advances in Pattern Recognition*, Lecture Notes in Computer Science 1876, Springer-Verlag, 2000, S. 173–182.

[Bau01]   C. Bauckhage: *Haskell script 'plansem.hls'*, http://www.techfak.uni-bielefeld.de/~cbauckha/plansemantics/plansem.lh s, 2001.

[Bla98]   M. J. Black, A. D. Jepson: *A probabilistic framework for matching temporal trajectories: CONDENSATION-based recognition of gestures and expressions*, Lecture Notes in Computer Science, Bd. 1406, 1998, S. 909–924.

[Boe98]   H.-J. Boehme, A. Brakensiek, U.-D. Braumann, M. Krabbes, H.-M. Gross: *Neural Architecture for Gesture-Based Human-Machine-Interaction*, in I. Wachsmuth, M. Fröhlich (Hrsg.): *Gesture and Sign Languge in Human-Computer Interaction*, Springer, Bielefeld, 1998, S. 219–232.

[BP99]   H. Brandt-Pook, G. A. Fink, S. Wachsmuth, G. Sagerer: *Integrated Recognition and Interpretation of Speech for a Construction Task Domain*, in *Proc. of the International Conference on Human Computer Interaction (HCI)*, Bd. 1, 1999, S. 550–554.

[Cha97]   S. Chakrabarty, J. Wolter: *A Structure-Oriented Approach to Assembly Sequence Planning*, *IEEE Transactions on Robotics and Automation*, Bd. 13, Nr. 1, 1997, S. 14–29.

[Com97]   D. Comaniciu, P. Meer: *Robust Analysis of Feature Space: Color Image Segmentation*, in *Proc. IEEE Conf. CVPR, Puerto Rico*, 1997, S. 750–755.

[Cos95]   M. Costa, L. Shapiro: *Analysis of Scenes Containing Multiple Non-Polyhedral Objects*, in C. Braccini, L. D. Floriani, G. Vernazza (Hrsg.): *Image Analysis and Processing*, Bd. 974 von *Lecture Notes in Computer Science*, Springer-Verlag, 1995.

[De 99]   T. De Fazio, S. Rhee, D. Whitney: *Design-Specific Approach to Design fo Assembly (DFA) for Complex Mechanical Assemblies*, *IEEE Transactions on Robotics and Automation*, Bd. 15, Nr. 5, 1999, S. 869–881.

[Fau93]   O. Faugeras: *Three-Dimensional Computer Vision*, MIT Press, Cambridge, Mass., 1993.

[Fin99]   G. A. Fink: *Developing HMM-based Recognizers with ESMERALDA*, in V. Matoušek, P. Mautner, J. Ocelíková, P. Sojka (Hrsg.): *Lecture Notes in Artificial Intelligence*, Bd. 1692, Springer, Berlin, 1999, S. 229–234.

[Fri00]   J. Fritsch, F. Loemker, M. Wienecke, G. Sagerer: *Erkennung von Konstruktionshandlungen aus Bildfolgen*, in *Mustererkennung 2000, 22. DAGM-Symposium Kiel*, Informatik aktuell, Springer-Verlag, 2000, S. 389–396.

[Fur87]   G. Furnas, T. Landauer, L. Gomez, S. Dumais: *The Vobabulary Problem in Human-System Communication*, *Communications of ACM*, Bd. 30, Nr. 11, 1987.

[Hei96]   G. Heidemann, F. Kummert, H. Ritter, G. Sagerer: *A Hybrid Object Recognition Architecture*, in C. von der Malsburg, W. von Seelen, J. Vorbrüggen, B. Sendhoff (Hrsg.): *Artificial Neural Networks – ICANN 96, 16.-19. July*, Springer-Verlag, Berlin, 1996, S. 305–310.

[Hei98]   G. Heidemann: *Ein flexibel einsetzbares Objekterkennungssystem auf der Basis neuronaler Netze*, Bd. 190 von *Dissertationen zur Künstlichen Intelligenz*, Infix, Sankt Augustin, 1998.

[Hom90]   L. Homem de Mello, A. Sanderson: *AND/OR Graph Representation of Assembly Plans*, *IEEE Transactions on Robotics and Automation*, Bd. 6, Nr. 2, 1990, S. 188–199.

[Hun94] M. Hunke, A. Waibel: *Face Locating and Tracking for Human-Computer Interaction*, in *Twenty-Eight Asilomar Conference on Signals, Systems & Computers*, Monterey, California, Nov 1994.

[Ike94] K. Ikeuchi, T. Suehiro: *Towards an Assembly Plan from Observation Part I: Task Recognition with Polyhedral Objects* , IEEE Transactions on Robotics and Automation, Bd. 10, Nr. 3, 1994, S. 368–385.

[Isa96] M. Isard, A. Blake: *Contour tracking by stochastic propagation of conditional density*, Lecture Notes in Computer Science, Bd. 1064, 1996, S. 343–356.

[Jac92] R. Jackendoff: *Languages of the Mind*, The MIT Press, 1992.

[Jun98] B. Jung, M. Hoffhenke, I. Wachsmuth: *Virtual Assembly with Construction Kits*, in *Proc. ASME Design for Engineering Technical Conferences*, 1998.

[Kum98] F. Kummert, G. A. Fink, G. Sagerer, E. Braun: *Hybrid object recognition in image sequences*, in *14th ICPR*, Bd. II, Brisbane, 1998, S. 1165–1170.

[Llo99] J. Lloyd, J. Beis, D. Pai, D. Lowe: *Programming Contact Tasks Using a Reality-Based Virtual Environment Integrated with Vision*, IEEE Transactions on Robotics and Automation, Bd. 15, Nr. 3, 1999, S. 423–434.

[Maß97] A. Maßmann, S. Posch, G. Sagerer, D. Schlüter: *Using Markov Random Fields for Contour-Based Grouping*, in *Proceedings International Conference on Image Processing*, Bd. II, IEEE, 1997, S. 207–210.

[Mes98] B. Messmer, H. Bunke: *A New Algorithm for Error-Tolerant Subgraph Isomorphism Detection*, IEEE Transactions on Pattern Analysis and Machine Intelligence, Bd. 20, Nr. 5, 1998, S. 493–504.

[Miu98] J. Miura, K. Ikeuchi: *Task Oriented Generation of Visual Sensing Strategies in Assembly Tasks*, IEEE Transactions on Pattern Analysis and Machine Intelligence, Bd. 20, Nr. 2, 1998, S. 126–138.

[Muk97] A. Mukerjee: *Neat vs Scruffy: A review of Computational Models for Spatial Expressions*, in P. Olivier, K.-P. Gapp (Hrsg.): *Representation and processing of spatial expressions*, Lawrence Erlbaum Associates, 1997.

[Par94] B. Parhami: *Voting Algorithm*, IEEE Trans. on Reliability, Bd. 43, Nr. 4, 1994, S. 617–629.

[Pea89] J. Pearl: *Probabilstic reasoning in intelligent systems: networks of plausible inference.*, Morgan Kaufmann, 1989.

[Rab96] M. Rabemanantsoa, S. Pierre: *An artificial intelligence approach for generating assembly sequences in CAD/CAM*, Artificial Intelligence in Engineering, Bd. 10, 1996, S. 97–107.

[Sag01] G. Sagerer, C. Bauckhage, E. Braun, G. Heidemann, F. Kummert, H. Ritter, D. Schlüter: *Integrating Recognition Paradigms in a Multiple-path Architecture*, in *International Conference on Advances in Pattern Recognition, Rio de Janeiro*, 2001, to appear.

[Sch96] J. Schürmann: *Pattern classification : a unified view of statistical and neural approaches*, Wiley, New York, 1996.

[Sid99] K. Siddiqi, A. Shokoufandeh, S. Dickinson: *Shock Graphs and Shape Matching*, Int. Journal of Computer Vision, Bd. 35, Nr. 1, 1999, S. 13–32.

[Soc98] G. Socher, G. Sagerer, P. Perona: *Baysian Reasoning on Qualitative Descriptions from Images and Speech*, in H. Buxton, A. Mukerjee (Hrsg.): *ICCV'98 Workshop on Conceptual Description of Images*, Bombay, India, 1998.

[Son93] M. Sonka, V. Hlavac, R. Boyle: *Image Processing, Analysis and Machine Vision*, Chapman & Hall, London, 1993.

[Stö99] M. Störring, H. J. Andersen, E. Granum: *Skin colour detection under changing lighting conditions*, in H. Araújo, J. Dias (Hrsg.): *SIRS'99 Proc. 7th Int. Symposium on Intelligent Robotic Systems*, July 1999, S. 187–195.

[Tho96]   J. Thomas, N. Nissanke: *An algebra for modelling assembly tasks*, *Mathematics And Computers In Simulation*, Bd. 41, 1996, S. 639–659.

[Ung82]   L. Ungerleider, M. Mishkin: *Two cortical visual systems*, in *Analysis of Visual Behaviour*, The MIT Press, 1982, S. 549–586.

[Wac96]   I. Wachsmuth, B. Jung: *Dynamic Conceptualization in a Mechanical-Object Assembly Environment*, *Artificial Intelligence Review*, Bd. 10, Nr. 3-4, 1996, S. 345–368.

[Wac98]   S. Wachsmuth, G. A. Fink, G. Sagerer: *Integration of Parsing and Incremental Speech Recognition*, in *Proceedings of the European Signal Processing Conference (EUSIPCO-98)*, Bd. 1, Rhodes, Sep. 1998, S. 371–375.

[Wac99]   S. Wachsmuth, H. Brandt-Pook, F. Kummert, G. Socher, G. Sagerer: *Integration of Vision and Speech Understanding using Bayesian Networks*, *Videre: A Journal of Computer Vision Research*, Bd. 1, Nr. 4, 1999, S. 62–83.

[Wac00]   S. Wachsmuth, G. A. Fink, F. Kummert, G. Sagerer: *Using Speech in Visual Object Recognition*, in *Proc. of DAGM-2000*, 2000.

[Wei99]   G. Wei, I. K. Sethi, N. Dimitrova: *Face Detection for Image Annotation*, in *Pattern Recognition Letters, special issue about "Pattern Recognition in Practice VI"*, Vlieland, June 1999.

Autonomous Fast Learning in a Mobile Robot[*]

Wolfgang Maass[1], Gerald Steinbauer[1,2], and Roland Koholka[1,3]

[1] Institute for Theoretical Computer Science
Technische Universität Graz, Austria
maass@igi.tu-graz.ac.at
[2] Roche Diagnostics, Site Graz
gerald.steinbauer@roche.com
[3] Knapp Logistik Automation GmbH, Graz
koholka@knapp.com

Abstract. We discuss a task for mobile robots that can only be solved by robots that are able to learn fast in a completely autonomous fashion. Furthermore we present technical details of a rather inexpensive robot that solves this task. Further details and videos of the robot are available from http://www.igi.tugraz.at/maass/robotik/oskar.

1 Introduction

We will discuss a task for an autonomous robot that *requires* learning insofar as a solution of this task by a non-learning robot is inconceivable (at least on the basis of budget constraints and currently available technology). On the other hand this task is also too difficult to be solved by a human. People can learn, but they do not have the mechanical skills which this task requires.

The task had been posed in the form of a student competition (Robotik 2000[1]) at the Technische Universität Graz. It can be outlined as follows. A 2×5 m white platform – surrounded by a black wall – had been divided by a black line into a *release zone* of about 1 m length and a *target zone* of about 4 m length (see Figure 1).

For each instance of the task one out of a large variety of green colored hills was placed at an arbitrary position – but at least 40 cm away from all walls – into the target zone. The hills were formed out of different kinds of hardening or non-hardening resins. These hills had a relatively rough surface and all kinds of odd shapes (diameters 30-60 cm), but they all had a round dip on the top of about 2 cm depth, with a diameters ranging from 8 to 12 cm, see Figure 2. The task was to accelerate and release a red billiard ball (diameter 5 cm, weight 142 g) in the release zone on the left part of the platform so that it comes to rest in the dip on top of the hill. To solve this task the ball has to be released with just the right speed v and angle α. For most hill positions the set of parameters

---

[*] This research was partially supported by the Land Steiermark (Austria), project Nr. P12153-MAT of the FWF (Austria), and the NeuroCOLT project of the EC.

[1] detailed information about this competition can be found online under http://www.igi.tugraz.at/maass/robotik

G.D. Hager et al. (Eds.): Sensor Based Intelligent Robots, LNCS 2238, pp. 345–356, 2002.

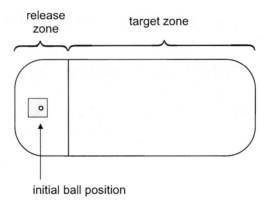

Fig. 1. Layout of the environment for the task.

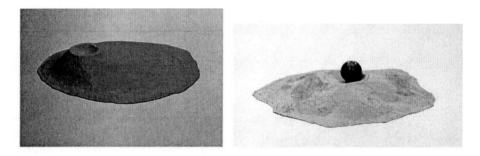

Fig. 2. Two of the hills that were used in the competition

$\langle \alpha, v \rangle$ that solved the task was so small that even an experienced human typically needed 40 or more trials before the first success. The number of trials needed for a second successful shot was not significantly smaller, indicating that in spite of all trying a human can solve this task essentially just by chance.

After any unsuccessful trial the robots in the competition had to find the ball, move it back to the release zone, and initiate the next shot. All of this – just as the trials themselves – had to be done completely autonomously, using only on-board sensors, computing devices and power supply (the latter had to last for at least 45 minutes). The only signals or interventions from the outside that the robots were allowed to receive were the *start-signal* at the beginning (by pushing a button) when the ball had been placed on the initial ball position marked in Figure 1, and a *repeat-signal* after a successful trial. The repeat-signal was given by pushing another button on the robot, and it signaled to the robot that the ball had been removed by a person from the top of the hill and had been placed again on the initial ball position, whereas the hill and hill position had been left unchanged[2]. The performance measure for the competition was

---

[2] We did not encourage that the robots themselves remove the ball from the top of the hill after a successful trial in order to protect our hills.

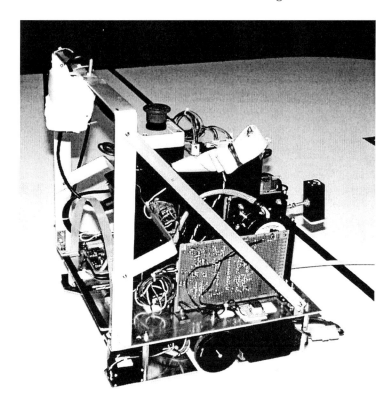

Fig. 3. This is the fast learning robot Oskar. The raised "lasso" for recapturing the ball is visible on his left. His hammer (visible on his right) has been pulled up, indicating that Oskar is about to shoot the ball from his ball-holding-bay underneath.

the total time needed by a robot until he had succeeded three times for the current hill and hill-position, averaged over a fair number of different hills and hill-positions.

A direct computation of the proper release values $\langle \alpha, v \rangle$ from measurements and other sensory data appears to be not feasible for an autonomous robot based on presently available technology[3]. Therefore a robot that was to be successful in this competition had no choice but to *learn* from his preceding trials in order to avoid a time consuming reliance on mere luck.

The winner of the competition was the robot Oskar[4] (see Fig. 3 and Fig. 4), designed and built by the second and the third author of this article. For technical details see Appendix B. Oskar usually needs 1-8 trials until the first successful shot, with a very high success rate for subsequent trials for the same hill in the same position. Oskar moves via two wheels powered by step motors, with

---

[3] More precisely: based on technology that can be afforded within the budget limit of 4000 Euro for each robot in the competition.

[4] see his homepage http://www.igi.tugraz.at/maass/robotik/oskar

Fig. 4. The robot Oskar shown while recapturing the ball (which is visible just in front of the robot). The camera in the white housing suspended from the square metal frame is used for controlling the recapture operation. Oskar's main camera, which is used for localization, shot planning, and learning is visible close to the top of the raised "lasso".

two Castor wheels in the back. For the control architecture see Figure 7. After recapturing the ball Oskar repositions himself in the release zone with a precision of 1 mm and an angular precision of 0.2 degrees, using the visual markers on the field for localization. As one camera looks othogonally down onto the lines of the platform from 30 cm above, the accuracy of the repositioning is only limited by the mechanical precision. This accuracy in repositioning is essential in order to turn experience from preceding trials into useful advice for the next trial. Oskar accelerates the ball by hitting it with a hammer that can be pulled up to variable heights, thereby allowing control over the initial speed v of the ball. Oskar catches the ball by lowering a flexible "lasso". He transports the ball back into the bay in front of the hammer through a suitable body movement initiated by his two wheels ("hip swing"). Oskar receives sensory input from 2 on-board cameras and 3 photo sensors. Processing the input of both cameras requires the whole CPU time. Therefore the management of all actors was outsourced to a micro controler board, which communicates with the PC via serial interface. Fig. 7 shows a diagram of Oskar's control system.

Videos that show the performance of Oskar and competing robots can be downloaded from the homepage of the robot competition http://www.igi.tugraz.at/maass/robotik/.

2 Oskar's Learning Strategies

Oskar uses both a longterm- and a shortterm learning algorithm. In order to compute the initial speed v for a new instance of the task he uses *longterm*

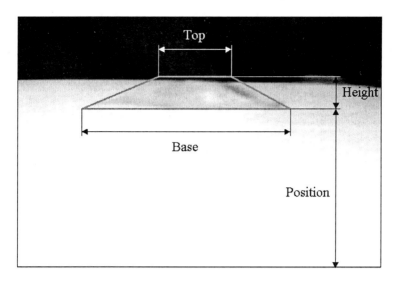

Fig. 5. A typical hill, as seen by the onboard camera of Oskar. Oskar approximates the hill by a trapezoid as shown and feeds its 4 corner points into a neural network that estimates a promising value for the speed v of the first shot.

learning via a sigmoidal neural network (MLP) with 4 hidden units. The inputs to the neural network are the coordinates of the 4 corner points of a trapezoid (see Fig. 5) that approximates the segmented video image of the hill, recorded from the starting position.

The neural network is trained via backprop, with training examples provided by preceding successful trials. The neural network started to make useful predictions after it was trained with data from 60 successful trials (error on training set: 1.2%). The other parameter α for the first trial on a new hill is simply computed by aiming at the center of the upper horizontal line segment ("Top") in the trapezoid that approximates the hill.

Since the first trial of Oskar is likely to succeed only for very simple hills (e.g., for small hills with a fairly deep dip at the top), the key point of his operation is his capability to learn via *shortterm learning* autonomously from preceding unsuccessful trials for the same instance of the task. For that purpose Oskar records for each shot the trajectory of the ball, and extrapolates from it an estimated sequence of positions of the *center* of the ball (see Fig. 6 for a few examples)[5]. This sequence of points is placed into one of 6 classes that characterize specific types of failures for a shot:

1. ball went too much to the right
2. ball went too much to the left
3. ball reached the dip, but went through it

---

[5] Oskar records 12 frames per second, yielding typically 3 to 25 frames for the period when the ball is on the hill.

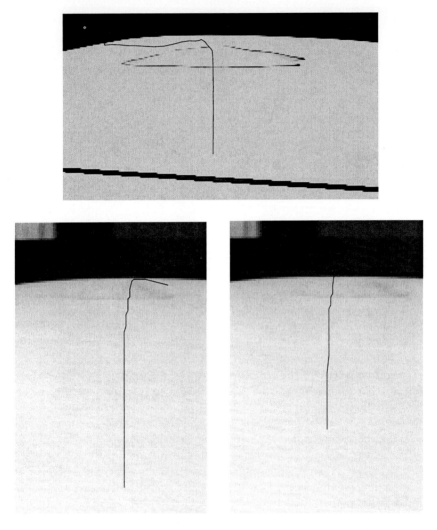

Fig. 6. Typical trajectories of the (estimated) center of the ball for unsuccessful trials, as computed by Oskar from his video recording. The trajectories shown were classified by Oskar into the classes 1,2, and 3. The characteristic feature of a trajectory in class 3 is that it goes up, down, and up again in the immediate vicinity of the hill.

4. ball rolled back from the hill
5. success
6. anything not fitting into classes 1-5

If the trial was unsuccessful, Oskar computes with the help of his learning algorithm QUICK-LEARN the parameters $\langle \alpha, v \rangle$ for the next trial from the classifications of the ball trajectories for the last 4 shots and a quantitative measure for their deviations to the left or right.

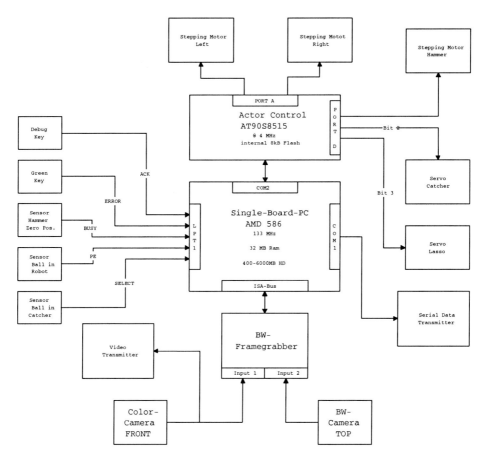

Fig. 7. Block diagram of Oskar's control system. It was designed to be simple and inexpensive. The main part is a low power Single Board PC. All control tasks except actor control are implemented on this PC. The actor control was sourced out to a microcontroller, to release the PC. The main sensors are two cameras connected to a 4-channel frame grabber on the ISA-Bus. To avoid an extra I/O-Module, the 3 photo sensors are read in via the parallel port.

There are no theoretical results known to us which could guide the design of such learning algorithm. There exists some theoretical work on binary search with errors (see for example [1]; pointers to more recent work can be found in [2]). However these studies focus on a worst case analysis under the assumption that the total number of erroneous answers lies below some known bound. In contrast to that, we need here a learning algorithm that converges very fast *on average* in the presence of stochastically distributed errors. Furthermore in our case the probability that an answer is erroneous is highly non-uniform: it is quite large if one is already close to the target, but it drops quickly with the distance between the current query and the unknown target value. To the best of our knowledge, no theoretical analysis of binary search with errors is known that can be applied to this more realistic case. In addition, our target vector lies in a *2-dimensional* space. Since it can be shown empirically that for most instances of the task the solution set does not form a rectangle in this 2-dimensional space, the search problem cannot be reduced to two independend binary search procedures.

The algorithm QUICK-LEARN merges two simultaneous binary searches, for the right angle α and the right speed v, where each of these binary search procedures receives just partially reliable feedback through the classification of previous trials into the classes 1 to 6. The algorithm QUICK-LEARN proceeds in all cases except one on the assumption that the parameters α and v can be determined independently.

The parameters α and v are left unchanged if the last trial was classified into class 6. Classifications of failures into the classes 1 and 2 are used as feedback for the binary search for α, and classifications into the classes 3 and 4 are interpreted as feedback for the binary search for v. This feedback assignment is only partially justified, since for example a ball that is a little bit too fast and a little bit too far to the right frequently runs along the distant inner rim of the dip on the hill and rolls back from the *left* side of the hill. Such misinterpretation, along with various other sources of noise and imprecision, causes errors in the binary search procedures for α and v.

When a trial yields for the first time a trajectory belonging to a class in $\{1,2\}$ or $\{3,4\}$, the corresponding parameter α or v is changed by some minimal step size in the obvious direction. Special precautions are built into QUICK-LEARN that allow it to recover quickly from erroneous feedback. The step size of the relevant binary search procedure is halved - like in the classical binary search procedure - only if the last two trials have produced trajectories that belong to classes that reflect complementary errors for the parameter in question, like the classes 1 and 2 for α, or the classes 3 and 4 for v. Whenever the trajectory switches between the groups of classes $\{1,2\}$ and $\{3,4\}$, the binary search is switched to the other parameter (with a step size equal to the last value when that other binary search had been interrupted). The step size for binary search is left unchanged if the trajectories for the last 2 or 3 trials all fall into the same class. If the last 4 trials have produced trajectories that all fall into the same class in $\{1,2,3,4\}$, the corresponding parameter is not only moved into the appropriate direction, but in addition the step size is doubled. Furthermore after 4 or more trials that yielded trajectories in one of the classes 3 or 4, not only the speed v is changed, but also the angle α is changed by some minimal step size in the direction which is suggested by a quantitative measure dev that evaluates left/right deviations of the ball that were caused by the hill (for details see the pseudocode of the algorithm QUICK-LEARN given in Appendix A).

3 Discussion

In contrast to many other applications of machine learning in robotics, where robots learn to avoid obstacles or other functions that can also be solved without learning, the robot Oskar that we have discussed in this article is the result of an evolutionary challenge where learning was indispensable to solve the given task. This given task happened to be one which is too difficult for humans in spite of their learning skills, because they are inferior to robots with regard to their precision and reproducibility of movements. Hence one may argue that Oskar solves a task that can neither be solved by non-learning robots nor by humans, thereby demonstrating the tremendous potential of autonomous adaptive robots.

On a more abstract level, the task that is solved by Oskar requires a very fast search in a multi-dimensional space for a parameter vector that provides a solution of the current instance of the task. This search has to be carried out without any human supervision, by computing autonomously suitable feedback for learning from the sensory input of the robot. Tasks of a similar nature arise in many application areas where one needs to procede by trial und error, for example in the area of construction, repair, or rescue. Many of these tasks arise in environments that are inaccessible or too hostile for humans, and therefore are interesting application areas for robotics. In the area of medical robotics miniature robots that learn fast to solve a task of skill in an autonomous fashion may potentially be useful for carrying out surgical procedures inside a human body that are so delicate that human supervision is too unreliable.

The robot Oskar that we have discussed in this article has become one of the longest serving autonomous adaptive robots that are known to exist. He has been running continuously 9 hours per day (for permanently changing hills and hill-positions) during the 6 months of a major exhibition[6] in Graz (Austria), during which Oskar delivered 31.108 shots (6.127 of which were successful). Hence this robot demonstrates that autonomous learning capabilities can be implemented in a mobile robot in a stable and reliable manner, at a very low cost[7].

Acknowledgement

We would like to thank Prof. A. Pinz, Prof. R. Braunstingl, Prof. G. Wiesspeiner, Dr. Thomas Natschläger, Oliver Friedl, and Dipl.-Ing. Harald Burgsteiner for their helpful advice.

References

1. Rivest, R. L., Meyer, A. R., Kleitman, D. J., and Winklmann, K.: Coping with errors in binary search procedures. Journal of Computer and System Sciences 20 (1980) 396–404
2. Sereno, M.:Binary search with errors and variable cost queries. Information Processing Letters 68 (1998) 261–270

---

[6] Steiermärkische Landesausstellung 2000, entitled comm.gr2000az, curated by H. Konrad and R. Kriesche. Oskar is scheduled to move in the summer of 2001 to the permanent collection of the Ars Electronica Center in Linz (Austria), see http://www.aec.at.

[7] The total cost of the components of Oskar lies below 3000 Euro. The only replacement parts that were needed during his 6 months of continuous performance was a new set of tires.

Appendix A:
Pseudocode for the Learning Algorithm QUICK-LEARN

Definitions:

last:	array with the classifications for the last 4 trajectories
last[0]:	classification of the immediately preceding trajectory
dev:	horizontal deviation at the hill of the last trajectory[1]

delta_a: current step size in binary search for angle
(initial value = 1/6 of the width of the hill at the base)

delta_v: current step size in binary search for velocity
(initial value = 1/20 of the value suggested by the neural network)

a:	parameter angle
v:	parameter velocity

a_min:	minimal step size in search for angle[2]
v_min:	minimal step size in search for velocity[3]

delta_a is never allowed to assume values less than a_min.
delta_v is never allowed to assume values less than v_min.

```
QUICK_LEARN(classes last[4], dev)
switch(last[0])
{
class_1:
if (all last == class_1)              // last 4 trajectories too far right:
delta_a *= 2                          // double step size for angle
else

if (last[1] == class_2)               // last 2 trajectories in classes 1
delta_a /= 2                          // and 2: halve step size for angle
delta_a = max(delta_a, a_min)         // step size not smaller than minimum
                                      // step size

if (last[1] == class_5)               // failure (of class 1) after success
a -= a_min⁴                           // (class 5): minimal change for
                                      // angle
else
a -= delta_a

class_2:
if (all last == class_2)              // last 4 trajectories too far left:
delta_a *= 2                          // double step size for angle
else

if (last[1] == class_1)               // last 2 trajectories in classes 1 and 2:
delta_a /= 2                          // halve step size for angle
delta_a = max(delta_a, a_min)         // stepsize not smaller than minimum step size

if (last[1] == class_5)               // failure (of class 2) after success:
a += a_min                            // minimal change for angle
else
a += delta_a
```

```
class_3:
if (all last == class_3)          // last 4 shots too fast (class 3):
a += sign(dev) * a_min            // additional minimal change of angle⁵

if (last[1] == class_4)           // last 2 trajectories in classes 3 and 4:
delta_v /= 2                      // halve step size for velocity
delta_v = max(delta_v, v_min)     // step size not smaller than minimum step size

if (last[1] == class_5)           // failure (of class 3) after success:
v -= v_min                        // minimal change of velocity
else
v -= delta_v                      // for all other classes: change with stepsize


class_4:
if (all last == class_4)          // last 4 shots too slow (class4):
a += sign(dev) * a_min            // additional minimal change of angle⁵

if (last[1] == class_3)           // last 2 trajectories in classes 3 and 4:
delta_v /= 2                      // halve step size for velocity
delta_v = max(delta_v, v_min)     // step size not smaller than minimum step size

if (last[1] == class_5)           // failure (of class 4) after success:
v += v_min                        // minimal change of velocity

else
v += delta_v                      // for all other classes:
                                  // change with step size


class_5:
a = a                             // last shot was successful:
v = v                             // leave parameters unchanged
store_example                     // save angle, velocity and hill
                                  // features as training example
                                  // for the longterm neural
                                  // network learning algorithm


class_6:
a = a                             // leave parameters unchanged,
v = v                             // no information extracted

}
```

Remarks:

[1] dev is the average of the horizontal deviations of points on the trajectory while the ball was on the hill, from a linear extrapolation of the initial segment of the trajectory defined by the points of the trajectory just before the ball reached the hill. Thus dev essentially measures the horizontal deviation of the ball caused by the hill.

[2] The minimum step size for the angle is one step of one of the two driving motors. The direction of the change determines which motor has to move.

[3] The minimal step size for the velocity is one step of the stepping motor that pulls up the hammer.

[4] Angles and angle changes to the left are counted negativ, angles and angle changes to the right are counted positiv. Angle 0 is defined by the straight line to the middle of the top of the hill (see Fig. 5).

[5] This rule prevents the algorithm to get stuck in just changing the velocity. Because of this rule one cannot view the algorithm as an implementation of 2 independent binary search procedures.

Appendix B:
Technical Data of the Learning Robot "Oskar"

Processor board
Single-Board-PC 586LCD/S by Inside Technology AMD 586 @ 133 MHz, 32 MB Ram, ISA and PC 104 Bus runs under DOS and the DOS-Extender DOS4GW by Tenberry
2,5 Laptop-Harddisc with 400 MB memory

Image processing
b/w-framegrabber pcGrabber-2plus by phytec with four camera inputs, ISA Bus color-CCD-camera VCAM 003 by phytec for ball/hill-analysis, no-name-wide-angle b/w-camera for navigation and obstacle avoidance

Motor/servo-steering
Risc-Microcontroller AT90S8515 by Atmel,
8kB integrated flash-program-memory
running under operating system AvrX by Larry Barello

Driving
two 1,8° stepping motors in differential drive

Shooting
one 0,9° stepping motor + hammer (250g)

Power Supply
High Performance NiMH-Accu with 2,6Ah capacity and 12V. Two DC/DC Converters provides the also needed 5V and -12V.

Exploiting Context in Function-Based Reasoning

Melanie A. Sutton[1], Louise Stark[2], and Ken Hughes[2]

[1] University of West Florida, Pensacola, FL 32514, USA
[2] University of the Pacific, Stockton, CA 95211, USA

Abstract. This paper presents the framework of the new context-based reasoning components of the **GRUFF** (**G**eneric **R**ecognition **U**sing **F**orm and **F**unction) system. This is a generic object recognition system which reasons about and generates plans for understanding 3-D scenes of objects. A range image is generated from a stereo image pair and is provided as input to a multi-stage recognition system. A 3-D model of the scene, extracted from the range image, is processed to identify evidence of potential functionality directed by contextual cues. This recognition process considers the shape-suggested functionality by applying concepts of physics and causation to label an object's potential functionality. The methodology for context-based reasoning relies on determining the significance of the accumulated functional evidence derived from the scene. For example, functional evidence for a chair or multiple chairs along with a table, in set configurations, is used to infer the existence of scene concepts such as "office" or "meeting room space." Results of this work are presented for scene understanding derived from both simulated and real sensors positioned in typical office and meeting room environments.

1 Introduction

Over the last decade, the **GRUFF** project has explored the extent to which representing and reasoning about object functionality can be used as a means for a system to recognize novel objects and produce plans to interact with the world. The reasoning behind this approach is that an object's function is often a geometric function. The function of a room is to be an enclosing volume. The function of a chair, desk or table is to be a flat surface at a comfortable height for sitting, writing, eating, etc. The recognition system can search for the proper geometric information that fulfills a pre-defined functional requirement. For example, find a *sittable surface* so I can rest.

The basic approach is as follows. **GRUFF** analyzes an object shape to 1) confirm the existence of required surfaces according to the required function (e.g. a surface or group of essentially co-planar surfaces of the proper size to support a sitting person); 2) confirm the relationship of required surfaces according to the required function (e.g. the handle of a hammer must be *below* and *essentially orthogonal* to the potential striking surface); and 3) confirm the accessibility or

G.D. Hager et al. (Eds.): Sensor Based Intelligent Robots, LNCS 2238, pp. 357–373, 2002.

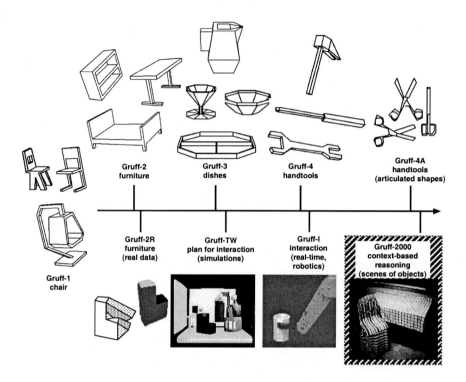

Fig. 1. The **GRUFF** project has encompassed the study of function-based reasoning applied to isolated rigid and articulated complete 3-D shapes, as well as partial 3-D shapes acquired from range and stereo systems. The area of expansion studied in this paper is highlighted in the bottom right corner.

clearance of potential functional elements[1] (e.g. the rim of the glass or cup must be accessible in order to use for drinking).

It has been argued that the function-based model is merely a structural model, specifying the required structure to pass the functional requirement tests. One can consider the function-based description as a weak structural model, which specifies not only the minimal required elements of structure and their relationships, but also the required **absence** of structure. The wide range of allowable structure can be best realized through example. Consider the chairs depicted in the left side of Figure 1. Each is confirmed to have an accessible sittable surface which can be stably supported essentially parallel to the ground plane within a certain height range. An accessible back supporting surface which is essentially orthogonal to the seat is also confirmed. The structure that exists between the sittable surface and the ground is of no interest to the recognition system, as long as it in some way provides the support and does not interfere with the usefulness of the object as a chair. All three of the chairs were identified

---

[1] Functional elements are defined as portions of the object structure identified as possibly fulfilling the functional requirement.

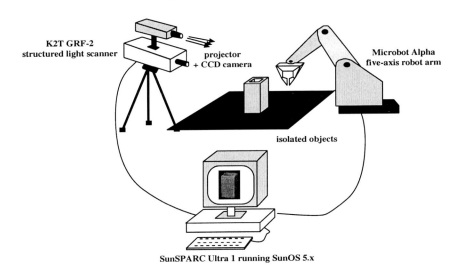

Fig. 2. The mounted workspace version of the **GRUFF** system was used to design and test interaction-based reasoning modules. At the time of its development, the overall cost of this system was roughly $15,000.

as straight back chairs by **GRUFF** despite the fact that one chair is supported by four legs, another by two legs and the third by essentially no legs (i.e. the hanging basket chair). What is important to remember is that all were recognized using the same representation.

The **GRUFF** approach has been methodical, in that initial work confirmed, using the category chair as the test case, the feasibility of a function-based representation for object recognition. Ideal (complete 3-D models) of chair and non-chair objects were tested, with promising results [26]. Over the years the knowledge base of domains has been extended to include a number of categories of furniture, dishes and handtools [26, 35].

As shown in Figure 1, the **GRUFF** project began with the study of isolated 3-D objects in simulated environments. The project has evolved to include analysis of articulated objects such as handtools and objects in real-world scenes using visual and haptic sensors. Our most recent work included the incorporation of an Interaction-based Reasoning Subsystem [33]. As shown in Figure 2, this version of the system was based on individual objects located in a static environment, utilizing a K2T GRF-2 structured light scanner and Microbot Alpha five-axis robot arm with workstation-based connectivity. It should be noted that all versions of **GRUFF** shown in Figure 1 have operated under an "expectancy paradigm" where functional evidence derived from bottom-up 3-D shape analysis has instantiated top-down exploratory modules. Although the systems have always been generative (i.e. capable of recognizing previously unencountered artifacts), the scope of the sensor-based version has previously been limited to a mounted workspace environment. This limited its utility to dynamic environments which mimic the types of applications we envision for the future.

The next obvious step in the evolution of the **GRUFF** project has therefore been to work with scenes rather than isolated objects. Through evaluation of a scene, the gathering of evidence is guided by cues provided by contextual information. Allowing multiple objects in a scene implies a high likelihood of occlusion. It is for this reason we must relax the **GRUFF** recognition process to simply *find evidence*, rather than to require the confirmation of functionality. The original **GRUFF** system discontinued probing for evidence of a back support if a sittable surface could not be identified. This requirement must be relaxed when considering models composed of multiple objects, as the surface may be occluded by a table top. Similarly, the requirement of surface accessibility must also be relaxed.

The key is that the functional evidence that can be observed must exist in the proper configuration to be considered. Humans do this all the time. When you walk into a meeting room where chairs are pulled up to the table, you still infer that it is a table surrounded by chairs. You see the backs of the chairs but no sittable surfaces to confirm the objects are actually chairs. You use the contextual information of knowing you are in an office building and expect a table surrounded by chairs as a common configuration. **GRUFF** exploits context using this same reasoning.

The layout of the remainder of the paper is as follows. Sections 2 and 3 explain related research in function, interaction, and context-based object recognition. Section 4 provides an overview of the experimental platform and equipment upon which the **GRUFF** system is based. Section 5 summarizes the evaluation techniques and preliminary experimental results of the implemented system. Finally, Section 6 provides a conclusion and directions for future research.

2   Background

The idea of using function to represent object categories for recognition purposes is not new. Binford and Minsky, among other researchers, have argued that object categories are better defined by focusing on the intended functionality of objects [4, 22]. The concept of how the use of function could be integrated into a computer vision system has matured over the years.

One body of work has looked at problems for which the assumed input is a complete symbolic description (e.g., a semantic network) of an object and its functionally relevant parts [38, 39]. This well-known work by Winston et al. explores reasoning by analogy between parts and functional properties of objects. Rivlin et al. also integrated function and object parts "by combining a set of functional primitives and their relations with a set of abstract volumetric shape primitives and their relations" [23].

Another body of work concentrated on producing and recognizing a function-oriented symbolic description of the object through reasoning about the 'raw' shape description rather than decomposing the object into parts. It was assumed that the input to the problem is a complete 3-D shape description (e.g., a boundary representation) of an object [8, 27, 28, 35, 18]. This work bypasses the 'real' vision problem, at least to some extent, by assuming complete shape models as input.

Functional information gained through an initial evaluation of an object can be used to help guide interaction with the object. Interaction can then be performed based on hypothesized areas of functional significance [6, 17, 25, 34]. A recent extension to the **GRUFF** research project also uses interaction for the purpose of verifying a functionality that was suggested by the object's shape [33]. This differs from the previously mentioned work in that the **GRUFF** system performs shape-based functional object labeling first, as a means to guide later interaction.

Functionality in object recognition has continued to gain attention as an alternative to the strict model-based approach. A broad sampling of other work that falls into this category can be found in the proceedings of the 1993 and 1994 AAAI Workshops on reasoning about object function [1, 2] and the special issue of *Computer Vision and Image Understanding* on functionality in object recognition [5, 7, 12, 13, 24, 23].

Our goal in this work is to further extend the function-based recognition capabilities of the **GRUFF** system to scenes of objects. This extension undoubtedly increases the complexity of the task. Cues that aid the process of disregarding irrelevant information in the scene (where relevance is defined according to the current task) are exploited to narrow the search space. This paper presents our approach to how context, defined in terms of required functional properties for a task, can be used to focus processing to verify whether specific geometric relationships exist that can fulfill the function.

3   Functional Context in Vision

Context is defined as both prior knowledge or background and the present situation or environment that is relevant to a particular event or action [36]. It is widely accepted that context can play an important role in the process of recognition.

The VISIONS image-understanding system, developed at the University of Massachusetts, was one of the early applications of context-based reasoning to resolve perceptual ambiguities in a scene [20]. This work incorporated *schemas* to capture object descriptions and contextual constraints relevant to prototypical outdoor and road scenes.

The work by Jensen, Christensen and Nielsen incorporate causal probability networks (CPN) along with Bayesian methods to model and interpret contexts [16]. The concept of *agents-with-instincts* was introduced as their control paradigm. Two agents, an image processor and an interpreter using a CPN, were used. It was found that "CPNs are very good for modeling scenes consisting of parts whose interpretation is context dependent." It was stated that this context dependence was "fairly easy to represent as conditional probabilities."

Fischler and Strat note that context was not being used as effectively as it could be by recognition systems [11]. The focus of their work was in recognizing objects in a natural environment using context. Their goal was to design a "perception system that not only recovers shape but also achieves a physical and semantic understanding of the scene" [11].

The focus of the work discussed in this paper is recognition of human artifacts instead of natural objects. However, the goal is essentially the same as the above mentioned work. The desire is to exploit context in the process of achieving a symbolic labeling of objects in a scene.

There are many different types of context available at different levels of processing. For example, in the area of image understanding, Hoogs [14] enumerates the following types of context that can be exploited:

1. image context (e.g. image resolution, image intensity, viewpoint)
2. geometric context (e.g. models of buildings or objects)
3. temporal context (e.g. time of day, image history)
4. radiometric context (e.g. orientation of the sun, material properties)
5. functionality context (e.g. orientation of vehicles in a parking lot, airport and railyards have an expected geometry stored as a model along with a description of the functionality)

Each of these types of context provide cues that aid the system in *understanding* the complex image data that must be interpreted. A broad sampling of other work that falls into this category can be found in the *Proceedings of the Workshop on Context-Based Vision* [3].

Our research effort is specifically concerned with demonstrating how *functional context* can be effectively exploited in the task of object recognition. It will be shown how functional context strongly influences which geometric relations are relevant. One of the major problems encountered in computer vision is discriminating what is really important and what is *extraneous* information. As discussed earlier, if the task at hand is to find a chair on which to sit and rest, the number of legs that support the chair should not be relevant, as long as the seat is stable. One can also ignore the fancy woodwork that decorates the back and legs of the chair. That information is not relevant in deciding if the object can *function* properly.

For the work being presented here, the task changes slightly. We may desire to find a meeting room that can accommodate up to ten people around a table. The first cue would be to find a room that is large enough to hold ten people which has a large table surface, normally, but not always, situated in the center of the room. There must be *evidence* of chairs in the room. Once again, the normal configuration would be that the chairs are arranged about the table. Supporting evidence for the chair functional requirement would be sittable surfaces and/or back supports.

The desire is to be able to capture this type of functional context of scenes in a representation that can be used in a vision system. In the next section the function-based recognition system **GRUFF**-2000, which is an implementation of such a representation, will be described.

4   Functionality in the Large: The Context-Based Framework and Platform

Another way to view the expansion of the **GRUFF** system is to consider it as the study of the implementation of a system which can attempt to do scene understanding by reasoning about "functionality in the large." The system was also designed with cost in mind, and a desire to test the platform in more realistic, potentially dynamically changing environments. An overview of the flow of control in the overall system, which includes both the older interaction-based modules and the new context-based analysis is provided in Figure 3. The system is composed of separate subsystems, as follows:

1. *Model Building* - the subsystem which builds 3-D world models from range images derived from stereo images of the scene
2. *Shape-based Reasoning* - the subsystem which performs static shape-based labeling of the 3-D scene model to identify functionally significant areas in the space, using scene probes (also called procedural knowledge primitives or PKPs) which assess clearance; stability; the proximity, dimensions, or relative orientation of surfaces; and/or the existence of enclosable areas
3. *Interaction-based Reasoning* - the subsystem which normally uses the visual and robotic components to perform dynamic confirmation of the 3-D model, by interacting with the object in the scene and analyzing subsequent 2-D intensity images
4. *Context-based Reasoning* - the subsystem which determines and reasons about the significance of the accumulated functional evidence derived from the scene

More specifically, the overall flow of the Context-based Reasoning Subsystem is shown in Figure 4. As an example, the functional label for *provides potential seating* can be realized by the acquisition of functional evidence for sittable areas and/or back support. In this way, partially occluded scene areas do not terminate scene processing. It should be noted that processing will terminate under two conditions; when no evidence can be found or the measure of the evidence found is below a specified threshold. Measures are returned by the procedural knowledge primitives. These measurements are combined for each functional requirement. For a more detailed description of how evaluation measures are combined see [26].

To get a better understanding of how the **GRUFF** recognition system would work, consider the scene depicted in Figure 4 of a table with nine chairs. It is actually four tables, but since the work surfaces are co-planar, the system would merge this into a single work surface. What draws our attention is the large workspace provided by the tables. **GRUFF** would label this area as *providing potential work surface*. From there, evidence of chairs would be investigated. **GRUFF** will try to identify object surfaces that could *provide potential seating*. Contextual cues thus drive the system to investigate the area near the work surface. The best evidence of a chair would be a sittable surface. For this example, four potential sittable surfaces are observable.

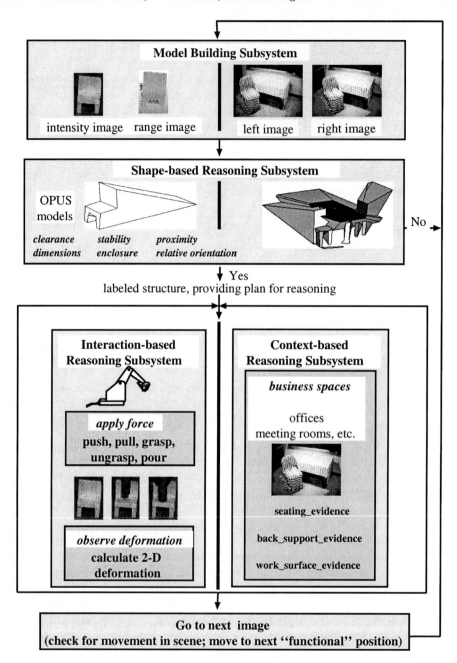

Fig. 3. The **GRUFF** system is based on a set of modules which label and reason about functionally significant areas in the scene. (The example OPUS (Object Plus Unseen Space) 3-D models shown in the Shape-based Reasoning box include real surfaces seen in a given view plus surfaces which bound the occluded areas in the scene [15, 21])

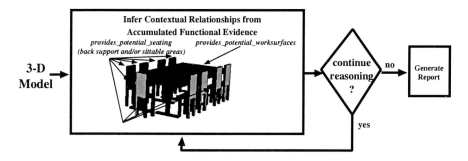

Fig. 4. The Context-based Reasoning Subsystem performs scene understanding by inferring and reasoning about contextual relationships of functional areas in the world model.

Processing continues by using the sittable surface evidence to find corresponding back support. Only one of the chairs (front left in the scene) has both a back and seat partially visible. Recognizing this, at this point the system would have strong evidence of one chair along with weaker evidence for three other chairs. Processing once again continues, looking for back support areas that are essentially orthogonal to the work surface and *close* such that the potential back of the chair is oriented toward the table. This would lend evidence toward a chair which is situated at the table with its sittable surface hidden under the table. Three other potential chairs are identified in this way.

Looking at the *big picture*, the system has identified, to some degree, seven of the chairs around the table. It should be noted that at this time, **GRUFF** restricts the labeling of a potential back support to surface areas which are in the proper orientation in relationship to a potential sittable surface and surface areas that are *facing* the table surface. Therefore, in this scene, two of the chairs on the right side of the table are not identified.

The recognition processing is guided by traversal of the function-based representation captured in the category definition tree depicted in Figure 5. As shown in Figure 5, incorporating context into the **GRUFF** structure involved augmenting the hierarchically stored world information which included base-level categories such as chairs, tables, etc. (see right side of Figure 5) with functional definitions of scene concepts such as "office" or "meeting room space" (see left side of Figure 5). The original **GRUFF** representation for base-level categories (see right side of Figure 5) required the identification of the key *functional elements* such as the *stable support* and *sittable surface* of the chair. The extension of the system to include functional definitions of scene concepts (see left side of Figure 5) allows more flexibility. Now the system need only identify *potential work surfaces* and *potential seating evidence* in the proper configuration to build contextual evidence of an office space. **GRUFF** can use both left and right function-based representations; however, the focus of our discussion will be on traversal of the functional definitions of the scene concept. The confirmation of this information involves the processing steps summarized in Figure 6.

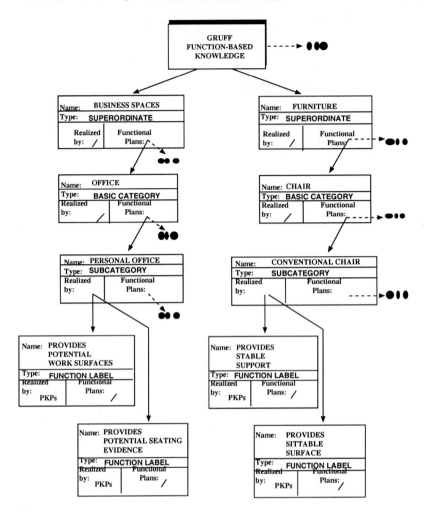

Fig. 5. The incorporation of scene concepts such as "business space" or "office" involves providing a hierarchical definition derived from category concept theory in humans. The procedural knowledge primitives (PKPs) probe the scene and measure the level of functional evidence for the category labels at the leaves of this world knowledge tree.

The next section summarizes initial experimental results using both simulated and real data.

5   Experimental Results

In order to design and test the Context-based Reasoning Subsystem, images derived from both simulated and real sensors positioned in typical office and meeting room scenes were used. Simulated data utilized to test the system was derived in two ways:

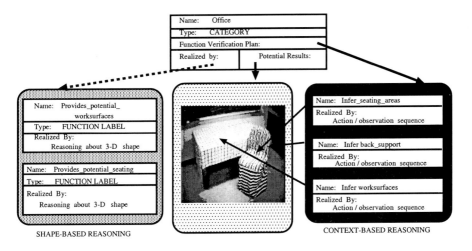

Fig. 6. The verification plan for an office space in the world.

1. Complete 3-D models, where all faces were considered "real", with no occlusion surfaces.
2. Partial 3-D models, derived from utilizing a laser range finder (LRF) simulation tool (lrfsim). These models contain both "real" and occlusion surfaces.

The items in (1) would be similar to those types of models which could be built if the system could move around a scene and combine information from multiple views. In practical terms, even with this type of combination of data, a completely accurate model, including all possible occluded surfaces such as the underside of the table, is unlikely to be recovered. The items in (2) are more typical of the types of models we expect systems can generate from a single viewpoint. Observed or real surfaces can be reconstructed by the system as a first step. The system can then make reasonable estimates of occluded areas, dependent upon the viewpoint and viewing direction.

Figures 7 and 8 provide examples of such type (2) data. Although the range image in Figure 7 is derived assuming a perfect sensor, not all derived segmentations are useful for model-building. In this example, pixels associated with the floor are erroneously merged with object pixels belonging to the chairs. This subsequently caused the Model-Building Subsystem to reject the current viewpoint, as a valid 3-D could not be created. When this occurs, no further scene analysis is possible until additional parameter-seeking heuristics can be instantiated.

Alternatively, Figure 8 demonstrates a recovered model for a simpler office space. In this case, the segmentation produces a reasonable 3-D model for additional analysis. From this model, the system finds evidence of a workspace area and several cues to possible sittable areas (two back support surfaces and one seat). From the category definition tree for this example, the accumulated functional evidence supports the recognition of the scene being a possible *office*.

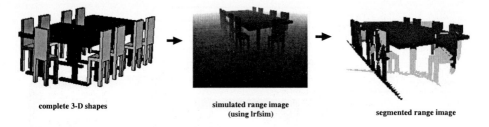

complete 3-D shapes   simulated range image   segmented range image
                                       (using lrfsim)

Fig. 7. Partially successful system result utilizing simulated sensor for a complex meeting room space.

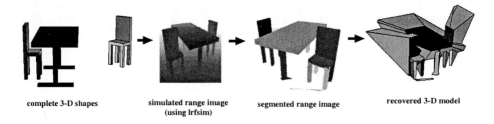

complete 3-D shapes   simulated range image   segmented range image   recovered 3-D model
                               (using lrfsim)

Fig. 8. System result utilizing simulated sensor for a simpler office space.

As stated earlier, the methodology of the **GRUFF** research plan is to first test the feasibility of extensions to the system, such as the scene evaluation extension, using a controlled (simulated) environment. Our long term goal is to be able to capture the required 3-D data as well as maneuver about the environment, gathering additional information as required. Where to look next would be driven by the results of the function-based analysis of the scene along with contextual cues. Such a system requires a stereo head and mobile base.

The newer setup we envision (shown pictorially in Figure 9) is currently PC-based, utilizing a low cost stereo head. We expect the stereo system depicted in Figure 9 to be mounted on the mobile base shown in the lower left image. We are in the process of developing such a platform, but in the interim have been able to acquire stereo images for processing assuming a stationary base. The real sensor data included here was therefore captured using of pair of wireless standalone cameras. The cameras were aligned such that the image planes of the two cameras were essentially co-planar (cameras level and viewing directions parallel). An example of real data acquired using this stereo pair is depicted in Figures 10 and 11.

SRI's (Stanford Research Institute's) Small Vision System software was then used to process the stereo pairs in these figures to obtain disparity maps, which subsequently produced the range images[2]. One of the characteristics of stereo vision is that without texture it is difficult if not impossible to extract correspondences to obtain a good disparity map. A trip to our local fabric store

---

[2] Software was acquired from Videre Design.

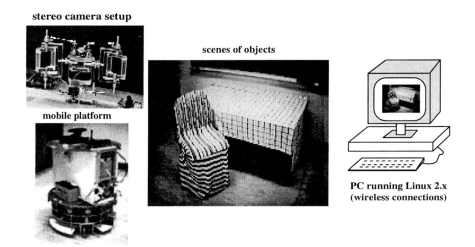

Fig. 9. The newer "world-based" version of the **GRUFF**-2000 system allows dynamic movement around a world space. This system is being developed at a more practical cost of under $3,000.

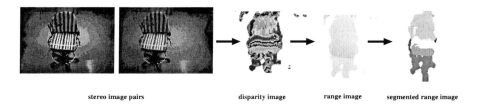

Fig. 10. System result based on actual stereo images taken of an isolated object in an office space. This example shows a system failure due to the missing data around the bottom portion of the seat back.

transformed our *office chair* and *common office scene* into a rather wild, but highly textured scene. This is one of those situations where, as computer vision researchers, we acknowledge the shortcomings of the current approach but can say that this approach helped us to test what we were really trying to test - context in function-based reasoning. Figure 10 represents how lack of texture affects the segmenter. In this case, there is an area where the back and seat surfaces meet which lacks texture. This area is ultimately merged with the floor area (which also lacks texture), subsequently causing a model-building error.

Figure 11 shows examples of more promising segmentation results derived from a textured office scene. In this case, the generated OPUS (Object Plus Unseen Space) [15, 21] model recovered from the sensor data includes both visible and occlusion surfaces. The visible surfaces are the ones representing the areas of the tabletop, back, and seat. In this case, a model can be recovered, but it ultimately fails to provide any usable surfaces for additional processing.

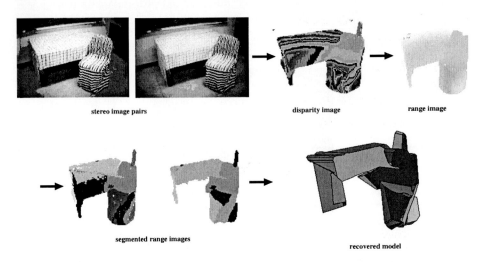

Fig. 11. Partially successful system result based on actual stereo images taken of a more complex scene with multiple objects in an office space.

6   Discussion and Future Work

This paper has summarized the new directions we will be taking the **GRUFF** project. We have augmented our knowledge base with context-based categories such as *office* and *meeting room*. Preliminary results with simulated sensors indicate we can derive functional evidence from scenes representative of these categories. Results utilizing real sensors are less successful at this point, due to lack of texture or surface fitting errors during the model-building stage. Since a closed solid model is ultimately not needed for processing, we are working on a new methodology in the Model-Building Subsystem which would allow us to relax the constraints for deriving a model. The alternative approach we envision would allow a coarse description of the 3-D surfaces in view to be processed.

Ultimately the final scene labeling acquired from context-based reasoning is to be incorporated into a mobile version of the interaction-based component of the **GRUFF-I** (**G**eneric **R**ecognition **U**sing **F**orm, **F**unction and **I**nteraction) system. One contextual cue which could potentially prove useful is determining the size of a room or space surrounding the robot. The mobile base shown in Figure 9 has its own collection of sensors (ultrasonics, encoders, etc) which are used for navigation, localization and environment mapping. The locations of these sensors are such that it is difficult to get the "larger picture" of a space without the robot actually moving about within the space. For example, the ultrasonics are arranged to detect objects near the floor in order for the robot to avoid obstacles while moving. They are therefore incapable of discerning the relative distance between the walls of a room, unless the room is devoid of objects on the floor. We plan to investigate the selection and positioning of sensors for the specific purpose of estimating room size by detecting walls. One approach is to mount ultrasonic sensors at "eye level", above tables and chairs, and to limit the "line of sight" of the sensors so that only objects within a narrow range of heights are detected. Although the robot would need to move about to gain a more accurate estimate of the room size, it should not need to traverse the room as much as a robot with only navigational ultrasonics.

Acknowledgements

The authors wish to thank Adam Hoover for his model building software and for his helpful comments.

References

1. Working notes of AAAI-93 Workshop on Reasoning About Function, Washington, D.C., (July 12, 1993)
2. Working notes of AAAI-94 Workshop on Representing and Reasoning About Device Function, Seattle, Washington, (August 3, 1994)
3. *Proceedings of the Workshop on Context-Based Vision*, Cambridge, Massachusetts, (1995)
4. Binford, T.O.: Survey of model-based image analysis systems, *Int. J. of Robotics Research*, **1**, (1982) 18–64
5. Bogoni, L. and Bajcsy, R.: Interactive Recognition and Representation of Functionality, *Computer Vision and Image Understanding*, special issue on Functionality in Object Recognition, Vol. 62, No. 2, (1995) 194–214
6. Connell, J.H.: Get Me That Screwdriver! Developing a Sensory-action Vocabulary for Fetch-and-Carry Tasks, *IBM Cyber Journal Research Report*, RC 19473 (April 1994)
7. Cooper, P., Birnbaum, L., and Brand, E.: Causal Scene Understanding, *Computer Vision and Image Understanding*, special issue on Functionality in Object Recognition, Vol. 62, No. 2, (1995) 215–231
8. Di Manzo, M., Trucco, E., Giunchiglia, F., Ricci, F.: FUR: Understanding FUnctional Reasoning, *Int. J. of Intelligent Systems*, **4**, (1989) 431–457
9. Doermann, D., Rivlin, E. and Rosenfeld, A.: The Function of Documents, *Int. J. on Computer Vision*, **16**, (1998) 799–814
10. Duric, Z., Fayman, J.A., and Rivlin, E.: Function from Motion, *IEEE Transactions on Pattern Analysis and Machine Intelligence*, Vol. 18, No. 6, (1996) 579–591
11. Fischler, M. A. and Strat, T. M.: Recognizing objects in a natural environment: a contextual vision system (CVS), *Proceedings: Image Understanding Workshop*, (May 1989) 774–788
12. Green, K., Eggert, D., Stark, L. and Bowyer, K.: Generic Recognition of Articulated Objects through Reasoning about Potential Function, *Computer Vision and Image Understanding*, special issue on Functionality in Object Recognition, Vol. 62, No. 2, (1995) 177–193
13. Hodges, J.: Functional and Physical Object Characteristics and Object Recognition in Improvisation, *Computer Vision and Image Understanding*, special issue on Functionality in Object Recognition, Vol. 62, No. 2, (1995) 147–163
14. Hoogs, A. and Hackett, D.: Model-Supported Exploitation as a Framework for Image Understanding, *Proceedings of the ARPA IU Workshop*, (November 1994) 265–268

15. Hoover, A., Goldgof, D. and Bowyer, K.W.: Building a B-rep from a segmented range image, *IEEE Second CAD-Based Vision Workshop*, Champion, Pennsylvania (February 1994), 74–81
16. Jensen, F., Christensen, H. and Nielsen, J.: Bayesian Methods for Interpretation and Control in Multi-agent Vision Systems, in *Proceedings of SPIE Conference on Application of AI X: Machine Vision and Robotics*, Kevin W. Bowyer, editor, Vol. 1708, (1992) 536-548
17. Kim, D. and Nevatia, R.: A Method for Recognition and Localization of Generic Objects for Indoor Navigation, *IEEE Workshop on Applications of Computer Vision*, (1994), 280–288
18. Kise, K., Hattori, H., Kitahashi, T., and Fukunaga, K.: Representing and Recognizing Simple Hand-tools Based on Their Functions, *Asian Conference on Computer Vision*, Osaka, Japan (November, 1993), 656–659
19. Konolige, K., Beymer, D.: SRI small vision system user's manual (software version 1.4). Stanford Research Institute. (December 1999)
20. Hanson, A. and Riseman, E.: *The VISIONS Image-Understanding System Advances in Computer Vision (2 vols)*, Erlbaum, Vol. 1, (1988), 1-114
21. Stark, L., Hoover, A.W., Goldgof, D.B. and Bowyer, K.W.: Function-based object recognition from incomplete knowledge of object shape, *IEEE Workshop on Qualitative Vision*, New York (June 1993), 11–22
22. Minsky, M.: The Society of Mind, Simon and Shuster, New York, (1985)
23. Rivlin, E., Dickinson, S. and Rosenfeld, A.: Recognition by Functional Parts, *Computer Vision and Image Understanding*, special issue on Functionality in Object Recognition, Vol. 62, No. 2, (1995) 164–176
24. Rivlin, E. and Rosenfeld, A.: Navigational Functionalities, *Computer Vision and Image Understanding*, special issue on Functionality in Object Recognition, Vol. 62, No. 2, (1995) 232–244
25. Rivlin, E., Rosenfeld, A. and Perlis, D.: Recognition of Object Functionality in Goal-Directed Robotics, in Working Notes on Reasoning About Function, (1993) 126–130
26. Stark, L. and Bowyer, K.: Generic Object Recognition using Form and Function, Series in Machine Perception Artificial Intelligence, Vol. 10, World Scientific, New York, (1996)
27. Stark, L., and Bowyer, K.W.: Achieving generalized object recognition through reasoning about association of function to structure, *IEEE Transactions on Pattern Analysis and Machine Intelligence*, Vol. 3, No. 10, (1991) 1097–1104
28. Stark, L., and Bowyer, K.W.: Indexing function-based categories for generic object recognition, *Computer Vision and Pattern Recognition (CVPR '92)*, Champaign, Illinois (June 1992) 795–797
29. Stark, L. and Bowyer, K.: Function-based generic recognition for multiple object categories, *CVGIP: Image Understanding*, **59** (1), (January 1994) 1–21
30. Stark, L., Hall, L.O. and Bowyer, K.W.: An investigation of methods of combining functional evidence for 3-D object recognition, *International Journal of Pattern Recognition and Artificial Intelligence* **7** (3), (June 1993) 573–594
31. Stark, L., and Bowyer, K.W.: Functional Context in Vision, *Proceedings of the Workshop on Context-Based Vision*, Cambridge, Massachusetts, (1995) 63–74
32. Strat, T.M. and Fischler, M.A.: The Role of Context in Computer Vision, *Proceedings of the Workshop on Context-Based Vision*, Cambridge, Massachusetts, (1995) 2–12

33. Sutton, M., Stark, L., Bowyer, K.W.: Function from visual analysis and physical interaction: A methodology for recognition of generic classes of objects. Image and Vision Computing. **16** (11) (August 1998) 745–763
34. Stansfield, R.A.: Robotic grasping of unknown objects: A knowledge-based approach, *Int. Journal of Robotics Research*, **10**, (1991) 314–326
35. Sutton, M., Stark, L. and Bowyer, K.W.: Function-based generic recognition for multiple object categories, in *Three-dimensional Object Recognition Systems*, A.K. Jain and P.J. Flynn, editors, Elsevier Science Publishers, (1993) 447–470
36. Webster's New World Dictionary, Second College Edition, Prentice Hall Press, (1986)
37. Weisbin, C.R., et al.: Autonomous mobile robot navigation and learning, *IEEE Computer*, Vol. 22, (1989) 29–35
38. Winston, P., Binford, T., Katz, B., and Lowry, M.: Learning Physical Description from Functional Definitions, Examples, and Precedents, *AAAI '83* (1983) 433–439
39. Winston, P. and Rao, S.: Repairing learned knowledge using experience, in *AI at MIT: Expanding Frontiers*, P.H. Winston and S.A. Shellard, editors, MIT Press (1990) 363–379

Author Index

Lecture Notes in Computer Science

For information about Vols. 1–2216
please contact your bookseller or Springer-Verlag

Vol. 2253: T. Terano, T. Nishida, A. Namatame, S. Tsumoto, Y. Ohsawa, T. Washio (Eds.), New Frontiers in Artificial Intelligence. Proceedings, 2001. XXVII, 553 pages. 2001. (Subseries LNAI).

Vol. 2254: M.R. Little, L. Nigay (Eds.), Engineering for Human-Computer Interaction. Proceedings, 2001. XI, 359 pages. 2001.

Vol. 2255: J. Dean, A. Gravel (Eds.), COTS-Based Software Systems. Proceedings, 2002. XIV, 257 pages. 2002.

Vol. 2256: M. Stumptner, D. Corbett, M. Brooks (Eds.), AI 2001: Advances in Artificial Intelligence. Proceedings, 2001. XII, 666 pages. 2001. (Subseries LNAI).

Vol. 2257: S. Krishnamurthi, C.R. Ramakrishnan (Eds.), Practical Aspects of Declarative Languages. Proceedings, 2002. VIII, 351 pages. 2002.

Vol. 2258: P. Brazdil, A. Jorge (Eds.), Progress in Artificial Intelligence. Proceedings, 2001. XII, 418 pages. 2001. (Subseries LNAI).

Vol. 2259: S. Vaudenay, A.M. Youssef (Eds.), Selected Areas in Cryptography. Proceedings, 2001. XI, 359 pages. 2001.

Vol. 2260: B. Honary (Ed.), Cryptography and Coding. Proceedings, 2001. IX, 416 pages. 2001.

Vol. 2261: F. Naumann, Quality-Driven Query Answering for Integrated Information Systems. X, 166 pages. 2002.

Vol. 2262: P. Müller, Modular Specification and Verification of Object-Oriented Programs. XIV, 292 pages. 2002.

Vol. 2263: T. Clark, J. Warmer (Eds.), Object Modeling with the OCL. VIII, 281 pages. 2002.

Vol. 2264: K. Steinhöfel (Ed.), Stochastic Algorithms: Foundations and Applications. Proceedings, 2001. VIII, 203 pages. 2001.

Vol. 2265: P. Mutzel, M. Jünger, S. Leipert (Eds.), Graph Drawing. Proceedings, 2001. XV, 524 pages. 2002.

Vol. 2266: S. Reich, M.T. Tzagarakis, P.M.E. De Bra (Eds.), Hypermedia: Openness, Structural Awareness, and Adaptivity. Proceedings, 2001. X, 335 pages. 2002.

Vol. 2267: M. Cerioli, G. Reggio (Eds.), Recent Trends in Algebraic Development Techniques. Proceedings, 2001. X, 345 pages. 2001.

Vol. 2268: E.F. Deprettere, J. Teich, S. Vassiliadis (Eds.), Embedded Processor Design Challenges. VIII, 327 pages. 2002.

Vol. 2270: M. Pflanz, On-line Error Detection and Fast Recover Techniques for Dependable Embedded Processors. XII, 126 pages. 2002.

Vol. 2271: B. Preneel (Ed.), Topics in Cryptology – CT-RSA 2002. Proceedings, 2002. X, 311 pages. 2002.

Vol. 2272: D. Bert, J.P. Bowen, M.C. Henson, K. Robinson (Eds.), ZB 2002: Formal Specification and Development in Z and B. Proceedings, 2002. XII, 535 pages. 2002.

Vol. 2273: A.R. Coden, E.W. Brown, S. Srinivasan (Eds.), Information Retrieval Techniques for Speech Applications. XI, 109 pages. 2002.

Vol. 2274: D. Naccache, P. Paillier (Eds.), Public Key Cryptography. Proceedings, 2002. XI, 385 pages. 2002.

Vol. 2275: N.R. Pal, M. Sugeno (Eds.), Advances in Soft Computing – AFSS 2002. Proceedings, 2002. XVI, 536 pages. 2002. (Subseries LNAI).

Vol. 2276: A. Gelbukh (Ed.), Computational Linguistics and Intelligent Text Processing. Proceedings, 2002. XIII, 444 pages. 2002.

Vol. 2277: P. Callaghan, Z. Luo, J. McKinna, R. Pollack (Eds.), Types for Proofs and Programs. Proceedings, 2000. VIII, 243 pages. 2002.

Vol. 2281: S. Arikawa, A. Shinohara (Eds.), Progress in Discovery Science. XIV, 684 pages. 2002. (Subseries LNAI).

Vol. 2282: D. Ursino, Extraction and Exploitation of Intensional Knowledge from Heterogeneous Information Sources. XXVI, 289 pages. 2002.

Vol. 2284: T. Eiter, K.-D. Schewe (Eds.), Foundations of Information and Knowledge Systems. Proceedings, 2002. X, 289 pages. 2002.

Vol. 2285: H. Alt, A. Ferreira (Eds.), STACS 2002. Proceedings, 2002. XIV, 660 pages. 2002.

Vol. 2286: S. Rajsbaum (Ed.), LATIN 2002: Theoretical Informatics. Proceedings, 2002. XIII, 630 pages. 2002.

Vol. 2287: C.S. Jensen, K.G. Jeffery, J. Pokorny, Saltenis, E. Bertino, K. Böhm, M. Jarke (Eds.), Advances in Database Technology – EDBT 2002. Proceedings, 2002. XVI, 776 pages. 2002.

Vol. 2288: K. Kim (Ed.), Information Security and Cryptology – ICISC 2001. Proceedings, 2001. XIII, 457 pages. 2002.

Vol. 2289: C.J. Tomlin, M.R. Greenstreet (Eds.), Hybrid Systems: Computation and Control. Proceedings, 2002. XIII, 480 pages. 2002.

Vol. 2291: F. Crestani, M. Girolami, C.J. van Rijsbergen (Eds.), Advances in Information Retrieval. Proceedings, 2002. XIII, 363 pages. 2002.

Vol. 2292: G.B. Khosrovshahi, A. Shokoufandeh, A. Shokrollahi (Eds.), Theoretical Aspects of Computer Science. IX, 221 pages. 2002.

Vol. 2293: J. Renz, Qualitative Spatial Reasoning with Topological Information. XVI, 207 pages. 2002. (Subseries LNAI).

Vol. 2296: B. Dunin-Kęplicz, E. Nawarecki (Eds.), From Theory to Practice in Multi-Agent Systems. Proceedings, 2001. IX, 341 pages. 2002. (Subseries LNAI).

Vol. 2300: W. Brauer, H. Ehrig, J. Karhumäki, A. Salomaa (Eds.), Formal and Natural Computing. XXXVI, 431 pages. 2002.

Vol. 2301: A. Braquelaire, J.-O. Lachaud, A. Vialard (Eds.), Discrete Geometry for Computer Imagery. Proceedings, 2002. XI, 439 pages. 2002.

Vol. 2302: C. Schulte, Programming Constraint Services. XII, 176 pages. 2002. (Subseries LNAI).

Vol. 2305: D. Le Métayer (Ed.), Programming Languages and Systems. Proceedings, 2002. XII, 331 pages. 2002.

Vol. 2309: A. Armando (Ed.), Frontiers of Combining Systems. Proceedings, 2002. VIII, 255 pages. 2002. (Subseries LNAI).

Vol. 2314: S.-K. Chang, Z. Chen, S.-Y. Lee (Eds.), Recent Advances in Visual Information Systems. Proceedings, 2002. XI, 323 pages. 2002.